T0143044

Studies in Computational Intelligence

Volume 544

Series editor   .

Janusz Kacprzyk, Polish Academy of Sciences, Warsaw, Poland
e-mail: kacprzyk@ibspan.waw.pl

For further volumes:
http://www.springer.com/series/7092

About this Series

The series "Studies in Computational Intelligence" (SCI) publishes new developments and advances in the various areas of computational intelligence—quickly and with a high quality. The intent is to cover the theory, applications, and design methods of computational intelligence, as embedded in the fields of engineering, computer science, physics and life sciences, as well as the methodologies behind them. The series contains monographs, lecture notes and edited volumes in computational intelligence spanning the areas of neural networks, connectionist systems, genetic algorithms, evolutionary computation, artificial intelligence, cellular automata, self-organizing systems, soft computing, fuzzy systems, and hybrid intelligent systems. Of particular value to both the contributors and the readership are the short publication timeframe and the world-wide distribution, which enable both wide and rapid dissemination of research output.

Theodor Borangiu · Damien Trentesaux
André Thomas

Editors

Service Orientation in Holonic and Multi-Agent Manufacturing and Robotics

 Springer

Editors

Theodor Borangiu
Faculty of Automatic Control and Computer
  Science (Automatica)
University Politehnica of Bucharest
Bucharest
Romania

Damien Trentesaux
PSI/TEMPO Lab
Universite de Valenciennes
Valenciennes cedex 9
France

André Thomas
ENSTIB - Ecole Nationale Supérieure des
  Technologies et Industries du Bois
Centre de Recherche en Automatique de
  Nancy
Epinal cedex 9
France

ISSN 1860-949X   ISSN 1860-9503   (electronic)
ISBN 978-3-319-34990-9   ISBN 978-3-319-04735-5   (eBook)
DOI 10.1007/978-3-319-04735-5
Springer Cham Heidelberg New York Dordrecht London

Printed on acid-free paper

Springer is part of Springer Science+Business Media (www.springer.com)

Foreword

Today, optimization and improvement efforts within established organizational borders suffer from diminishing returns (i.e. the proverbial low-hanging fruits have been picked already) whereas society is experiencing difficulties when attempting to coordinate activities crossing these traditional borders at unprecedented scales. Such large-scale cross-border coordination is urgently needed to answer the challenges that our society is facing. For instance, sustainable and energy-aware production cannot afford to compartmentalize its systems and operate them in uncoordinated manners.

In this regard, note that these traditional borders have been defined by the capabilities of established technology and solutions. In other words, the manner in which these borders emerged implies with almost mathematical certainty (by a so-called reduction *ad absurdum*) that the above challenges require innovative solutions. Indeed, if established solutions would have been able to manage across such borders, these borders had disappeared already whenever the competitive advantage of superior coordination pushes the old borders into non-existence. Improvements based on established designs will continue to improve incrementally by translating and benefiting from technological progress in ICT as well as the wider spreading of ICT skills but the own contribution of enhancements to established designs will be minor when developers remain within their comfort zones.

Novel solutions must change the manner in which systems are coupled and interacting. *Distributed intelligence* is a first aspect of such novel approaches. It is necessary if only because in a multi-organization setting, organizations will not accept to yield control over their own data and information processing. *Service orientation* provides answers because it decouples while shielding the internal aspects and it generalizes what can be employed, activated and recruited amongst different organizational entities. It also makes functionalities (services) available in smaller compose-able and orchestrate-able chunks. In contrast, flexible equipment such as robots allow service providers capable to offer generic services, allowing to build complete production chains based on available service offerings (less gaps to be filled by yet-to-be-developed systems).

Furthermore, the direct interaction and coordination of enterprises (resources) is unlikely to scale in a dynamic and demanding environment. It requires too much knowledge beyond *knowing one-self*, which makes such system designs short-lived in a fast-changing context almost by definition. In the real world products *connect* the resources and, consequently, intelligent products interacting with intelligent resources, offering intelligent services, are superior at mirroring a coherent and consistent reality especially when this reality is dynamic. Both intelligent products and intelligent resources need to be first-class citizens and should not be reduced to data formats. Moreover, neither shall act on behalf of the other; a single source of truth design is needed. In this context, *intelligent products* driving their own production enables the scaling and coordination across organizational borders, provided they do not make the mistake of reducing the resources to data in some document formats.

Finally, insights in *complex-adaptive systems* need to be accounted for. For instance, the insights from Herbert Simon lead to holonic systems. Equally important, the need for a critical user mass relative to the size and complexity of the (intelligent) artifacts that we develop has to be acknowledged. In particular, software has reached levels of complexity that render the whole manufacturing community as too small to deliver critical mass. Adopting, adapting and completing mainstream technologies from the IT and the Telecom communities are rapidly becoming unavoidable. The scenario of computer graphics in CAD – benefiting from computer games – will become the norm. Our scientific community needs to learn which mainstream technologies to select, adopt and adapt, and make our current habit of developing too many manufacturing automation only systems a part of history. The parts of the world that will do this first will hurt the continents that stay behind.

In view of the above, this book contributes to the investigation, development, understanding and validation of such innovative solutions. It does not deliver final answers but documents the efforts that lead toward solutions that can make a difference beyond the incremental.

December 2013   Paul Valckenaers
                                    Professor, Catholic University of Leuven

Preface

This volume gathers the peer reviewed papers which were presented at the third edition of the International Workshop "Service Orientation in Holonic and Multi-agent Manufacturing and Robotics – SOHOMA'13" organized on June 20--22, 2013 by the Centre of Research in Computer Integrated Manufacturing and Robotics – CIMR Bucharest, and hosted by the University of Valenciennes, France.

The SOHOMA'13 scientific event was organized in the framework of the European project no. 264207 ERRIC, the objective of which is to foster innovation in manufacturing control through intelligent IT and in this context to empower excellence in research at the faculty of Automatic Control and Computer Science within the University Politehnica of Bucharest.

The book is structured in five parts, each one covering a specific research domain which represents a trend for modern manufacturing control: *Distributed Intelligence for Sustainable Manufacturing* (Part 1), *Holonic and Multi-Agent Technologies for Manufacturing Planning and Control* (Part 2), *Service Orientation in Manufacturing Management and Control* (Part 3), *Intelligent Products and Product-driven Automation* (Part 4) and *Robotics for Manufacturing and Services* (Part 5).

These five evolution lines have in common concepts related to *service orientation* in a distributed planning and control agent-based industrial environment; today it is generally recognized that the Service Oriented Enterprise Architecture paradigm has been looked upon as a suitable and effective approach for industrial automation and management of manufacturing enterprises.

Manufacturing systems that use *robot-vision workstations* are amongst the most complex and demanding artefacts in modern society but also amongst the most valuable ones. The challenges include coping with their heterogeneous nature and their on-line interactive nature in combination with competitive pressures. Off-line plans are known to become invalid within minutes after arriving on the factory floor. Therefore, researchers are looking into matching technologies which are able to answer these challenges. *Holonic systems* are, actually by definition, targeting such challenges. *Agent technologies* focus on interactive and decentralized aspects. In particular, developments aim to deliver open systems and system components, as well as infrastructure and infrastructural components rather than closed systems. This open nature implies that developments will not solve industrial problem on their own but rather contribute while avoiding the unnecessary constraining of an overall solution.

Technological advances in wireless sensor networks are enabling new levels of distributed intelligence in several forms such as active products that interact with the working environment and smart metering for monitoring the history of products over their entire life cycle and the behaviour of resources. These distributed intelligences offer new opportunities for developing techniques to reduce myopic decision making in manufacturing control systems thereby potentially enhancing their sustainability. Control architecture could itself switch modes of operation to adapt to severe disruptions. *Manufacturing sustainability* is addressed in this special issue with respect to: circular economy paradigm, fault-tolerance to disturbances; energy efficiency at resource and shop floor level; balancing resource usage; cost efficiency and in line quality control of products. Innovative services will be enablers and drivers of growth of next generation of manufacturing enterprises that are competitive and sustainable. Finally, the debate was open on the question: Are Intelligent Manufacturing and Services Systems sustainable?

Several frameworks are proposed for classifying, analysing initiatives and potentially developing distributed intelligent automation systems. These frameworks will be referred to in the book as the *Distributed Intelligent Automation Systems Grid*. In particular we are interested in systems in which the planning or execution of tasks normally associated with a centralized operational level are reassigned to be carried out instead by a number of units at a different level. Or conversely, a task normally using information from a single source makes use of data spread across a range of operations – and potentially a range of organisations.

The book defines and explains the main ways to implement *intelligent products*: by putting intelligence at the object (Intelligent Embedded Systems) or through the computing network (using Automatic or Biometric Identification and Data Capture technology attached to the product to allow it to be identified by a computer system). These technologies enable the automated identification of objects, the collection of data about them, and the storage of that data directly into computer systems.

The service-oriented multi-agent systems (SoMAS) approach discussed in the book is characterized by the use of a set of distributed autonomous and cooperative agents (embedded in smart control components) that use the SOA principles, i.e. oriented by the offer and request of services, in order to fulfil industrial and production systems goals. This approach is different from the traditional Multi-agent Systems (MAS) mainly because agents are service-oriented, i.e. individual goals of agents may be complemented by services provided by other agents, and the internal functionalities of agents can be offered as services to others agents (note that these service-oriented agents do not only share services as their major form of communication, but also complement their own goals with different types of external provided services).

Special attention is paid in the book to the framework for manufacturing integration, which matches plant floor solutions with business systems and suppliers. This solution focuses on achieving flexibility by enabling a low coupling design of the entire enterprise system through leveraging of Service Oriented Architecture (SOA) and Manufacturing Service Bus (MSB) as best practices.

The *Manufacturing Service Bus* (MSB) integration model described in some papers included in this volume is an adaptation of ESB for manufacturing enterprises and introduces the concept of bus communication for the manufacturing systems. The MSB acts as an intermediary (middle-man) for the data flows, assuring loose coupling between modules at shop floor level.

The book offers a new integrated vision combining complementary emergent technologies which allow reaching control structures with distributed intelligence supporting the enterprise integration (vertical and horizontal dimensions) and running in truly distributed and ubiquitous environments. Additionally, the enrichment of these distributed systems with mechanisms inspired by biology supports the dynamic structure reconfiguration, thus handling more effectively with condition changes and unexpected disturbances, and minimizing their effects. As an example, the integration of service-oriented principles with MAS allows to combine the best of the two worlds, and in this way to overcome some limitations associated to multi-agent systems, such as interoperability.

A brief description of the book chapters follows.

Part 1 reports recent advances and on-going research in sustainable manufacturing based on distributed approaches such as Holonic Manufacturing Execution Systems. Distributed intelligences offer new opportunities for developing techniques to reduce myopic decision making in manufacturing control systems thereby potentially enhancing their sustainability. There are contributions analysing the concept of Intelligent Product and related techniques for Product-driven Automation. The rapid development of this concept is mainly due to the fact that, over the last decade, the increasing growth of embedded technologies (e.g., RFID, smart cards, wireless communication), associated with the concepts of ambient intelligence and machine-to-machine intelligence, has allowed the development of products that are fully able to interact in an intelligent mode with their environment. Also, working on the closed-loop PLM (Product Life Cycle Management), interoperability and traceability topics leads to some relevant specifications that can be applied using an "intelligent product" approach, from the product's design to its recycling.

Part 2 is devoted to holonic and multi-agent technologies for manufacturing planning and control. The demand for large-scale systems running in complex and even chaotic environments requires the consideration of new paradigms and technologies that provide flexibility, robustness, agility and responsiveness. Holonic systems are, actually by definition, targeting challenges that include coping with the heterogeneous nature of industrial systems and their on-line interactive nature in combination with competitive pressures. Multi-agents systems is considered as a suitable approach to address these challenge by offering an alternative way to design control systems, based on the decentralization of control functions over distributed autonomous and cooperative entities. This part of the book gathers contributions on on-line simulation and on benchmarks aiming at delivering open systems which feature agility, optimization in hybrid structures, myopia decrease and robustness.

Part 3 approaches the trend of service orientation in the management and control of processes in manufacturing enterprises. The service orientation is emerging at multiple organizational levels in enterprise business, and leverages technology in response to the growing need for greater business integration, flexibility and agility of manufacturing enterprises. Close related to IT infrastructures of Web services, the Service Oriented Architecture represents a technical architecture, a business modelling concept, an integration source and a new way of viewing units of automation within the enterprise. Business and process information systems integration and interoperability at enterprise level are feasible by considering the customized product as "active controller" of the enterprise resources – thus providing consistency between the material and informational flows within the enterprise. Service orientation in the manufacturing domain is not limited to just Web services, or technology and technical infrastructure either; instead, it reflects a new way of thinking about processes that reinforce the value of commoditization, reuse, semantics and information, and create business value. The unifying approach of the contributions for this third part of the book relies on the methodology and practice of disaggregating siloed, tightly coupled business processes at manufacturing enterprise level into loosely coupled services and mapping them to IT services, sequencing, synchronizing and automating the execution of processes which encapsulate the software description of such complex business processes related to agile production by means of distributed information systems.

Part 4 analyses, develops solutions and describes applications in the domain of intelligent products and product-driven automation. The rise of embedded technologies (e.g. RFID, smart cards, wireless communication), as well as research in the field of ambient intelligence, have enabled the development of "intelligent" products that can fully interact with their environment. Various typologies of intelligent products are reviewed: "Individual", focusing on the product as an individual entity classified either according to a unique criterion-their level of intelligence or to multiple criteria such as: sensory capacities, location of intelligence, level of intelligence, aggregation level of intelligence; "Collective", characterizing the types of interactions which exist in a collective of "intelligent" products. After reviewing different IP categories, limitations are outlined; among them, the difficulty in describing the dynamics of a collective of "intelligent" products. In general, the majority of the typologies proposed do not deal with the interactions that exist in a collective of products.

Following a brief presentation of the notion of "activeness", an analysis framework is proposed to describe the "activeness" of a product throughout its life cycle. This framework focuses on the analysis of the interaction situation between a collective of "active" products and the support system for a given function. The framework contains two views (functional and organic), which can be divided into several sections that describe the different facets of the interaction situation. Via an increase in its informational, communicational and decisional capacities, an "active" product is considered as an entity capable of interacting with the different support systems (e.g. manufacturing system, supply chain) during the successive phases of its life cycle.

This part also discusses new challenges and opportunities which arise with concepts such as Internet of Things (IoT), where objects of the real world are linked with the virtual world, thus enabling connectivity anywhere, anytime and for anything. The rapid expansion of the IoT and the web technology enabled the development of new Business-to-Business (B2B) infrastructures based on the concept of Service Oriented Architecture (SOA). The technical provision of services is not a uniform activity but is skewed by prevalent technological solutions, which have often a lack of interfaces required by users or are designed in an isolated faction that limit their openness. Such a limitation has a direct impact on organizational infrastructures that, in today's world, need more than ever to exchange product information in an appropriate manner (e.g. to provide the right service or information, whenever it is needed, wherever it is needed, by whoever needs it). Quantum Lifecycle Management (QLM) messaging standards are proposed as a standard application-level interface that would provide flexible and a wide range of properties/interfaces, which aim to increase the SOA scope as well as the data exchange interoperability in the IoT. Applications of Intelligent and Active Products in the general IoT framework are proposed in manufacturing, transport, and precision agriculture.

Part 5 reports advances in robot control and integration in manufacturing tasks and services. In order to adapt themselves to the environment and characteristics of material flows, robot systems are equipped with vision systems. Vision-guided robot motion using visual servoing methods provide best performances in the generation of accurate, task-oriented motion patterns. Integrating Visual Quality Control (VQC) services in the manufacturing environment is described as product traceability means. In the context of agent-based manufacturing reconfiguring, this section of the book also describes planning and tracking of cooperative activities in robot teams.

The *service value creation model* at enterprise level consists into using a Service Component Architecture (SCA) for business process applications, based on entities which handle (provide, ask for, monitor) services. In this componentization view, a service is a piece of software encapsulating the business control logic or resource functionality of an entity that exhibits an individual competence and responds to a specific request to fulfil a local (product operation, verification) or global objective (batch production).

If SOA is the conceptual framework for service orientation of manufacturing enterprise processes, then **Service Oriented Computing** (SOC) represents the methodology and implementing framework for embedded monitoring and control systems in *Service Oriented Enterprise Architectures*, and **Service and Computing Oriented Manufacturing** (SCOM) unifies existing advanced manufacturing models by centring them on internet/network, cooperative work and resource sharing, which creates the premises for *Digital Manufacturing* of the future.

All these aspects are treated in the present book, which we hope you will find useful reading.

November 2013

The Editors
Theodor Borangiu
Damien Trentesaux
André Thomas

Contents

Part II: Holonic and Multi-Agent Technologies for Manufacturing Planning and Control

Extraction of Priority Rules for Boolean Induction in Distributed Manufacturing Control 127
Nassima Aissani, Baghdad Atmani, Damien Trentesaux, Beldjilali Bouziane

Supply Chain Management Using Multi-Agent Systems in the Agri-Food Industry 145
Ait Si Larbi El Yasmine, Bekrar Abdel Ghani, Damien Trentesaux, Beldjilali Bouziane

Part III: Service Orientation in Manufacturing Management and Control

A Generic Service System Activity Model with Event-Driven Operation Reconfiguring Capability **159**
Theodor Borangiu, Monica Drăgoicea, Virginia Ecaterina Oltean, Iulia Iacob

Product Specification for Flexible Workflow Orchestrations in Service Oriented Holonic Manufacturing Systems
Francisco Gamboa Quintanilla, Olivier Cardin, Pierre Castagna

A Multi-Agent Architecture for Compensating Unforeseen Failures on Field Control Level **195**
Christoph Legat, Birgit Vogel-Heuser

Part IV: Intelligent Products and Product-Driven Automation

QLM Messaging Standards: Introduction and Comparison with Existing Messaging Protocols

Part I
Distributed Intelligence
for Sustainable Manufacturing

Are Intelligent Manufacturing Systems Sustainable?

André Thomas[1] and Damien Trentesaux[2]

[1] Research Centre for Automatic Control (CRAN), CNRS (UMR 7029), Lorraine University,
ENSTIB 27, rue Philippe Seguin, 88000 Epinal, France
`Andre.Thomas@cran.uhp-nancy.fr`
[2] Université Lille Nord de France, UVHC, Tempo-Lab., F-59313 Valenciennes,
cedex 9, France
`Damien.Trentesaux@univ-valenciennes.fr`

Abstract. This paper introduces and opens the debate on the role of "product-driven control" and, on a broader level, Intelligent Manufacturing and Services Systems in sustainable development and circular economy. First, the concept of IMS and perspectives relating to future industrial and economic systems, as well as their expected advantages and the challenges to be addressed are introduced. Different interpretations of sustainability are then described. Third, our vision concerning the possible impacts of IMS on a sustainable world is presented. A set of challenging prospects and illustrative examples conclude this paper.

Keywords: intelligent manufacturing systems, product-driven control, holonic control, intelligent product, sustainable development, sustainable manufacturing, C2C.

1 Introduction

In today's world, we are invaded by a wide variety of smart appliances: smart-phones, smart-homes, smart-cars and so on. Physical objects are becoming more and more intelligent. Internet of Things (IoT), Ubiquitous or Pervasive Computing are now common terms. The industrial and services sectors are also affected. Computer Integrated Manufacturing has evolved from classical top-down approaches to more bottom-up approaches where intelligence is distributed in products and resources (supply chains (SC) and production systems), leading to Intelligent Manufacturing or Services Systems.

However, the international community is increasingly sensitive, aspiring for greater sustainability worldwide. Concepts such as sustainable development, circular economy, industrial ecology and sustainable manufacturing are becoming ever more popular and it is now becoming necessary and urgent to implement them in today's economic activities.

Thus, the aim of this article is to open the debate on the role of "product-driven control" and, on a broader level, Intelligent Manufacturing Systems (IMS) including services, in sustainable development and circular economy. A short state of the art on

T. Borangiu et al. (eds.), *Service Orientation in Holonic and Multi-Agent Manufacturing and Robotics*, Studies in Computational Intelligence 544,
DOI: 10.1007/978-3-319-04735-5_1, © Springer International Publishing Switzerland 2014

IMS is presented and the interpretations of sustainable approaches described. Finally, some challenging prospects are proposed according to critical sustainable development indicators.

2   Towards a Ubiquitous World and Intelligent Manufacturing

New opportunities are arising with the deployment of concepts such as Internet of Things, and Ubiquitous or Pervasive Computing. Through these concepts, objects in the real world are linked with the virtual world, thus people as well as things can be connected anytime, anyplace. Such concepts refer to a world where physical things or even people, as well as virtual environments, interact with each other [1], [2]. A huge quantity of applications exist today, such as mobile phones, smart homes, smart cars, in various sectors including services: medical sector [3], textiles industry [4], military applications [5], home automation [6], dismantling operations [7] or transportation [8].

It seems to us that production systems, supply chains, hospitals, and services in a broader sense, are no exception to this evolution. The industrial, healthcare or services sectors are today also affected. Initially, centralized approaches for control and decision-making in such domains were used. These approaches were successful, mostly due to their ability to provide long-term, global optimization of production planning and scheduling, given a relatively stable operational context. In the 1990s, Production and Supply Chain Systems changed from the traditional mass production to mass customization in order to deal with the increasing competition in global markets.

Industrial requirements have clearly evolved from the traditional performance criteria, described in terms of static optimality or near-optimality, towards new performance criteria, described in terms of reactivity, adaptability and visibility. A growing number of industrialists now want control systems that provide satisfactory, adaptable and robust solutions rather than optimal solutions that require several hard assumptions to be met [9], [10]. Information technology has gradually improved, giving physical system entities (e.g. parts, resources) some decision-making capabilities and information-carrying capacities. These improvements could be a new way of dealing with the as yet unsolved problem concerning the de-synchronization of material and informational flows [11]. The concepts of Holonic Manufacturing Systems, Product-Driven Control Systems, Smart Product, Intelligent Products, Intelligent Manufacturing Systems, and Agent-Based Manufacturing, to name a few, have been proposed to design these future manufacturing systems. These concepts advocate that the products, and also more globally, all the production resources, can be modelled as an association between two parts (i.e., a physical part and an informational part) to become a holon interacting with human operators.

The French community of automatic control, at the request of the CNRS (Centre National de la Recherche Scientifique – French National Centre for Scientific Research), recently issued its report on "manufacturing of the future" (ARP FuturProd Report). The main objective of this research team was to highlight some strategic and tactical prospects and design a global vision of product and services production systems for 2020/2030. The main keywords that emerged were "innovation" and "sustainability". Behind these words and from a more comprehensive point of view, it seems obvious that "distributed intelligence" will be implemented in a large variety of applications. The residual question will then be how can they be in harmony with new sustainable development trends?

3   Sustainable Development and Related Concepts

The "Ecotechnic" concept concerns all techniques that help preserve the environment and get the best out of the ecological resilience of "ecosystems". In this sense, an "ecosystem" is a balanced, stable and complex life system. The use of materials is optimal, each type of waste is reused by other organisms and energy needs will be met using renewable sources such as solar energy.

The most frequently quoted definition of "Sustainable development" is from *Our Common Future*, also known as the Brundtland Report [12]:

"Sustainable development is development that meets the needs of the present without compromising the ability of future generations to meet their own needs. It contains within it two key concepts:

- *the concept of needs, in particular the essential needs of the world's poor, to which overriding priority should be given; and*
- *the idea of limitations imposed by the state of technology and social organization on the environment's ability to meet present and future needs."*

To meet these targets, policies based on a viable, global approach which is socially, environmentally and economically structured will have to be implemented, leading to the proposal of three pillars of sustainability (Fig. 1) suggesting that both the economy and society are constrained by environmental limits.

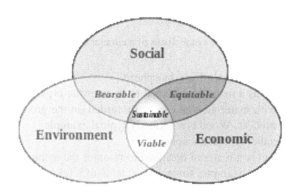

Fig. 1. The three pillars of *Sustainable Development* (SD)

The main aim of a circular economy is to optimize energy and material flows at system level, which could be a plant, an industrial park, a country, etc... To limit energy consumption and waste, the aim is to achieve quasi-cyclic functioning in the same way as an eco-system

In the approach presented above, the system is constituted of two metabolisms. Fig. 2, proposed by the Ellen MacArthur foundation, provides an example of such a system.

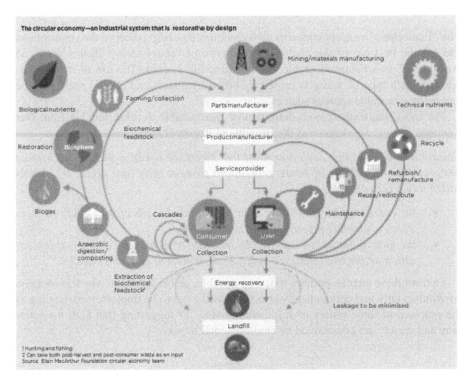

Fig. 2. The two metabolisms of a circular economy [13]

The biological/organic metabolism: biosphere.

An organic nutrient is a material designed to re-enter the organic cycle. The idea is to design products with materials that can be discarded on the ground and return to a safe biodegradable process. Wood is a perfectly typical example.

The technical metabolism: technosphere.

A technical nutrient is a material designed to re-enter the technical/industrial cycle from which it came. For example, some components of a television set may be interesting materials/components that could be used for other purposes.

So, to control such systems according to the three pillars of sustainable development, a system of indicators (Fig. 3) will help develop new actions and establish new fields of research. The famous "Cradle to Cradle" (C2C) book describes some implementations of such a circular economy and recommends the development of indicators and relevant accreditations such as C2C.

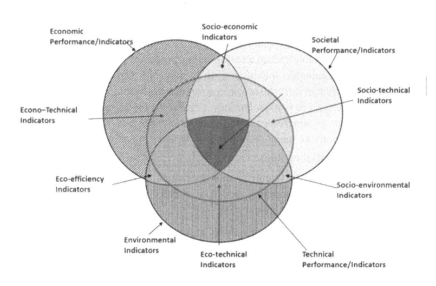

Fig. 3. Economic, ecological, social and technical dimensions of a sustainable product [14]

Fifteen high level indicators have been associated with the 9 challenges of the French "Stratégie Nationale de Développement Durable" (SNDD) 2010-2013 [National Strategy for Sustainable Development]. Four are related to economic and social contexts. These high level indicators are augmented with thirty detailed indicators. According to the challenges, perspectives and opportunities relative to IMS, it would be interesting to highlight the following indicators:

Challenge 1: Sustainable consumption and production (excerpts)

      Material productivity

      Evolution in production waste

      Industrial efficiency

      Employment

Challenge 4: Climate change and energy

      Total greenhouse gas emissions

      Carbon footprint

Challenge 5: Sustainable transport/mobility

      Transport energy consumption

Challenge 7: Public health and risk management

      Life expectancy

      Occupational accident

The main objective of this article is to highlight how IMS could be compatible with sustainable development. Consequently, the questions are: how can the target values of these indicators be reached in the context of IMS and how can IMS help improve the value of these indicators?

4   Challenges for IMS in a Sustainable World

The recognition that sustainable natural eco-systems result from cyclic functioning led to the concept of sustainable manufacturing being proposed [15]. One of the most recent definitions of sustainable manufacturing is given by Garetti and Taisch [16] as "*the ability to smartly use natural resources for manufacturing, by creating products and solutions that, thanks to new technology, regulatory measures and coherent social behaviours, are able to satisfy economic, environmental and social objectives, thus preserving the environment, while continuing to improve the quality of human life*".
Consequently, in industrial ecology:

- *ecology* is relative to scientific ecology, which is the study of eco-systems.
- *industrial* concerns modern industrial society (production, distribution systems, public services and so on).

By seeking to copy rules and principles from nature, sustainable manufacturing could also have scientific approaches, such as biomimetic. Some research already exists in the fields of stigmergy and viable systems [17].

From the definition of sustainable manufacturing presented previously, it is interesting to note the use of the adverb "smartly" enables us to link the "IMS" concept to the "sustainable manufacturing" concept, in the sense that IMS can be seen as the "smart use of natural resources", in conjunction with Challenge 1 introduced previously ("Sustainable consumption and production: Industrial efficiency"). For example, opportunistic energy management at shop floor level may be naturally addressed using smart distributed devices. A typical illustration of this approach is proposed by [18]. The authors studied the integration of an energy monitoring function in the entities of a manufacturing system (e.g., programmable logic controller, enterprise resource planning, and embedded devices). According to their models, the future resources are expected to auto-manage their energy consumption. These resources will switch to a low energy consumption state if no task is currently being executed on them.

In this context, the key point that favours to the use of more "bottom-up" and more intelligent approaches rather than classical top-down approaches is related to the fact that in job-shop and flow-shop problems, the scheduling problem with no attention paid to energy is already known to be NP-Hard. Consequently, the resolution time of centralized methods is very long. Adding another constraint (e.g., energy consumption) or adding another objective in the objective function (e.g., minimize the energy) will make the resolution even more difficult. This could make such methods unable to react to uncertainty (e.g., changes or perturbations in a short time, dynamic energy pricing in the smart grid, etc.), and scarcely compatible with the reactivity required in a dynamic context.

But the "intelligent systems" approach can also be concerned indirectly, on a more global level, beyond the classical scope of the manufacturing phase, by proposing solutions to increase links among the different phases of product and resource life cycles, including better traceability, quality management and contributing to the optimization of recycling. Indeed, embedding or distributing intelligence among products and resources enables the information processing capabilities to be physically closer to the products and systems, and even connect this information to the systems, thus reducing decisional loops and increasing information accuracy. For example, if a smart system is able to memorize the way it has been used and its history of parts that have been replaced, this information can be useful for maintenance or dismantling and recycling issues [19].

According to the indicators introduced previously, we can illustrate some interests and contributions of IMS thanks to the frameworks presented in table 1.

Table 1. The decision level framework

	Sustainable development indicators (extracts)									
	Challenge 1 = Sustainable consumption and production				Challenge 4 = Climate change and energy			Challenge 5 = Sustainable transport/mobility		Challenge 7 = Public health and risk management
	Material productivity	Evolution in Production wastes	Employment	Industrial efficiency	Total greenhouse gas emissions	Carbon footprint	Energy consumption	Transport energy consumption	Greenhouse gas emissions	Life expectancy
Strategic		Diversificatio	Perpetuation of activities		(Short) Supply Chain Design					
Tactical	Inventory and expenditure minimization	Inventory management and traceability	Resource optimization	Inventory management and traceability	Resource and vehicle routing optimization – lead-time and inventory minimization					Inventory management and traceability in hospitals
Operational	Material yield	Control of physical		Physical flows control	Energy and short distance road transport flow optimization and control					Drug dispensing
Real time	"on-line"	physical								

This first framework highlights some of the challenges chosen according to the decision-making and industrial/structural levels. These are explained and presented in more detail below.

For the first challenge:

- The first SD indicator, *material productivity*, concerns materials from a global point of view, but more specifically the use of natural resources. The goal is to obtain the best yield and to avoid material waste. This problem is not generally a strategic challenge (except in some specific industries which use rare and very expensive raw materials). From a tactical viewpoint and because the material is an important factor in Sales, Inventory & Operations Planning process, inventory and expenditure minimization models have to be developed and proposed to balance budget and Supply Chain objectives. Obviously, from an operational and very short term point of view, material waste needs to be avoided through "on-line" optimization of material yield.

- Concerning the evolution of the production waste indicator, the main idea is to exploit the production and supply chain systems in the most efficient way. Physical system and quality process control is the most common approach. Utilization and Efficiency are the main very short-term indicators for physical flows and production system control. Products, sustainable objectives and customer satisfaction have to be tracked by means of traceability processes and so tactical inventory management models have to be implemented to ensure the best equilibrium between funds, demand, expenditure on energy and resources, and inventory. Finally, diversification models could be a very good way to minimize and optimize production and SC resources, thus smoothing the demand.

- Employment has to be saved to contribute to the social pillar of SD. Obviously company continuity and the perpetuation of activities become objectives. Labour being resources, as previously, control and optimization models have to be proposed.
- Controlling physical systems, work centers, labour, inventories and so on, according to customer demand can also help improve industrial efficiency.

Challenges 4 and 5 concern the global impact of energy consumption and activities. Concerning greenhouse gas emissions, at a strategic level short Supply Chain design becomes the most promising way to implement B2B or B2C models that can take such SD indicators into account because they have a direct impact on carbon footprint and energy consumption.

At tactical level and in the very short term, the well-known resources and ve-hicle routing models must be adapted with new parameters and constraints relative to SD.

Table 2. The conceptual level framework

	Sustainable development indicators (extracts)									
	Challenge 1 = Sustainable consumption and production				Challenge 4 = Climate change and energy			Challenge 5 = Sustainable transport/mobility		Challenge 7 = Public health and risk management
	Material productivity	Evolution in Production wastes	Employment	Industrial efficiency	Total greenhouse gas emissions	Carbon footprint	Energy consumption	Transport energy consumption	Greenhouse gas emissions	Life expectancy
Context	Consumption and material waste minimization	Co-products usages	Local employment	Low prices for economic sustainability	Sustainable Suppy Chains design			Short distance roads		Hospital
Concept	Defects, products real time vision	Co-products need global knowledge	Employment robustness	Decision making at various levels	Routing and material resource usage minimization					Quality of service and availability
Logical	Mono-multi criteria			Management and lot sizing rules	Mono-multi criteria - Heterogeneous models			Mono-multi criteria		Mono-multi criteria
Physical	Marking			RFID, barcodes	RFID, GPS, WSN			RFID, GPS		RFID, Barcodes
Detail	knapsack algorithms			Optimization, simulation				PL, PLNE, ...		

A second framework highlights the same challenges according to system/Zachman abstraction levels (Table 2). In the same way as for the previous framework, the re-search perspectives relative to some challenges and SD indicators are presented.

For the first challenge:

- In terms of detail and to address Material yield and productivity, bin-packing and/or knapsack algorithms are useful tools. Mono or multi criteria models are used for these kinds of algorithms. A lot of research has been published by the OR community on this subject.
- Marking is the easiest way to track and trace the material throughout the SC. But biometric algorithms associated with vision tools could be also used. A lot of research has also been proposed in this area.
- Concerning the use of natural resources/materials, such as wood, vision tools (X-rays, 2D colour scanners, micro-waves or micro-sounds) could be used to highlight/solve defects and quality problems using "on-line processes". Classification models that often use fuzzy logic could lead to the automation

of the evaluation process by a human expert. In doing so, it is possible to propose new ways to minimize material consumption and reduce material waste.

For challenges 4 and 5:

- Linear programming (and its extensions such as Mixed Integer Linear Programming MILP) is a widely used, comprehensive tool for solving vehicle routing problems taking into account greenhouse gas emissions and carbon footprint with the objective of minimizing routes and resource consumption in short and local Supply Chains. Multi-agent bio-algorithms are often implemented for distributed systems as well.
- New information technologies such as RFID, WSN, GPS and so on are commonly used today.

5 How Can IMS Help Improve Sustainability in Manufacturing?

Unquestionably, IMS technologies, architectures, management and control approaches will be useful for sustainable industries. A lot of successful experiments already exist [20], especially at production activity control level, but in a restrictive SD sense. Indeed, classical research concerning IMS applications relating to shop-floors and supply chains resulting in greater industrial efficiency are relevant. But, they only address the economic pillar of SD. Furthermore, in this domain, performance is measured according the lowest level of the 3 pillars (economic, environmental and societal pillars, see Fig. 3)! Finally, we have observed that very few studies in the IMS community actually concern SD in industry with a will to pay equal attention to the three pillars. Some production management research communities, such as the IFIP, pay attention to SD but little consideration is yet given to linking this with IMS approaches (see for example the APMS conference series). On the contrary, IEEE and IFAC production management communities pay a lot of attention to IMS approaches, but little attention to linking this with SD. Maybe the time has come to merge these perceptions. Typically, the concept of PSS (Product – Service System) could be the missing link in the context studied.

It seems obvious that real-time identification, tracking or tracing could improve system efficiency and control, but some interest could also be highlighted at higher levels (strategic and tactical), which concern supply chain design or industrial systems specification, leading to shorter routings in transportation, more economic robustness and so on. On the other hand, from a conceptual point of view and thanks to the capacity of IMS to provide global knowledge of the system state, some other targets could be addressed, such as economic or quality of service robustness.

As concrete examples, we propose the following set of research areas where IMS can provide solutions to improve SD in manufacturing:

During the manufacturing phase, IMS could help handle energy-related issues, namely:

- Controlling the energy in an opportunistic and reactive way,
- Adapting scheduling dynamically to unpredictable energy prices,
- Limiting peak consumption of energy in real time,

- Adapting to the variety and variability of energy sources and power according to the weather conditions (sun, wind).

Throughout the product life cycle, IMS could also help:

- Improve product traceability and maintenance,
- Contribute to the optimization of recycling and re-manufacturing,
- Optimize the re-design of future versions of products (next generation),

It can also be noted that these IMS opportunities could be implemented in the building sector. IMS tools, especially digitalization possibilities, could become a very interesting way of managing products such as buildings in the production/installation phase as well as the usage phase. BIM (Building Information Modelling) or Building Numerical Engineering is a promising concept for IMS applications.

To conclude this chapter, in the next two sections, we provide two illustrative examples dealing with two of the research areas introduced previously.

6   Effective, Energy-Aware Control of a FMS

In this example, SD is concerned with the energy savings during manufacturing in Flexible Manufacturing Systems (FMS). IMS principles are used with the integration of intelligence into both products and resources to dynamically solve the routing and scheduling problem. The merging of the two concepts (IMS and SD) gives birth to a control system able to effectively schedule and route in real time products while at the same time, save energy in an opportunistic way by shut downing energy supply of resources if products do not intend to reach them within a short term window.

More precisely, in this example, routing (by products) and shut down (by resources) decisions are based upon emission and reception of potential fields (PF). Resources emit attractive PFs, which are scanned by the products when a decision must be made. Each resource owns an initial attractiveness value, which is amplified, propagated and altered by the environment between the product and the emitting resource. To allow energy savings, idle resources should shut down their power supply if there is no product around that concerns them within a time window. If a product selects a resource and comes near it, the resource should turn on the power supply. However, to make this behaviour consistent, resources need information about the products around them. This is handled by products emitting a second PF type: when a product has chosen a resource to obtain a service, it emits an intention PF for this resource. A resource can sense all the PFs emitted by the products that have chosen this resource. As a result, the resource decides dynamically to manage the power supply.

First results are promising: energy savings are about of 50% on average for simple cases, from 20% to 40% in large simulations and 32% in experiments with a limited increase in completion times values (3% on average for simple cases, 2% in large simulations and 6% in experiments). For more information about this example, refer to [21].

7   Energy Optimization in Complex Building

In this example, SD is concerned (because other SD indicators are also concerned: material productivity or industrial efficiency, etc...) with energy savings during the manufacturing, erection on site and usage phases. Concerning the manufacturing and erection phases, we face the same challenges as those introduced previously in FMS applications. Concerning the usage phase, IMS principles are used with the integration of intelligence in complex products (buildings), critical resources (cranes, etc...), but also the CAD-based digital mock-ups, to solve problems relating to energy consumption minimization and user service quality. In this context, merging the two concepts (IMS and SD) provides a control system implemented in the mock-up that can schedule the activation of the energy resources according to the local and conjectural usage and conditions. For example, on a university campus, if we consider a building where practical training computer labs exist alongside professors' offices, we can imagine that with 20 students working on 20 computers the energy generated increases the temperature thus causing the boilers to stop, and thus lowering the temperature in the professors' offices!

So, new models are needed which take into account heterogeneous parameters (temperature, humidity, time for increasing or lowering, thermal intertie, CO_2...) with specific constraints and costs. Implementing them in the mock-up could be useful for the building company but also for the company that will manage the building during its usage phase. We are today only at the start of such developments, but nevertheless, some test benches, databases and platforms already exist.

8   Conclusion

In this chapter, the objective was to open the debate on the role of Intelligent Manufacturing and Services Systems in sustainable development and circular economy. We discussed IMS and the interpretations of sustainable approaches in this context. We proposed a framework describing the role of some challenging prospects according to critical sustainable development indicators. This chapter was completed with a set of research challenges which show what IMS can bring to SD and provides two illustrative examples pointing out the potential benefits of using IMS in SD.

Again, it is unquestionable that IMS technologies, architectures, management and control approaches will be useful, in our foreseeable future, for sustainable industries. Thereby, the proposed frameworks could be a first tentative to highlight some channels of innovation and to structure future research. We will present a more detailed analysis in the near future.

Acknowledgments. The authors gratefully acknowledge the financial support of the CPER 2007-2013 "Structuration du Pôle de Compétitivité Fibres Grand'Est" (Fibre Competitiveness Cluster), through local (Vosges General Council), regional (the Lorraine Region), national (DRRT and FNADT) and European (FEDER) funds.

References

1. Sundmaeker, H., Guillemin, P., Friess, P., Woelfflé, S.: Vision and challenges for realising the Internet of Things. Cluster of European Research Projects on the Internet of Things, European Commission (2010)

2. Ley, D., Becta, I.: Ubiquitous Computing. Emerging Technology 2, 64–79 (2007)
3. Bargagli, R., Wynn-Williams, D., Bersan, F., Cavacini, P., Ertz, S., Frati, F., Freckman, D., Lewis-Smith, R., Russell, N., Smith, A.: Field report, Biotex 1: first BIOTAS expedition. Newsletter of the Italian Biological Research in Antarctica 1, 42–58 (1995-1996) (1997)
4. Schwarz, A., Van Langenhove, L.: Report on State-of-the-art on intelligent textiles, Development of a strategic Master plan for the transformation of the traditional textile and clothing into a knowledge driven industrial sector by 2015, Tech. Rep., Clevertex (2006)
5. Park, S., Mackenzie, K., Jayaraman, S.: The wearable motherboard: a framework for personalized mobile information processing (PMIP). In: ACM/IEEE 39th Design Automation Conference, pp. 170–174 (2002)
6. Augusto, J.: Ambient Intelligence: The confluence of ubiquitous/ pervasive computing and artificial intelligence. In: Intelligent Computing Everywhere, pp. 213–234 (2007)
7. Wong, C.Y., McFarlane, D., Zaharudin, A.A., Agarwal, V.: The Intelligent Product Driven Supply Chain. In: International Conference on Systems, Man and Cybernetics, pp. 4–6 (2002)
8. Le Mortellec, A., Clarhaut, J., Sallez, Y., Berger, T., Trentesaux, D.: Embedded Holonic Fault Diagnosis of Complex Transportation Systems. Engineering Applications of Artificial Intelligence 26(1), 227–240 (2013)
9. Thomas, A., Trentesaux, D., Valckenaers, P.: Intelligent distributed production control. Journal of Intelligent Manufacturing 23(6), 2507–2512 (2012)
10. Trentesaux, D.: Distributed control of production systems. Engineering Applications of Artificial Intelligence 22(7), 971–978 (2012)
11. Plossl, W.G.: La nouvelle donne de la gestion de la production – Afnor gestion, Paris (1993)
12. WCED - World Commission on Environment and Development: Our common future, p. 43. Oxford University Press, Oxford (1987)
13. Webster, K., Johnson, K.: Sense & sustainability. Ellen McArthur Foundation (2012)
14. Kiritsuis, D.: Semantic technologies for engineering asset lifecycle management. IJPR 50 (2013)
15. Ehrenfeld, J.: Can Industrial Ecology be the Science of Sustainability? Journal of Industrial Ecology 8, editorial, 1–3 (2004), doi:10.1162/1088198041269364
16. Garetti, M., Taisch, M.: Sustainable manufacturing: trends and research challenges. Production Planning and Control 23(2-3), 83–104 (2012)
17. Herrera, C., Berraf, S.B., Thomas, A.: Viable System Model Approach for Holonic Product Driven Manufacturing Systems. In: Borangiu, T., Thomas, A., Trentesaux, D. (eds.) SOHOMA 2011. SCI, vol. 402, pp. 169–182. Springer, Heidelberg (2012)
18. Karnouskos, S., Colombo, A.W., Lastra, J.L.M., Popescu, C.: Towards the energy efficient future factory. In: 7th IEEE Int Conf on Industrial Informatics, pp. 367–371 (2009)
19. Sallez, Y., Berger, T., Deneux, D., Trentesaux, D.: The Life Cycle of Active and Intelligent Products: The Augmentation concept. International Journal of Computer Integrated Manufacturing 23(10), 905–924 (2010)
20. Pannequin, R., Morel, G., Thomas, A.: The performance of product-driven manufacturing control: An emulation-based benchmarking study. Computers in Industry, 195–203 (2009)
21. Pach, C., Berger, T., Sallez, Y., Trentesaux, D.: Effective, energy-aware control of a production system: a potential fields approach. In: 11th IFAC Workshop on Intelligent Manufacturing Systems, São Paulo, Brazil, pp. 130–135 (2013)

Distributed Feedback Control for Production, Inventory, and CO_2 Emissions in an Assemble-To-Order System

Seokgi Lee and Vittaldas Prabhu

Harold and Inge Marcus Department of Industrial and Manufacturing Engineering,
The Pennsylvania State University, University Park, PA 16802 USA
sul201@psu.edu, prabhu@engr.psu.edu

Abstract. We study continuous variable feedback control of an assemble-to-order system with multiple components and multiple workstations to analyse interrelationships among the production system and corresponding CO_2 emissions. The proposed dynamic models are designed by proportional and integral control laws, and represent assembly job arrivals along with the component consumption rates at each workstation for controlling finished-goods assembly schedule, and component production rate for controlling component stock levels, respectively. We also develop the unified feedback control algorithm whose objective is to minimize due date deviation from the customer-requested final assembly due date and component inventory discrepancy in respect to the optimum inventory level, simultaneously. Using numerical simulations, we show how dynamics between the inventory and production systems are interrelated, and resulting CO_2 emission variation by the production system.

Keywords: continuous variable feedback control, CO_2 emissions, production schedule, machine capacity, inventory control.

1 Introduction

The importance of identifying and quantifying sources of greenhouse gases (GHG) has increased dramatically in view of the need to actively address recent and future climate change. It was shown that the primary GHG emitted by many human activities is carbon dioxide (CO_2), which accounts for approximately 83.6% (53 billion tons) of total GHG emissions in the United States. The main source of CO_2 is fossil fuel combustion, which accounted for 94.4%, along with other sources, including natural gas systems, iron and steel production, petrochemical production [1]. Especially in the industrial sector, which accounts for 14.4% of fossil fuel combustion, a huge effort has been made to reduce CO_2 emissions. This has been done through energy-saving measures, such as applying energy-saving equipment focusing on motive-power facilities, increasing efficiencies through revised manufacturing processes and management changes, promoting the measurement and visualization of energy consumption and proactive use of that data, and using natural energy sources, such as solar power [2].

T. Borangiu et al. (eds.), *Service Orientation in Holonic and Multi-Agent Manufacturing and Robotics*, Studies in Computational Intelligence 544,
DOI: 10.1007/978-3-319-04735-5_2, © Springer International Publishing Switzerland 2014

Although various GHG reduction initiatives have been developed by industries and governments, research from a daily or hourly operational perspective is still limited and has just begun to receive attention in several areas: production, inventory, and transportation. In inventory management, for example, the optimal order quantity regarding impacts of trading and emitting CO_2 were analytically examined [3] and the environmental importance of logistics and inventory management was explained [4]. Regarding both production and inventory system, the effects of the introduction of an emission trading program, i.e., an environmental license, on the production-inventory strategy of a firm were analysed [5]. Also, Jaber et al. [6] proposed a two-level, vendor-buyer supply chain, mathematical model to optimize the joint production-inventory policy regarding CO_2 emissions cost, penalties for exceeding emissions limits, and inventory-related costs. Most of the previous research has focused on production and inventory policy or planning perspectives, but there is as yet no literature considering daily or hourly controls, except for a few specific industry or operations areas, including energy-saving computing and green transportation scheduling systems [7][8].

In this research, the continuous-time variable feedback control model for real-time control of a production schedule, machine capacity, and an inventory level in a unified manner is proposed, and the relationship between production performance and the cost of CO_2 emissions is explained. The control criteria of production and inventory systems are minimizing customer demanded due-date deviations and guaranteeing safety stocks of inventories for demand uncertainty. Dynamics among those three functions are analytically examined, and production schedules with corresponding production rates and inventory levels, considering CO_2 emissions, are suggested. The organization of this research is as follows: in section 2, dynamic models of production and inventory system are developed. In section 3, dynamics of proposed models are presented in detail. Lastly, performance of production and inventory controls is examined in section 4 using numerical examples.

2   Problem Definition

We consider Assemble-to-Order (ATO) production systems ranging from the component production line to the final assembly line. n finished goods are assembled from w different components that are produced on either a part of the assembly line or an outsourcing facility in a make-to-stock fashion. When the component production order of the component j is placed, the component production rate, θ_{Bj}, is also specified so as to maintain its optimum stock level θ_{Ij}^* in a component warehouse.

In the assembly line, finished goods have different assembly sequences according to their different component combinations requested by customers. We assume that each of the assembly workstations has a unique assembly job by different components, so w different workstations need w different components (or assembly jobs). Each of workstation j (for the component j's assembly job) is initialized to operate by the nominal assembly processing time, $p_{i,j}$, for the finished good i, that can be flexibly adjusted within a specific processing time range allowed by a workstation capacity limit. The finished good i has own due date, $d_{i,|J_i|}$, indicating the time when the final assembly has to be completed by the final workstation. Because of these assembly production characteristics, we assume that assembly production follows the job shop production system.

Based on this ATO production environment, the objective of the proposed dynamic models is to represent interrelationship among (i) a production order release time, (ii) a production rate, and (iii) a finished-goods-inventory (FGI) level, regarding a customer demanded due date, a desired throughput to guarantee a specific safety stock level, and CO_2 emissions.

The cost function of the proposed models consists of two factors: (i) penalty cost of due date deviation, ε (deviation unit/unit time), and (ii) the amount of CO_2 emissions per unit (emission unit/unit), which can be represented by the function of the production rate r (unit/unit time) such that

$$e_1 r^2 - e_2 r + e_3 \tag{1}$$

where e_1, e_2, and e_3 are emissions function parameters whose units are emission unit $\cdot$ (unit time)^2/unit^3 , emission unit $\cdot$ unit time/unit^2 , and emission unit/unit, respectively [6].

Consequent CO_2 emission cost per year (dollar/unit time) is calculated by

$$(e_1 r^2 - e_2 r + e_3) d C_{ec} \tag{2}$$

where d is demand rate (unit/unit time) and C_{ec} is emission tax (dollar/emission unit).

Furthermore, to assess performance of the ATO production system, the mean-squared due-date deviation (MSD) is measured for Just-in-Time (JIT) performance of assembly scheduling:

$$\text{MSD} = \sqrt{\frac{\sum_{i=1}^{n} \sum_{j=J_i^{[1]}}^{J_i^{[|J_i|]}} \left(d_{i,j} - c_{i,j}(t)\right)^2}{\sum_{i=1}^{n} |J_i|}}. \tag{3}$$

Finally, total cost of the proposed model is given by

$$\text{TC (dollar/unit time)} = \alpha \text{MSD} + (e_1 r^2 - e_2 r + e_3) d C_{ec} \tag{4}$$

where α is the penalty cost of due date deviation (dollar/MSD unit).

3   Dynamic Models

In this section, we propose a feedback control model for integrating the functions of (i) production scheduling, (ii) machine capacity, and (iii) inventory stock level. For production and inventory controllers, specifically, the integral controller is used to adjust the arrival time of production jobs at workstations by penalizing earliness and tardiness of jobs, while simultaneously controlling machine capacity and inventory levels. The following notations are used in models:

J	Set of workstations (or equivalently components or jobs), $J = \{1, \dots, w\}$
$\mathcal{N}$	Set of finished goods, $\mathcal{N} = \{1, \dots, n\}$
i	Index of a finished good, $i \in \mathcal{N}$
j	Index of a workstation (or equivalently components or jobs), $j \in J$
J_i	Subset of J for the finished good i, $J_i \subseteq J$
$J_i^{[u]}$	The uth assembly job (or component or workstation) to produce the finished good i, $J_i^{[u]} \in J_i$, $u = 1, 2, \dots, |J_i|$
$\mathcal{N}_j$	Subset of $\mathcal{N}$ that needs the component j (or job or workstation j) to be assembled, $\mathcal{N}_j \subseteq \mathcal{N}$
θ_{Ij}	On-hand component inventory of the component j
θ_{Ij}^*	Optimum component inventory level of the component j
ξ_j	Planned inventory time of the component j
θ_{Bj}	Component production rate of the component j (unit/unit time)
θ_{Pj}	Component consumption rate of the component j (unit/unit time)
$d_{i,j}$	Due date of the component j's assembly job of the finished good i
$a_{i,j}$	Arrival time of the component j's assembly job of the finished good i
$c_{i,j}$	Completion time of the component j's assembly job of the finished good i
$p_{i,j}$	Processing time of the component j's assembly job of the finished good i

3.1   Production Controller

The basic mechanism of the production controller is that $a_{i,j}(t)$ is adjusted in proportion to the size of the discrepancy between $d_{i,j}(t)$ and $c_{i,j}(t)$. Thus, the arrival time dynamic in the production controller is explained by

$$\dot{a}_{i,j}(t) = k_{P_{i,j}} \left(d_{i,j}(t) - c_{i,j}(t) \right) \tag{5}$$

where $k_{P_{i,j}}$ is the control gain.

The predicted job completion time of the job j of the finished good i is given by

$$c_{i,j}(t) = a_{i,j}(t) + q_{i,j}(t) + p_{i,j}(t) \tag{6}$$

where $q_{i,j}(t)$ is the queuing time which is calculated by $q_{i,j}(t) = a_{1,j}(t) + \sum_{s=1}^{i-1} p_{s,j}(t) - a_{i,j}(t)$ if $a_{i,j}(t) < a_{1,j}(t) + \sum_{s=1}^{i-1} p_{s,j}(t)$, otherwise zero.

Here, failure of assembly facilities and corresponding repair times are assumed to be not considered to keep the dynamic model simple.

Substituting Equation (7) into (6) yields

$$c_{i,j}(t) = a_{1,j}(t) + \sum_{s=1}^{i} p_{s,j}(t). \tag{7}$$

It should be emphasized that changing the sequence of arrival times incurs variations of the assembly sequence at a workstation due to FIFO discipline as we mentioned above, and finally discontinuous changes of the completion time. Thus, the completion time in Equation (7) is a highly nonlinear function of the arrival time with discontinuities, even though it appears to be affected only by the first arrival of a job. Using feedback in a local arrival time controller for each job, these nonlinearities can be ignored and aggregated in a local arrival time controller without any global information [9]. Hence, we use the following integral control law for arrival time control:

$$a_{i,j}(t) = k_{Pi,j} \int_0^t \left(d_{i,j}(\tau) - c_{i,j}(\tau) \right) d\tau + a_{i,j}(0), \tag{8}$$

where $a_{i,j}(0)$ is an arbitrary initial condition.

3.2   Machine Capacity Control

The objective of the capacity controller is to dynamically adjust the assembly processing time of a job at a workstation, considering a time constraint assigned to each assembly job. We call the control for adjusting component assembly processing time as the capacity control, because the maximum amount of finished goods to be produced is determined by the assembly processing time. Specifically, the capacity controller adjusts the assembly processing time from its nominal value in proportion to the amount of earliness or tardiness, i.e., due date deviation of the assembly job. Here, we assume that the assembly processing time includes only the actual assembly operation time of a component, and other time factors, such as the tool replace time, the machine repair time, and the setup time are not considered in the model.

Thus, the capacity controller for the job j of the finished good i is given by

$$\dot{p}_{i,j}(t) = k_{Ci,j} \left(d_{i,j}(t) - c_{i,j}(t) \right), \tag{9}$$

where the $d_{i,j}(t) - c_{i,j}(t)$ term represents the earliness or tardiness of the assembly job j of the finished good i, and $0 < k_{Ci,j} < 1$ is a given positive real number making $d_{i,j}(t) - c_{i,j}(t)$ less than unity.

It should be noted that, based on the control logic in Equation (9), the capacity controller decreases (or increases) the nominal assembly processing time when tardiness of the assembly job occurs, i.e., $d_{i,j}(t) < c_{i,j}(t)$, (respectively, when earliness occurs, i.e., $d_{i,j}(t) > c_{i,j}(t)$), and this capacity-controlled assembly processing time has an effect of reducing the total due date deviation in the production controller that will be explained in the next section.

3.3   Inventory Control

The inventory controller aims to keep the on-hand component inventory at its optimum level that is assumed to be given by demand forecast at the master schedule phase. The on-hand inventory level of the component j, $\theta_{Ij}(t)$, varies by the input and output flows of a component, that is, the component production rate, $\theta_{Bj}(t)$, and the component consumption rate, $\theta_{Pj}(t)$, controlled by the component production line and the assembly line, respectively.

In particular, $\theta_{Pj}(t)$ is directly decided by the aggregate assembly processing time for all finished goods that should be processed at the workstation j, such that

$$\frac{|\mathcal{N}_j|}{\sum_{i\in\mathcal{N}_j}\left(p_{i,j}(t)+q_{i,j}(t)\right)}, \tag{10}$$

where $|\mathcal{N}_j|$ is the total number of finished goods to be processed at the workstation j.

Assuming that the assembly schedule at the workstation j decided by the capacity and production controllers in (9) and (5) is repeated for the planned inventory time ξ_j, $\xi_j > \sum_{i\in\mathcal{N}_j}\left(p_{i,j}(t)+q_{i,j}(t)\right)$, the average component consumption rate over ξ_j is given by

$$\theta_{Pj}(t) = \frac{\xi_j|\mathcal{N}_j|}{\sum_{i\in\mathcal{N}_j}\left(p_{i,j}(t)+q_{i,j}(t)\right)}. \tag{11}$$

Here, ξ_j is defined as the waiting time of a finished good in a distribution center before shipping to a customer to protect against uncertain demand.

Now, we can model the inventory system as follows:

$$\theta_{Ij}(t) = \xi_j\left(\theta_{Bj}(t) - \theta_{Pj}(t)\right), \tag{12}$$

$$\dot{\theta}_{Bj}(t) = k_{Ij}\left(\theta_{Ij}^*(t) - \theta_{Ij}(t)\right). \tag{13}$$

where k_{Ij} is the control gain for the inventory control.

The average on-hand component inventory over time ξ_j is determined by the component input and output flows as represented in Equation (12). Equation (13) represents a rule of decision, specifying the component production rate as a function of the excess and deficiency of component inventory. If the average on-hand component inventory is expected to be less than the optimum level during ξ_j, the component production rate should increase to make up the inventory deficiency, and therefore, $\dot{\theta}_{Bj}(t) > 0$. The opposite case is also well established.

Furthermore, we choose the integral control law for controlling $\theta_{Bj}(t)$ that has the following integral form:

$$\theta_{Bj}(t) = k_{Ij}\int_0^t \left(\theta_{Ij}^*(\tau) - \theta_{Ij}(\tau)\right)d\tau + \theta_{Bj}(0), \tag{14}$$

where $\theta_{Bj}(0)$ is the arbitrary initial component production rate.

Unifying all the proposed controllers for inventory, production, and capacity is enabled by the unified feedback control algorithm. Detail procedures of the algorithm and analysis will be explained in section 4.

3.4   Capacity of Component Production and Assembly Systems

Controlling capacity of a workstation, i.e., the assembly processing time, adjusts the nominal processing time within the upper and lower limits that are predefined to a workstation, and noted as $p_{MINi,j}$ and $p_{MAXi,j}$, respectively. Similarly, the component production rate also needs to be capacitated by its upper and lower limits of a component production facility. Assuming that these values are basically given by the component production system, we represent the upper and lower limits of the component production rate as θ_{BMAXj} and θ_{BMINj}, respectively.

Substituting Equation (12) into (13) and solving the first order differential equation with respect to $\theta_{Bj}(t)$ yields the solution of $\theta_{Bj}(t)$ for each element j as follows:

$$\theta_{Bj}(t) = \frac{\theta_{Ij}^*(t) + \xi_j \theta_{Pj}(t)}{\xi_j}\left(1 - e^{-k_{Ij}\xi_j t}\right) + \theta_{Bj}(0)e^{-k_{Ij}\xi_j t}, \qquad (15)$$

where $\theta_{Bj}(0)$ is the arbitrary initial production rate of the component j.

If $t \to \infty$, the right-hand side of Equation (15) becomes

$$\theta_{Bj}(t) = \frac{\theta_{Ij}^*(t) + \xi_j \theta_{Pj}(t)}{\xi_j} = \theta_{Bj}^*(t) + \theta_{Pj}(t), \qquad (16)$$

where $\theta_{Bj}^*(t)$ is the average component production rate over ξ_j that can fulfill $\theta_{Ij}^*(t)$ when no output flow of a component exists. For example, if $\theta_{Ij}^*(t) = 300$ and $\xi_j = 10$, then $\theta_{Bj}^*(t)$ should be 30 to maintain the given optimum stock level.

Applying the upper and lower capacities of a component production facility yields the following relations

$$\theta_{BMINj} \leq \theta_{Bj}(t) = \theta_{Bj}^*(t) + \theta_{Pj}(t) \leq \theta_{BMAXj}, \qquad (17)$$

and therefore,

$$\theta_{BMINj} - \theta_{Bj}^*(t) \leq \theta_{Pj}(t) \leq \theta_{BMAXj} - \theta_{Bj}^*(t), \qquad (18)$$

indicating the upper and lower capacity limits of a workstation j.

Finally, by Equation (11), the aggregate assembly processing time at the workstation j over ξ_j should satisfy the following condition:

$$\frac{\xi_j |N_j|}{\theta_{BMAXj} - \theta_{Bj}^*(t)} \leq \sum_{i \in N_j} p_{i,j}(t) \leq \frac{\xi_j |N_j|}{\theta_{BMINj} - \theta_{Bj}^*(t)} \qquad (19)$$

that should be considered as the capacity condition of each assembly workstation.

4   Computational Experiments

In this section, we show interrelated dynamics among the capacity, production, and inventory control systems, and analyse how the unified control approach influences on system performance that can be evaluated by numerical simulation. As noted in section 3.3, a chronological sequence of the arrival times for assembly jobs decides a final assembly schedule; it can be obtained by the unified control algorithm, which is performed based on a discrete event simulation. For this simulation, the controllers in Equation (8), (9), and (14) are needed to be executed on a digital computer, and thereby being approximated by discretizing in time as follows:

$$p_{i,j}(T) = k_{Ci,j} p_{i,j}(m-1)\left(d_{i,j}(m-1) - c_{i,j}(m-1)\right) + p_{i,j}(m-1) \quad (20)$$

$$a_{i,j}(T) = k_{Pi,j}\bar{T}\left(d_{i,j}(m-1) - c_{i,j}(m-1)\right) + a_{i,j}(m-1) \quad (21)$$

$$\theta_{Bj}(T) = k_{Ij}\bar{T}\left(\theta_{Ij}^*(m-1) - \theta_{Ij}(m-1)\right) + \theta_{Bj}(m-1) \quad (22)$$

where $p_{i,j}(m-1)$, $a_{i,j}(m-1)$, $d_{i,j}(m-1)$, $c_{i,j}(m-1)$, and $\theta_{Bj}(m-1)$ are the assembly processing time, the arrival time, the due date, the assembly completion time, and the component production rate at the $(m-1)$th time step, respectively, and $\bar{T}$ is the time step for integration. The proposed algorithm is designed to perform all functions of capacity, production, inventory controls simultaneously, and simulates dynamics of all above discrete time controllers for n finished goods at w workstations. The high-level description of the algorithm is outlined as follows:

The unified feedback control algorithm

STEP 1:   Update arrival times, assembly processing times, and due date

for $j = w$ to 1

for $i = 1$ to n

if $j \in J_i$ then

$a_{i,j}(m) = k_{Pi,j}\bar{T}\left(d_{i,j}(m-1) - c_{i,j}(m-1)\right) + a_{i,j}(m-1)$

$p_{i,j}(m) = p_{i,j}(m-1) + k_{Ci,j}p_{i,j}(m-1)\left(d_{i,j}(m-1) - c_{i,j}(m-1)\right)$

$d_{i,j-1}(m) = a_{i,j}(m) - \pi_{j,j-1}(m)$

STEP 2:   Update component inventory production rate and inventory stock level

$\theta_{Pj}(m) = \xi_j |\mathcal{N}_j| / \sum_{i \in \mathcal{N}_j} p_{i,j}(m)$

$\theta_{Bj}(m) = k_{Ij}\bar{T}\left(\theta_{Ij}^*(m-1) - \theta_{Ij}(m-1)\right) + \theta_{Bj}(m-1)$

STEP 3:   Build new assembly schedule by sorting arrival times based on FCFS rule

STEP 4:   Compute completion times based on new schedule

STEP 5:   Calculate total cost

STEP 6:   Go to **STEP 1**

To illustrate dynamic interactions among three controllers and show performance of the unified feedback control algorithm above, the following data sets consisting of 2 finished goods and 3 workstations in job-shop manufacturing environment are tested to analyse dynamics among three controllers, and initial conditions are described in Fig. 1. Each of finished goods has different pre-defined processing routes that basically satisfy the master assembly sequence, $1\rightarrow2\rightarrow3$. The first and second finished goods have the assembly sequence $1\rightarrow3$ and $1\rightarrow2\rightarrow3$, respectively. As a result, there exist 5 different combinations of (i, j) for $i = 1,2$ and $j = 1,2,3$, and therefore we can obtain five different arrival, completion, and processing time variations.

Due dates of each finished good at the final workstation are set to 20.0 and 24.8, respectively. It should be noted that such a tight condition of due dates make the production system infeasible, because the assembly processing time at the workstation 3 $(p_{i,3})$ takes 5.0 unit time for both jobs, resulting in one of two assembly jobs always being not able to keep its final due date. In the arrival time system point of view, such infeasibility makes the arrival time vector in the $R^{|N_j|}$ space move and eventually remain in the dead-zone region where more than two jobs interact with each other, causing positive queuing times [9].

For the simulation, a total of 1000 iterations are performed and ξ_j is set to 100. Control gains of each controller are set as $k_{Pi,j} = 0.01$, $k_{Ij} = 0.001$, and $k_{Ci,j} = 0.005$, so that trajectories of the arrival time and the inventory stock level take a form of curvlinear and their steady states can be obtained at around 500 iterations. The capacity control is set to start after the 800th iteration and the capacity limits of both the component production and consumption rate are not specified in this example.

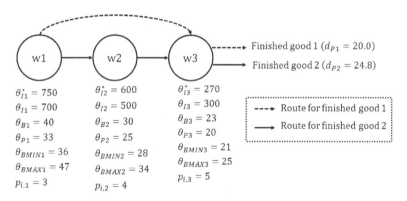

Fig. 1. 3-workstation and 2-finished good case

Fig. 2 illustrates variations of assembly processing times by the capacity controller, and corresponding changes of job arrival times. As mentioned above, the given problem is supposed to initially yield the infeasible production schedule due to the tight due-date condition at the final workstation, and therefore the job arrival times at the final workstation, i.e., $a_{1,3}$ and $a_{2,3}$ (red solid lines), conflict with each other at the beginning of the simulation as shown in Fig. 2(b). However, after the 800th iteration, the assembly processing times of each job at the final workstation, i.e., $p_{1,3}$ and $p_{2,3}$, gradually decrease from 5.0 to 4.5 as illustrated in Fig. 2(a), resulting in complete separation of $a_{1,3}$ and $a_{2,3}$ as shown in Fig. 2(b) and finally leading the conversion of the initially infeasible production schedule into the feasible schedule. These system variations can be more clearly understood by representing the movement of the arrival time vector in the arrival time space as shown in Fig. 2(c). The arrival time vector of $a_{1,3}$ and $a_{2,3}$ moves in a certain direction, but still remains in the dead-zone region (light red area) until the capacity controller is applied. After the 800th iteration, however, it changes its direction and finally moves out from the dead-zone region, meaning that the production schedule changes to be feasible. The final arrival time values obtained at the 3000th iteration for each workstation are shown in Table 1, indicating that the assembly sequence at the workstation 1 and 2 are decided as "assembly job of the first finished-good" → "assembly job of the second finished-good", identically.

Table 1. Arrival times of each job at each workstation

Workstation 1		Workstation 2	Workstation 3	
$a_{1,1}$	$a_{2,1}$	$a_{2,2}$	$a_{1,3}$	$a_{2,3}$
11.87	11.88	16.58	15.50	20.27

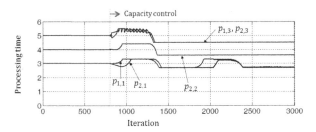

(a) Processing time variation by capacity control

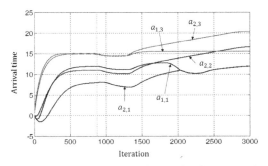

(b) Arrival time variation by production control

Fig. 2. Effect of capacity and production control

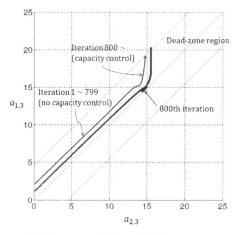

(c) Movement of the arrival time vector

Fig. 2. (*continued*)

In a component inventory control point of view, adjusting assembly processing times by the capacity controller causes variations of the component consumption rates as shown in Fig. 3(a), incurring unexpected changes of the component stock levels. At this moment, the inventory controller adaptively adjusts the component production rate, as illustrated in Fig. 3(b), according to excess and deficiency of component inventory resulted from variation of the component consumption rate, and consequently the component stock level maintains almost the optimum stock level (the red dot line) as described in Fig. 3(c).

Fig. 4 illustrates how the performance of the ATO production system improves in terms of MSD. As illustrated in Fig. 4(a), MSD converges into 0.01as the time-average of the due-date deviation becomes zero by the production controller. In case of the component inventory, there are several peak observed in the trajectory in Fig. 4(b) due to complementary system behaviours between the production and inventory systems, but very small average inventory discrepancy which is below 5 units of component can be eventually obtained by the unified feedback control algorithm.

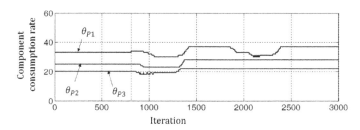

(a) Component consumption rate variation

Fig. 3. Variations of component production/consumption rates and inventory

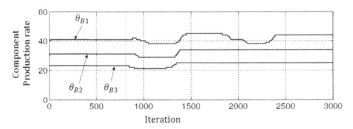

(b) Component production rate variation

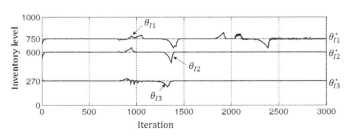

(c) Inventory variation

Fig. 3. (*continued*)

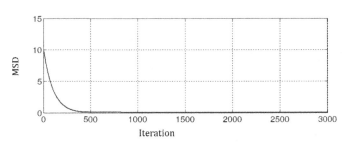

(a) MSD variation

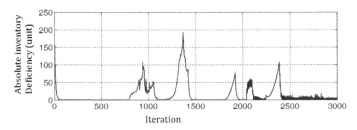

(b) Absolute inventory deficiency variation

Fig. 4. System performance

Lastly, we also investigate how the costs of due date deviation (JIT cost) and CO_2 emission vary over the simulation time. JIT cost and emission tax are set to $10.0 per MSD unit, $18.0 per ton, respectively. Demand rate is also set to 1000 units per year and last of parameters used in emission cost (2) are set as $e_1 = 3 \times 10^{-7}$, $e_2 = 0.012$, and $e_3 = 1.4$, respectively. The production rate is simply calculated by average of component consumption rates of three workstations.

Variations of the JIT cost and the CO_2 emission cost are illustrated in Fig. 5. Specifically, MSD cost reduction shown in Fig. 5(a) results in JIT cost reduction as shown in Fig. 5(a), and variation of CO_2 emission cost by dynamic adjustment of production rate to minimize MSD is illustrated in Fig. 5(b). Consequent total cost reduces in this example as shown in Fig. 5(c). Currently, the integral controller of the production system is designed to minimize the JIT cost, and variation of production rate is a corresponding result, meaning that a control function for reducing the CO_2 emission cost is not aggressively included in the feedback control algorithm. However, dynamics of MSD, production rate, and CO_2 emissions, and their interrelationships explained by dynamic models can help to finally control CO_2 emissions, for example, by flexibly limiting upper and lower ranges of production rate of a workstation.

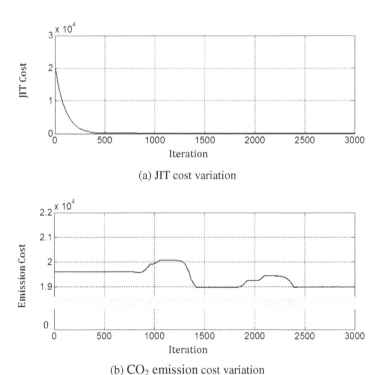

(a) JIT cost variation

(b) CO_2 emission cost variation

Fig. 5. Cost variation

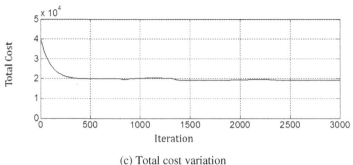

(c) Total cost variation

Fig. 5. (*continued*)

5   Conclusion

This work has shown dynamics of multi-functional control combining component inventory control at the manufacturing supply chain level, production scheduling at the shop floor level, and machinery capacity control at the machining level, regarding JIT performance and Eco-performance driven by CO_2 emissions. In a conventional manufacturing planning and control system, these functions are usually performed by different decision-making phases in a top down manner, and therefore frequently changed shop-floor environment and demands are hard to be reflected again on overall management of resource capacity and production planning. This decision-making structure causes the inaccuracy of production information, yielding unexpected costs on inventory and manufacturing execution. Thus, analysis of dynamics and the attempt to make an integrated decision over these functions have a significant role in improving production system as well as other sub-systems.

To do this, we developed continuous variable feedback control models representing assembly job arrivals along with the component consumption rates at each workstation for setting a finished-good assembly schedule, and the component production rates for managing component stock levels, respectively. Specifically, dynamics of each controller were designed by proportional and integral control laws that represent assembly job arrivals along with the component consumption rates at each workstation, and the component production rates, respectively. We also developed the unified feedback control algorithm to decide the assembly schedule, capacities of workstations, and the component production rates in such a way to minimize due date deviation of assembly jobs and CO_2 emissions by dynamically adjusting production rate, simultaneously.

Analysis of such multi-function dynamics has significant practical meaning for production managers, because the unified control approach provides the possible way to make an integrated decision considering shop-flow dynamics and production planning, simultaneously. Inaccuracy of inventory by frequently changed customer demands can be minimized by real-time information sharing and adaptive decision making by the proposed approach. Not only performance of production and inventory systems, green enabled production system by reducing CO2 emissions can make these systems more profitable from a long-term point of view. Extending this unified control concept, mathematical frameworks proposed in this paper need to be extended across transportation among distributors and delivery for customers in order to reflect real dynamic situations in complex production and supply chain networks. Furthermore, the corresponding predictive analytical models must be derived, so that more widely integrated decision-making with good global performance can be achieved.

References

1. U.S. EPA: Inventory of U.S. Greenhouse Gas Emissions and Sinks: 1990 to 2010. U.S. Environmental Protection Agency (2012)
2. Turner, V., Bigliani, R., Ingle, C.: Reducing Greenhouse Gases Through Intense Use of Information and Communication Technology: Part 1. IDC White Paper (2009)
3. Hua, G., Cheng, T.C.E., Wang, S.: Managing Carbon Footprints in Inventory Management. International Journal of Production Economics 132(2), 178–185 (2011)
4. Bonney, M., Jaber, M.Y.: Environmentally Responsible Inventory Models: Non-Classical Models for a Non-Classical Era. International Journal of Production Economics 133(1), 43–53 (2011)
5. Dobos, I.: Tradable Permits and Production-Inventory Strategies of the Firm. International Journal of Production Economics 108(1-2), 329–333 (2007)
6. Jaber, M.Y., Glock, C.H., El Saadany, A.M.A.: Supply Chain Coordination WithEmission Reduction Incentives. International Journal of Production Research, 1–14 (2012)
7. Truong, V.T.D., Sato, Y., Inoguchi, Y.: Performance Evaluation of a Green Scheduling Algorithm for Energy Savings in Cloud Computing. In: IEEE International Symposium on Parallel & Distributed Processing, Workshops and Ph.D. Forum (IPDPSW), pp. 1–8 (2010)
8. Lee, S., Prabhu, V.V.: Simulation-based Control for Green Transportation WithHigh Delivery Service. In: Winter Simulation Conference, pp. 2046–2056 (2010)
9. Prabhu, V.V., Duffie, N.A.: Nonlinear Dynamics in Distributed Arrival Time Control of Heterarchical Manufacturing Systems. IEEE Transactions on Control Systems Technology 17(6), 724–730 (1999)

Holonic Condition Monitoring and Fault-Recovery System for Sustainable Manufacturing Enterprises

Sobhi Mejjaouli[1] and Radu F. Babiceanu[2]

[1] Department of Systems Engineering
University of Arkansas at Little Rock, Little Rock, AR 72204, USA
sxmejjaouli@ualr.edu
[2] Department of Electrical, Computer, Software, and Systems Engineering
Embry-Riddle Aeronautical University, Daytona Beach, FL 32114, USA
babicear@erau.edu

Abstract. Relatively new technologies such as sensor networks and automated identification, and increased computational power available through distributed computing offer the advantage of real-time monitoring of manufacturing resources and production orders. This work integrates the distributed manufacturing concept, largely used previously in constructs such as agent-based and holonic manufacturing systems with the new sensor network and distributed computing technologies embedded within the manufacturing system for condition monitoring and fault-recovery purposes. Failure, mode, and effects analysis method is also considered in the model. Through this integration the simulated operations of the proposed manufacturing enterprise model receive more visibility, flexibility, and agility and, ultimately, provide a sustainable enterprise model able to address in real-time the uncertainties of the manufacturing environment for the entire lifetime of the manufactured products, as well as the life-cycle of the manufacturing equipment.

Keywords: real-time monitoring, distributed manufacturing, agent-based and holonic modelling and simulation, sustainable enterprise systems.

1 Introduction

The fierce global competition that manufacturing industry is facing in the last decade asks for new solutions to complete all orders in the shortest amount of time, with the lowest budget, and while maintain a high level of quality. Requirements such as strict deadlines, keeping low inventories, uncertain demand, standardization of manufacturing processes, products diversity, increased production control, and resource limitation changed the operational mode of manufacturing systems, with increased challenges for optimization of the operations within the manufacturing system and across the supply chain. Classic problems related to manufacturing optimization, such as production planning, job scheduling, and assignment of resources in the presence of different uncertain variables such as stochastic and diverse demand and resource unavailability (e.g., machine breakdowns) cannot be addressed efficiently anymore

T. Borangiu et al. (eds.), *Service Orientation in Holonic and Multi-Agent Manufacturing and Robotics*, Studies in Computational Intelligence 544,
DOI: 10.1007/978-3-319-04735-5_3, © Springer International Publishing Switzerland 2014

using analytical and mathematical models. Enhancing the manufacturing environment for more visibility and better control of the production processes by integrating new technologies presents opportunities for the manufacturing domain.

Distributed intelligent manufacturing literature abounds of multiple types of agent-based and holonic frameworks and constructs, which present the advantages as well as disadvantages of adopting such an operational mode. Though for large enterprise systems the proposed models outperform the classical centralized manufacturing schemes, the real-world has yet to see a significant penetration of such systems. This reduced adoption of distributed agent-based and holonic systems in manufacturing operations is a result of varied reasons, which mostly deal with the cost of replacing current shop-floor and control process equipment. Sensor networks are already used for condition monitoring for a large spectrum of applications. Individual sensors provide feedback control in manufacturing applications, by monitoring parameters such as tool temperature and vibration levels, as well as machine-tool monitoring. Undetected wear of tooling and of the machine tools leads to low surface quality, many times resulting in scrapping the raw material. Any potential safety issues can also be monitored by individual sensors deployed on machine tools.

This work continues the research started by the authors and reported in [1]. It complements the distributed intelligent manufacturing concept with relatively new technologies, such as sensor networks and distributed computing, in an attempt to add more value to the entire distributed manufacturing enterprise concept and raise the attention about its potential benefits in uncertain environments. A network of wireless sensors is proposed to be embedded within the physical resources of the manufacturing enterprise. Just as an agent-based or holonic manufacturing system, the sensor network collects, analyses data, and exchanges information with other entities in the system, using computational agent-based and holonic distributed schemes. The performance of the proposed holonic model, enhanced with preventive maintenance capabilities, featuring a real-time condition monitoring and fault-recovery mechanism, is compared with that of a holonic system that uses the traditional corrective maintenance approach. As a further enhancement of the preventive maintenance capabilities, the proposed real-time condition monitoring and fault-recovery mechanism also includes a Failure, Mode, and Effects Analysis (FMEA) module that provides data related to the likelihood and severity of faults.

2   Literature Review

Pereira and Carro [2] propose an agent-based manufacturing systems consisting of a set of autonomous, intelligent, and goal oriented units that can benefit from large-scale distributed real-time embedded systems. One of the systems considered is a wireless sensors network, where the sensor nodes cooperate effectively to reach their global objective by using coordinated decision-making. Flexible and agile manufacturing, which provide the ability to react to expected or unexpected changes and uncertainties is addressed by Stecke [3]. The author considers that flexible manufacturing problems are not restricted to scheduling, planning and design problems only, but they also include the control problems associated with the continuous monitoring of systems and tracking of production orders which must meet specified requirements and due dates.

The holonic manufacturing systems concept presented by van Brussel *et al.* [4] states that its goal is to obtain in manufacturing the benefits that holonic organization provides to living organisms and societies. Holonic systems consist of a set of entities, called holons, which are strongly connected with the physical level devices [5-7]. The holons are viewed as building blocks of the entire system architecture, consisting of an information processing part and a physical processing part. The holons execute all sorts of manufacturing operations ranging from material transformation processes, assembly processes, loading, unloading, and transportation processes, storing and inventory build-up processes, or just information processing related to the physical entities part of the system. A holon is characterized by the following two mandatory characteristics [4]:

- *Autonomy*, defined as the capability to make decisions and execute the selected plans; and,
- *Cooperation*, defined as the relationship with other holons in the system to develop mutual plans and execute them with respect to basic rules that limits their autonomy in order to achieve a goal or mutual objectives.

The literature review clearly identified these characteristics, on the one hand, as being of utmost importance for the operation of holonic manufacturing systems. On the other hand, the operation of holonic manufacturing systems, and of any other manufacturing systems for that matter, cannot be sustained without monitoring and control of the manufacturing operations. In the case of a holonic system, which attempts to provide real-time routing and scheduling of production orders and resources, the manufacturing and control system need to provide the information about the operational status of both production and resources and make it available in real-time for the decision-making process. Several frameworks that consider the implementation of a "real-time monitoring component" within the architecture of advanced manufacturing systems are proposed in the literature. For example, in one of the surveyed works, Zhang *et al.* [8], consider a real-time distributed control system composed of two components. The first component, named Execution Control, is responsible for planning for reconfiguration, fault monitoring and fault detection, while the second component, named Control Execution, is responsible for the basic monitoring and alerts, and for the handling of low-level fault recovery procedures. Koomsap *et al.* [9] reports a significant improvement in performance for distributed manufacturing environments when making process control decisions in conjunction with condition monitoring and machine maintenance decisions. One way to realize this combined decision making process is to link process planning and optimization systems with shop-floor control decision processes (Shaikh and Prabhu [10]).

One of the potential solutions which ensures real-time monitoring for large-scale manufacturing systems is the use of wireless sensor networks. These sensor networks have the required capabilities of sensing, computing and communication of data in real-time. Pereira and Carro [2] mention the use of data provided by sensors and computerized components embedded within the manufacturing resources and other equipment. The data can be used as the main input for implementing intelligent maintenance systems, since the manufacturing resources undergo the process of degradation before their failure, thus the data provided by the deployed sensors can be used for health estimation and failure prediction. The advantages of using sensor networks and distributed computational capabilities for process manufacturing and control are summarized in the work of Shaikh and Prabhu [10], as follows:

- Extend existing manufacturing and process control systems reliably;
- Improve asset management by continuous monitoring of critical equipment;
- Help identify inefficient operation or poorly performing equipment;
- Help automate data acquisition to reduce user intervention; and,
- Provide detailed data to improve preventive maintenance programs.

The review of the literature identified a few papers that consider the Failure, Mode, and Effects Analysis technique in conjunction with distributed control architectures [11-13]. For example, Enright *et al.* [11] provide a comprehensive review and devise an optimal FMEA solution for electronics manufacturing, while Lopez *et al.* [12] adapted the FMEA methodology for failure diagnosis for distributed architectures.

3   Holonic System Model without Condition Monitoring

Building on the holonic architecture described in previous works by Babiceanu *et al.* [5-7] consisting of five types of holons (Order Holon, three Resource Holons, and a Global Scheduler Holon), Babiceanu and Tudoreanu [15] transformed an initial passive system component (*System Monitoring and Database Holon*) into an active holon element. Furthermore, the authors added another component, the *Monitoring System Holon*, which includes a series of wireless sensor networks distributed across the manufacturing enterprise, for the purpose of achieving real-time monitoring and recovery from faults, as follows:

- *System Monitoring and Database Holon*, responsible for monitoring the jobs in the system and availability of resources; and,
- *Monitoring System Holon*, responsible for the preventive and corrective maintenance procedures within the manufacturing system, which include condition monitoring, fault diagnosis, and recovery processes.

The capabilities of these two constructs are further defined and refined in this current work for the proposed holonic system. The new holonic system architecture for condition monitoring and fault recovery, viewed from the enterprise level, is depicted in Fig. 1. The architecture was obtained using a Systems Modeling Language (SysML^{TM}) Block Definition Diagram. The proposed holonic system model consists of a manufacturing plant that, initially, does not include the infrastructure for real-time monitoring of manufacturing resources and processes. Specifically, the proposed manufacturing system model is composed of the holonic entities listed next.

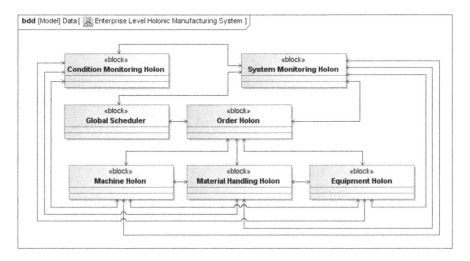

Fig. 1. Holonic enterprise architecture for condition monitoring and recovery

- *Order Holon*, which includes the data regarding the new jobs that enter the system; the *Order Holon* is responsible for scheduling the jobs to the available resources;
- *Resource Holons*, which include all the *Machine, Material Handling*, and any other *Equipment*, including storage equipment, such as Automated Storage and Retrieval Systems (AS/RS), and recycling capabilities, that are part of the manufacturing system; the *Resource Holon* is responsible for adding value to the orders processed within the system, and storing the finished parts; and,
- *System Monitoring Holon*, represented by the corresponding holonic instances of the *Purchasing, Quality Control, Maintenance*, and *Production Control* departments; the *System Monitoring Holon* is responsible for making raw materials available for processing, controlling the quality of the processed orders through a statistical sampling process, performing corrective maintenance of the manufacturing resources, and for any type of production control needs.

The operational instance of the holonic systems, depicted in Fig. 2, was obtained using a SysML^{TM} Activity Diagram. The system, as depicted, will serve its purpose well-enough for a certain time period. However, the system is expected to be effective for the entire lifecycle of the manufacturing resources. The degradation process that machine cells undergo is a continuous process that cannot be noticed in real-time. At some point in time, the machine cells will not be able anymore to provide the expected quality for the orders entering the system. This change in the quality of the throughput will be noticed by the *System Monitoring Holon*, through the *Quality Control* unit, which will conclude that the manufacturing resources (machine cells) entered a degradation process. Once the threshold of statistical sampling process is surpassed, the *Quality Control* unit implements the process described next.

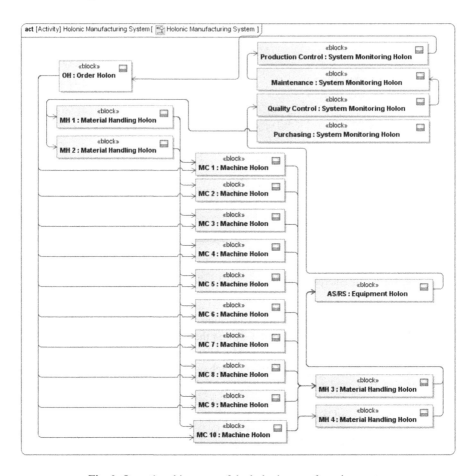

Fig. 2. Operational instance of the holonic manufacturing system

- The identified machine cells of the *Resource Holon* interrupt the execution of the assigned production orders;
- The new jobs corresponding to the rest of the order are re-directed to the *Order Holon* for re-assignment and re-scheduling through holonic allocation among the remaining machine cells;
- The *System Monitoring Holon* starts the fault diagnosis process by collecting data related to the operational variables of the machine cells and the type of faults identified; and,
- Corrective actions are recommended and executed, as a result of the fault diagnosis process.

4   Embedded Condition Monitoring Holonic System Model

4.1   Architecture of the Condition Monitoring Holonic System

The enhanced holonic model, depicted in Fig. 3, is designed to include real-time monitoring and control capabilities and considers the improvement that monitoring of manufacturing resources brings to quality control and responsiveness. The architecture of Fig. 3 was obtained using a SysML™ Activity Diagram.

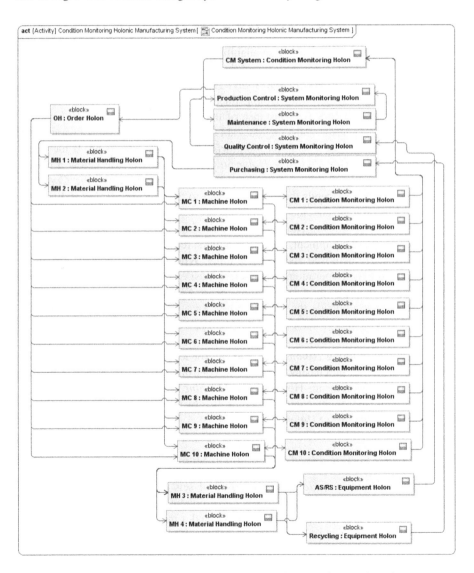

Fig. 3. Operational instance of the condition monitoring holonic manufacturing system

A wireless sensor network is deployed across the holonic manufacturing enterprise to every machine cell to monitor the condition of tooling and of the machines themselves. The holonic manufacturing system with real-time monitoring and control capabilities is composed of the *Order Holon, Resource Holons* (i.e., *Machine Holon, Material Handling Holon, Equipment Holon*), *System Monitoring Holon*, and *Condition Monitoring Holon*, which are described next.

- *Order Holon* is responsible for the real-time resource allocation and job scheduling. These processes are performed based on the holonic autonomy and cooperation processes together with the *Machine* instances of the *Resource Holons*. To accomplish its role efficiently, the *Order Holon* needs real-time data about the status of the machine cells (e.g., availability, reliability). These data are received from the *Production Control* instance of the *System Monitoring Holon*.
- The *Machine Holon* is part of the *Resource Holon* and is responsible for processing the jobs assigned by the *Order Holon*. Each *Machine* instance is monitored by a wireless sensors network (i.e., instance of the *Condition Monitoring Holon*) consisting of a set of sensors collecting data about the machine operational parameters. The data are sent to the *Condition Monitoring Holon* for processing. If the monitored parameters of the *Machine Holon* increase up or decrease below a certain threshold, which is already defined by the process requirements, the *Condition Monitoring Holon* suspends the order assigned to the *Machine Holon* and sets the machine status to unavailable.
- The *Condition Monitoring Holon* is the monitoring unit of the machine cells, and any other pieces of equipment, material handlers, recycling units, and facility units, when sensors are deployed also to these units. When the machine cells are in operational mode, their operational parameters must exhibit values in a defined range depending on the order processing requirements. When one or more of the parameters exhibit values outside the defined range for a specific machine cell, that machine cell entered a degradation process. Consequently, there is a high risk that the produced order is defective.
- The *Material Handling Holon* is responsible for loading/unloading and transporting items from the purchasing department, controlled by the *Purchasing* instance of the System Monitoring Holon to the *Machine Holons* on the shop-floor and from the shop-floor to the equipment controlled by the *AS/RS* and the *Recycling* instances of the *Equipment Holon*.

4.2   Detailed Functionality of the Condition Monitoring Holon

The *Condition Monitoring Holon* is composed of sets of deployed sensors that are measuring and then routing data to their corresponding sink nodes according to a predefined routing protocol, an FMEA module, and a monitoring computational system module. The deployed sensors associated with a *Machine* instance and its computational unit, in this case the sink node, form a deployed instance of the *Condition Monitoring Holon*. The *Condition Monitoring Holon* also exhibits instances associated with the operations on each of the shop-floor of the manufacturing enterprise, and, as mentioned above, *FMEA*, and *Condition Monitoring System* instances. Besides *Machine* instances, the instances associated with the shop-floor operations, could also include *Equipment*, *Material Handling*, *Recycling*, and *Facility* instances. The *FMEA* instance is discussed in the next section.

The *Condition Monitoring System* instances are responsible for processing the data received from the deployed instances of the *Condition Monitoring Holon*. If any of the physical resources of the system enters a degradation process, the *Condition Monitoring Holon* stops all activities performed by that physical resource, changes its status to unavailable and launches the fault-recovery process, with corrective actions starting to be implemented. The data received from the *Material Handling Holon* corresponding to the suspended order indicates the number of remaining items to be processed by the faulty machine cell. A corresponding new job will be started by the *Order Holon*.

4.3   FMEA Holonic Module

FMEA Background Information. Regardless of the application domain, when a system, product, process or project design includes some type of a risk, the FMEA process can be used as a tool to identify and correct design deficiencies by analyzing the failure modes, their mechanisms and effects, and to propose corrective actions. Yang [16] considers that FMEA can be described as a group of activities that lead to the recognition and evaluation of potential failures pertaining to a process or a product and their effect, in a first step, and the actions that attempt to eliminate the occurrence of those failures in a second step.

Even though, several improvements of the FMEA process were made after the initial release, that led to the development of extended versions, such as: Failure, Mode, Mechanisms, and Effects Analysis (FMMEA) or Failure, Mode, Effects, and Criticality Analysis (FMECA), the underlying concepts remained the same, and are summarized below [16-17].

- *Potential failure mode* is the manner in which a system, subsystem or a component fails to meet the requirements it was designed for;
- *Potential failure effect* is the impact of the failure or the consequences if a failure happens;
- *Potential failure causes* are the circumstances, during design, manufacturing or use that lead to the failure mode; and,
- *Failure mechanism* is the combination of stresses that induce failure, where a stress can be electrical, chemical, physical or mechanical.

Wessels [18] considers that criticality analysis attempts to evaluate the consequences of failures and establish a ranking method to determine criticality level. One of the most known approaches to evaluate the criticality of a failure mode is to compute the Risk Priority Number (*RPN*), which is obtained by multiplying three factors: severity, occurrence and detection, defined below, with all these three factors being associated with the same perceived risk.

- *Severity* factor is evaluated based on the effects of the failure mode, where these effects can be related to the system (e.g., damage, repairing costs, availability, etc.), personnel (e.g., death, injury, idle time, etc.) or regulations, which are punitive actions taken as a result of certain violations. Usually, severity is measured on a scale from 1 (least severe) to 10 (most severe);
- *Occurrence* factor is evaluated based on the likelihood of a failure to happen, from unlikely to frequent. Usually, the occurrence factor is measured on a scale from 1 (least likelihood of occurrence) to 10 (certainty of occurrence);
- *Detection* factor is evaluated based on the ability to detect the failure. Usually, failure modes range on a scale from 1 (obviously detectable) to 10 (not detectable).

By analysing the components of the *RPN*, it can be stated that the risk of a failure increases when *RPN* takes higher values. As a consequence, actions like design change, material upgrade and revision of test plans can be implemented to reduce the identified risk.

Implementation of the FMEA Holonic Module. Since the *Condition Monitoring Holon* is designed for real-time detection of failures and fault-recovery, the FMEA implementation, as part of the *Condition Monitoring Holon*, emphasizes potential failure causes and associated failure mechanisms. Through real-time monitoring of the system, the *Condition Monitoring Holon* collects the data that serves to identify under which circumstances and conditions failures are likely to occur. For example, a certain failure in raw materials can be caused by temperature cycling, random vibration or shock impact. In this case, if the temperature and vibration sensors are deployed, the data collected is used by the *Condition Monitoring Holon* to establish the ranges of values for the temperature and vibration values under which the failure is likely to occur. During operations, if any of these value ranges is detected, the *Condition Monitoring Holon* takes the appropriate action to flag the unavailability of the machine or equipment for future processing until the likelihood of failure is reduced to acceptable levels.

The detection, fault-isolation, and recovery process can be enhanced by adding an *FMEA Holon* to the *Condition Monitoring Holon*. The *Condition Monitoring Holon* uses the data received from the deployed condition monitoring units and the data retrieved from the FMEA unit to assess the status of the machines and decide about their availability and the preparation for the recovery process. The *FMEA Holon* serves as a storage and processor of data and considers the following parameters, performance measures, and actions related to the potential failures of the processing machines and equipment. Using a SysML™, Internal Block Diagram, the internal architecture of the *FMEA Holon* is presented in Fig. 4.

- *Failure modes*: the *FMEA Holon* identifies and stores different failure modes exhibited by the instances of the *Machine Holon*;
- *Failure effects*: the *FMEA Holon* acknowledges and addresses the potential failure effects on the *Machine Holon*;

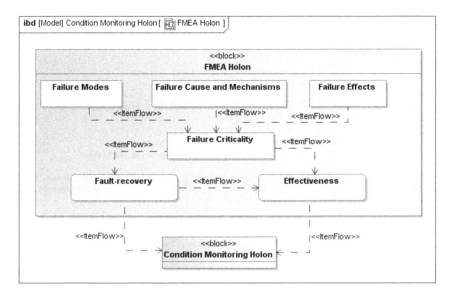

Fig. 4. The internal architecture of the FMEA Holon and data exchange

- *Failure cause and mechanism*: the *FMEA Holon* describes the potential causes or mechanisms (combination of stresses) that may induce the failure, and the different value ranges for the different mechanisms/causes that lead to the failure; as an example, if the failure cause is determined as being an elevated temperature, and the failure is likely to be induced if the temperature goes above 150°F, then the range is defined as [150, +infinity];
- *Failure criticality*: the *FMEA Holon* calculates the *RPN* number associated with the failure which includes the measures of severity, occurrence and detection;
- *Fault-recovery*: the *FMEA Holon* provides recommended actions for fault-recovery purposes, which increase the likelihood that the occurred failure is contained, and future failures are avoided;
- *Effectiveness*: the *FMEA Holon* measures the results of implementing the recommendations by re-calculating an updated *RPN* number.

Functionality of the FMEA Holonic Module. With the addition of the *FMEA Holon*, the *Condition Monitoring Holon* is able to retrieve data related to the failure cause and mechanism, which includes the defined range, and compares it with the real-time data received from the deployed instances of the *Condition Monitoring Holon*. Out of this interaction, two failure scenarios can result, for which the data flows and the overall functional architecture of the *Condition Monitoring Holon* are presented in Fig. 5. First scenario considers that the real-time data values received from the deployed instances of the *Condition Monitoring Holon* begin to cover the predefined range. In this specific scenario, a failure is likely to occur in the system, so the *FMEA Holon* recommends the following list of actions to be considered within the system.

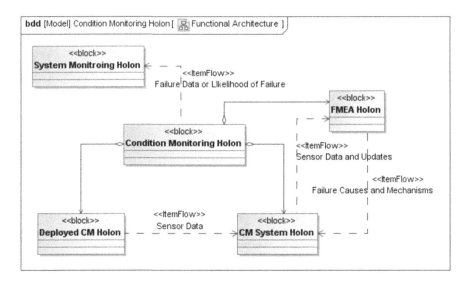

Fig. 5. Functional architecture of the Condition Monitoring Holon

- The *fault-recovery* process is to be launched with the implementation of the recommended actions.
- The *effectiveness* measure is updated; if the implemented actions were effective, then the occurrence of the corresponding failure mode should be reduced, which should result in a decrease of the *RPN* number.

The second scenario considers that a failure actually occurs in the system. This second scenario can be further organized in another two possible cases.

- If the failure mode was not found in the *FMEA Holon* database (i.e., the failure mode was not defined or occurred before), the database should be updated with the new failure mode along with its causes, mechanisms, effects, and criticality to the extent possible; also, the data collected by the deployed instances of the *Condition Monitoring Holon* is to be retrieved and will serve as an analysis basis for the failure causes and mechanism and for the definition of the pertaining value ranges;
- If the monitored condition ranges, pertaining to mechanisms, are not well defined, then an analysis of the data received from the deployed instances of the *Condition Monitoring Holon* will serve to establish new failure condition ranges, pertaining to mechanisms, and the *FMEA Holon* database will be updated.

5   Simulation Study

Two simulation models corresponding to the two holonic systems described above were implemented using the Arena® simulation environment. The *Machine Holon* included the use of 10 holon instances, each of them having associated 10 instances of *Condition Monitoring Holons*. There were four instances considered for the *Material*

Handling Holon, two for the *Equipment Holon*, another instance for the *Condition Monitoring Holon*, four different instances for the *System Monitoring Holon*, and a single instance for the *Order Holon*. The design of the simulation experiments for both models included 100 replications for two months of continuous operation simulation time. The other running parameters for the two simulation models were selected such that the systems worked at high utilization and all the potential events, such as resource failures, were sufficiently represented to obtain an accurate means of comparison between the performances of the two systems.

- Orders arrival was approximated by an exponential distribution with the mean of 30 minutes: EXPO(30);
- Number of items to be processed for each of the order was considered to be equal to 240;
- Failure rate of the physical resources was approximated by a Weibull distribution with parameters as follows, measured in hours: WEIBULL(8.5, 25.5);
- Job arrival rates for each of the machine cells was selected as coming from an exponential distribution with a mean of two minutes: EXPO(2);
- Processing time on the machine cells was approximated by a triangular distribution with the following times, expressed in minutes: TRIA(1.8, 2, 2.2);
- Time needed for performing the corrective maintenance actions, in minutes, were approximated by using exponential distributions, EXPO(40) for the first simulation model, and EXPO(20) for the second simulation model;
- Quality control sampling process used in the first simulation was able to identify the machine cells degradation process after producing 30 defective parts.

Overall, the simulation results show an improvement in the manufacturing plant metrics when real-time condition monitoring was implemented. First, the results show an increase in the utilization for the machine cells. For example, there are utilization increases of 1.52% and 1.40% for machine cell 3 and machine cell 10 respectively. Also, the overall utilization of the manufacturing plant was recorded to increase by 1.19%. The productivity of the machine cells also increased leading to an overall increase in the number of items processed by 2.09%, and the number of jobs executed by 4.23%. For example, the number of items processed increased by 2.02% and 2.38% for machine cells 6 and 10, respectively. These improvements in machine utilization and the productivity of the resources were obtained due to the fact that less time was necessary for performing fault diagnosis and identifying necessary corrective actions. In the case of the condition monitoring-enabled system, the data necessary for fault diagnosis is made available in real-time by the *Condition Monitoring Holon*, eliminating the time needed for performing the quality control sampling process. Fig. 6 presents the reported results above.

The increase in the number of jobs completed (for example, 4.23% and 4.65% more jobs completed for machine cells 3 and 7, respectively) is obtained due to the real-time job re-scheduling capabilities implemented when a certain machine cell, initially responsible for job processing, enters a degradation process (Fig. 7). Also, the results of simulating the proposed condition monitoring holonic manufacturing system showed a significant improvement of quality of the completed orders.

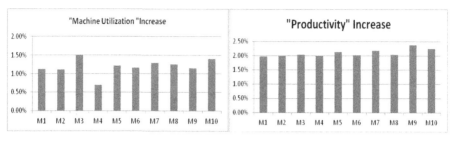

Fig. 6. Improvement in utilization and productivity when using real-time condition monitoring

The manufacturing plant defect rate, which was recorded in the first simulation model to be more than 4% (Fig. 8), decreased in the second model to a value of almost 0%. In fact, any job assigned to any machine cell will be cancelled and rescheduled immediately if the machine cell goes into a degradation process which prevents the processing of any part using a failed resource.

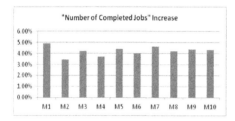

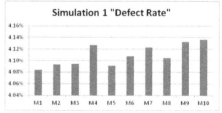

Fig. 7. Completed jobs performance   **Fig. 8.** The initial defect rate recorded

6   Conclusions and Future Research Directions

This work presents a holonic condition monitoring and recovery system for sustainable enterprises, which includes mechanisms for fault-detection, isolation, and recovery. The addition of an FMEA module is further enhancing the capabilities of the proposed system. The simulation results presented in the previous section show a significant improvement in terms of productivity, responsiveness, flexibility and quality when the condition monitoring and recovery system is employed. This improvement comes from the transition to a more advanced holonic manufacturing architecture that uses wireless sensor networks for real-time monitoring of resources. Sustainable enterprises need adequate models that take in consideration potential real-life scenarios and have the ability to respond with real-time solutions such that the operations of the facility are not disrupted. As future research directions, this work can be continued by investigating the impact of real-time monitoring on a more complex manufacturing environment presenting different types of resources executing jobs related to many products and facing several sources of disturbances and uncertainties such as demand, machine, equipment, and facility failures with combined FMEA scenarios associated with the processing machines, equipment, and facility.

References

1. Mejjaouli, S., Babiceanu, R.F.: Distributed intelligent real-time condition monitoring and recovery system for sustainable manufacturing enterprises. In: Borangiu, T. (ed.) Service Orientation in Holonic and Multi-Agent Manufacturing and Robotics, pp. 55–64. AGIR Publishing House (2013)
2. Pereira, C.E., Carro, L.: Distributed Real-Time Embedded Systems: Recent Advances, Future Trends and Their Impact on Manufacturing Plant Control. Annual Reviews in Control 31(1), 81–92 (2007)
3. Stecke, K.E.: Design, Planning, Scheduling, and Control Problems of Flexible Manufacturing Systems. Annals of Operations Research 3(1-4), 3–12 (1985)
4. Van Brussel, H., Wyns, J., Valckenaers, P., Bongaerts, L., Peeters, P.: Reference Architecture for Holonic Manufacturing Systems: PROSA. Computers and Industry 37(3), 255–274 (1998)
5. Babiceanu, R.F., Chen, F.F., Sturges, R.H.: Framework for the Control of Automated Material-Handling Systems Using the Holonic Manufacturing Approach. International Journal of Production Research 42(17), 3551–3564 (2004)
6. Babiceanu, R.F., Chen, F.F.: Development and Applications of Holonic Manufacturing Systems: a Survey. Journal of Intelligent Manufacturing 17(1), 111–131 (2006)
7. Babiceanu, R.F., Chen, F.F., Sturges, R.H.: Real-time Holonic Scheduling of Material Handling Operations in a Dynamic Manufacturing Environment. Robotics and Computer Integrated Manufacturing 21(4-5), 328–337 (2005)
8. Zhang, X., Balasubramanian, S., Brennan, R.W., Norrie, D.H.: Design and Implementation of a Real-time Holonic Control System for Manufacturing. Information Sciences 127(1-2), 23–44 (2000)
9. Koomsap, P., Shaikh, N.I., Prabhu, V.V.: Integrated Process Control and Condition Based Maintenance Scheduler for Distributed Manufacturing Control Systems. International Journal of Production Research 43(8), 1625–1641 (2005)
10. Shaikh, N.I., Prabhu, V.V.: Monitoring and Prioritizing Alerts for Exception Analytics. International Journal of Production Research 47(10), 2785–2804 (2009)
11. Enright, J., Lewis, H., Ryan, A.: FMEA as applied to electronics manufacturing: a revised approach to develop a more and robust and optimised solution. In: Proceedings of the 23rd International Conference on Flexible Automation and Intelligent Manufacturing, Porto, Portugal (2013)
12. Lopez, C., Subias, A., Combacau, M.: An adapted FMEA-based approach for failure diagnosis in distributed architectures. In: Proceedings of the 17th IMACS World Congress, Paris, France (2005)
13. Sankar, N.R., Prabhu, B.S.: Modified approach for prioritisation of failures in a system failure mode and effects analysis. International Journal of Quality and Reliability Management 18(3), 324–335 (2001)
14. Chin, K.S., Wang, Y.M., Poon, G., Yang, J.B.: Failure mode and effects analysis using a group-based evidential reasoning approach. Computers and Operations Research 36(6), 1768–1779 (2009)
15. Babiceanu, R.F., Tudoreanu, M.E.: Incorporation of Sensor Networks into Holonic Enterprise Systems for Life-cycle Process Monitoring. In: Johnson, A., Miller, J. (eds.) Proceedings of the 2010 Industrial Engineering Research Conference, Miami, FL (2010)

16. Yang, G.: Life Cycle Reliability Engineering. John Wiley & Sons, Inc., Hoboken (2007)
17. Pecht, M.: Product Reliability, Maintainability and Supportability Handbook, 2nd edn. Taylor & Francis Group, LLC, Boca Raton, FL (2009)
18. Wessels, W.R.: Practical Reliability Engineering and Analysis for System Design and Life-Cycle Sustainment. Taylor & Francis Group, LLC, Boca Raton, FL (2010)

Resource, Service and Product: Real-Time Monitoring Solution for Service Oriented Holonic Manufacturing Systems

Octavian Morariu[2], Cristina Morariu[1], and Theodor Borangiu[2]

[1] Cloud Computing Research Department, Cloud Troopers Intl.
Cluj Napoca, Romania
cristina@cloudtroopers.ro
[2] University Politehnica of Bucharest, Dept. of Automation and Applied Informatics
Bucharest, Romania, 060042
{octavian.morariu,theodor.borangiu}@cimr.pub.ro

Abstract. Service orientation of holonic manufacturing systems represents a major milestone in increasing efficiency, flexibility and standardization for manufacturing enterprises. SOA governance assures the capability for dynamic composition of services at runtime without human intervention, allowing the system to automatically align itself to the business drivers. In this context there is a need for accurate and real time monitoring of the shop floor activities during the manufacturing process. This paper presents a shop floor monitoring solution based on distributed multi-agent system architecture capable of real time data collection and presentation for production tracking. The solution provides a monitoring portal where system administrators can track key performance indicators in real time. The paper discusses the strategies for handing the monitoring data in real time and also long term, focusing on the consolidation of information in persistent data structures.

Keywords: Shop-floor monitoring, production tracking, simulation, monitoring portal, real time data, alerts, multi agent systems.

1 Introduction

Holonic manufacturing systems evolution in the last decade was driven by the technological advances in the underlying areas and by the business environment changes. This class of manufacturing systems has gained capabilities that allow real time decision making based on unpredictable events that provides a high degree of flexibility and adaptability to change as described by Borangiu [1][2][3]. In this context there are three inter-related vectors that define the effectiveness of the manufacturing system: robustness, composability and observability. In any complex manufacturing system there is a risk that a failure of a module has the potential of stopping the entire system. In other words, a partial system failure can lead to a complete failure if it cannot be detected, isolated and handled in a timely manner. In Service Oriented

T. Borangiu et al. (eds.), *Service Orientation in Holonic and Multi-Agent Manufacturing and Robotics*, Studies in Computational Intelligence 544,
DOI: 10.1007/978-3-319-04735-5_4, © Springer International Publishing Switzerland 2014

manufacturing systems, the impact of partial system failure is litigated by promoting a loosely coupled architecture, assuring isolation both between individual modules and the underlying runtime platform. In practice, there are two approaches that are used together for isolation of modules: data isolation and execution isolation. Data isolation refers to the internal state and information stored in a module. Execution isolation refers to the runtime platform independence for each module. Recent research done by Morariu et al. [4][5] has identified ways to assure this isolation by promoting loosely coupled systems, connected through service busses. Lack of one or both types of isolation can have catastrophic effects on the entire system even in case of a minor module failure. In order to achieve robustness, manufacturing systems need to become compositions of loosely coupled modules. This raises another important problem of how to locate and compose these modules into a functioning manufacturing system. Traditionally the composition was static and it was the responsibility of the system architect, due to intricacies of document and protocol standards and poor discovery and publishing services. With the emergence of standards in manufacturing domain like ISA 95/99, MIMOSA and others, the support for runtime and dynamic service composition has become available. Another critical aspect of any manufacturing system is the possibility to know what it is doing, the current state it is in, and the state history from the beginning of the execution to the current state. In other words, in order to know whether the manufacturing system is working correctly, it must provide means to be observed, or monitored in real time. However, the notion of observability as an abstract term in service oriented architecture (SOA) goes beyond just monitoring the system by inspecting the state of its services. For manufacturing systems, observability can have more concrete meanings, like energy consumption at shop floor level, resource utilization, time per operation, average make-span etc.

In 2010, Gartner Research published the Application Performance Monitoring (APM) Conceptual Framework [6] which defines five directions for APM: end-user experience, run-time application architecture, business transactions, component monitoring and reporting. While these directions are generally valid for generic APM solutions targeted for end user applications, for Manufacturing Systems Performance Monitoring (MSPM) a derived schema is more relevant. This is more evident if the manufacturing system design is based on SOA design patterns in conjunction with meta-heuristic algorithms or local intelligence for autonomous decision making. In these conditions the overall performance of the system is not a simple composition of the performance measured or estimated on each individual sub-module. For these reasons we propose a derived conceptual model for manufacturing systems performance monitoring, as illustrated in Fig. 1.

The MSPM framework has four main directions: service monitoring, resource monitoring, product monitoring and analytics/reporting.

Service Monitoring focuses on the actual web services exposed and used by the system sub-modules. The monitoring is done at the HTTP/SOAP protocol layer and the data consists in service status, response time for requests and overall workload.

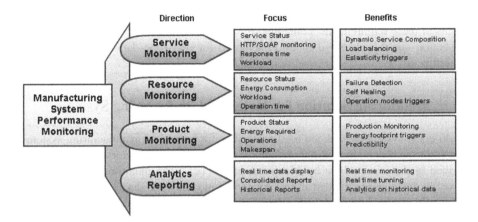

Fig. 1. Manufacturing System Performance Monitoring Directions

Using this information in real time allows the system to automatically reconfigure and re-compose complex services in case of an individual service failure. In case of an increased workload, a threshold mechanism can be used to balance the load to a secondary service provider. The data gathered here is used to monitor the information flow in the system. Resource Monitoring refers to the actual robots on the shop floor that are executing operations during the manufacturing process. The focus here is on the resource status, the energy consumption and on the resource utilization. These factors can generate triggers that can change the manufacturing system behaviour, as changing from a hierarchical operational mode to a heterachical operational mode or can change the scheduling strategy to avoid a defective resource. *Product Monitoring* is used to keep track of the products that are in production at any given time. The data collection focuses on energy consumption, real time status and make-span for each intelligent product. The data gathered here is used to monitor the material flow in the manufacturing system. The concept of intelligent product considered here was first described by McFarlane et al. [7] and developed further by Meyer [8]. Finally the *analytics and reporting* direction represent the mechanisms to consolidate the collected data around relevant key performance indicators (KPIs) for the manufacturing system. The most important KPIs and their impact on manufacturing system performance and integration were studied by Ahmad et al. [9], Cai et al. [10] and Zheng et al. [11]. The MSPM solution should be able to provide both real time reports that would allow dynamic tuning of the system and long term historical reports that would bring more predictability in the manufacturing processes.

Holonic manufacturing systems operating in an online scheduling mode, or also known as heterarchical operation mode, are affected by the risk of myopia, where intelligent products and agents have a limited information horizon. The implementation of advanced monitoring solutions can reduce this risk by augmenting the real time information available in the system with relevant real time KPIs.

This paper presents a MSPM solution for holonic manufacturing systems that aligns to these requirements, containing a MAS based architecture for real time data collection and a web based monitoring portal for real time data visualization and reporting.

2   Monitoring Solution for Holonic Manufacturing Systems

Multi agent systems have been used previously for control of manufacturing systems and implement communication and complex behaviour of the actors involved as described in the survey done by Monostori et. al [12]. The monitoring solution proposed in this paper is divided in two functional modules: the data collection module, responsible for gathering real time data from monitored targets and consolidate it in a persistent storage, and the monitoring portal which provides a user rich user interface for data visualization. The general architecture of the monitoring solution is illustrated in Fig. 2.

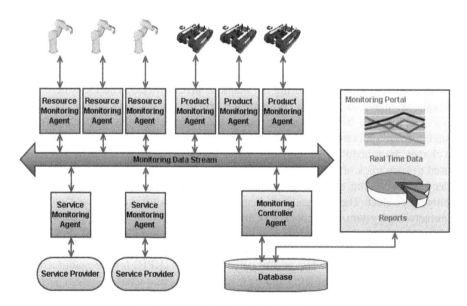

Fig. 2. Monitoring Solution Architecture

2.1   Data Collection Agents Design

There are four types of agents developed for this solution: resource monitoring agent, product monitoring agent, service monitoring agent and a controller agent. Along with these agents, there are three more components important in the architecture: the monitoring data stream, the monitoring controller agent and the monitoring portal. The following section provides some implementation details for each of these components.

2.1.1   Resource Monitoring Agent (RMA)

The resource monitoring agent (RMA) is implemented as a JADE agent with a cyclic behaviour. The agent does a cyclic poll at 5 seconds interval on the resource in order to gather information about the resource status, energy usage and the operations performed. The information is serialized and sent to the Controller Agent over the monitoring data stream.

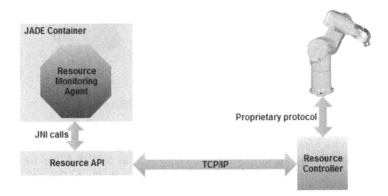

Fig. 3. Resource monitoring agent architecture

RMA typically interacts with the shop floor resource through a proprietary API provided by the resource vendor (Fig. 3). These APIs are provided most of the times in the form of native libraries, specific for each operating system and platform supported. For this reason, a generic implementation of the RMA cannot be provided, as the integration will differ for each resource type. However, the general architecture of the RMA, validated in the pilot implementation, consists of a JNI (Java Native Interface) interface that provides a wrapper over the native library and allows integration for the polling mechanism of the agent. The actual metrics available are highly dependent on the API provided by the vendor, but typically these include: resource real time status, resource power consumption, current operation and operation duration.

2.1.2   Product Monitoring Agent (PMA)

The product monitoring agent (PMA) is implemented as a JADE agent with a consumer behaviour. The agent runs in a shared JADE container and is notified after each operation performed. The agent records the operation data and sends these details on the monitoring data stream.

On the product pallet carrier (containing an Intelligent Embedded Device, IED) a small footprint monitoring agent runs, implemented specifically for the native embedded platform. This monitoring agent acts as a service client, creating and sending the notifications to the PMA over SOAP protocol using the build in WiFi network connection (Fig. 4). The monitoring data depends on the IED characteristics, but typically includes: operation execution time, time spent on the conveyor, cumulated travel time, current position and current action.

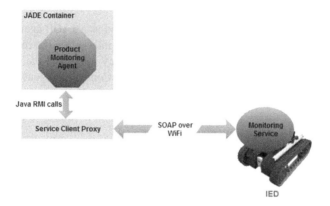

Fig. 4. Product monitoring agent architecture

2.1.3   Service Monitoring Agent

The service monitoring agent (SMA) is implemented as a Web Filter for the Web Container in the application server (Fig. 5). The filter calls the JADE API to send a FIPA INFORM message to the monitoring data stream including the timestamp, the URL of the service and the payload used to invoke it.

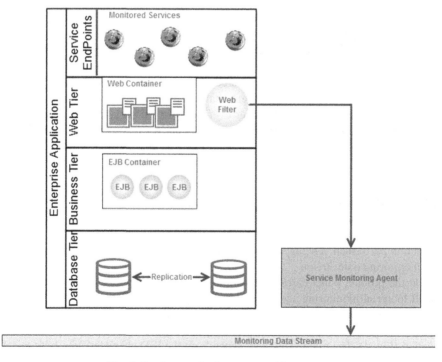

Fig. 5. Service monitoring agent architecture

Web Filter component definition was introduced in Java Servlet specification version 2.3. The filter intercepts requests and responses and has full access to the information contained in these requests or responses. Filters are useful for many scenarios where common processing logic can be encapsulated. Historically filters have been used for access management (blocking requests based on user identity), logging and tracking users of a web application, data compression, localization, XSLT transformations of content, encryption, caching, triggering resource related events, mime-type processing and many others. The implementation of a Web Filter is governed by the following interfaces: Filter, FilterChain, and FilterConfig in the javax.servlet package. The actual filter is a implementation of the Filter interface. The filters are invoked in a chained fashion by the servlet container. The Filter interface declares the doFilter method, which contains the actual processing of the request/response objects.

In our implementation, the doFilter method sends a message to the Monitoring Agent containing the service being invoked by the user. The information sent by the Web Filter to the monitoring agent has the following structure:

Table 1. Web Filter message structure

ID	The ID of the thread in which the web filter is called
Client ID	Client ID invoking the service
Service URL	URL of the service invoked
EndPoint	The service endpoint being invoked
Parameters	The complete payload used to invoke the service
Time	Time elapsed between request and response

2.1.4   Monitoring Data Stream, Controller Agent and Monitoring Portal

The *Monitoring Data Stream* is implemented as a message queue at JADE agent platform level. This queue allows asynchronous communication between the data producers, in this case the monitoring agents, and the data consumers: the controller agent and the monitoring portal.

The *Monitoring Controller Agent* is a JADE agent that implements a cyclic behaviour and consumes the monitoring messages sent to the monitoring data stream. The agent aggregates the data based on the monitoring target and saves it in the persistent storage. Monitoring Portal is a web based application that allows real time data display with AJAX and Partial Page Rendering, and report generation based on the data saved in the database. The integration and the dataflow between these components is illustrated in Fig. 2.

2.2   Data Storage Strategy

The strategy for the data storage is common for all the metrics recorded. For each metric collected by the target monitoring agents, the data is stored in two database tables using the following strategy:

* Short term storage (Staging): consists in a rolling table with a fixed number of rows, having a timestamp as a primary key. This table contains the "last N time intervals" of that particular metric and is used to display real time data in the web application.
* Long term storage: consists in a table containing averaged data for each metric for given time intervals (i.e. one hour). This data is generated from the short term storage, at each rollover by the controller agent. The long term storage table is used for reporting and analytics purposes (Fig. 6).

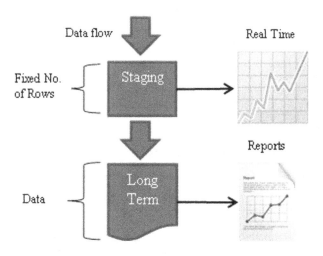

Fig. 6. Data Flow

The granularity in both short and long term storage is configurable globally and for each metric individually, allowing higher or lower history in real time views and detailed reporting. The agent uses an incoming message queue where messages from the other agents are placed in order to be processed. There are typically two types of messages for each metric. In case the target agent is sending real time data, the data will be stored in the staging table and the agent will rotate the staging table and promote aggregate metrics in the long term table. The other scenario is when the agent is sending directly aggregated metrics, which are added directly to the long term storage tables. Because of the large number of target agents and large amount of data to be managed, the monitoring controller agent (or main agent) is using a database connection pool where connections are re-used. This avoids the need to re-authenticate for each connection and improves the overall performance.

3   Agent Interaction and Scalability

The agent interaction is based on FIPA standard messaging protocol implemented by JADE [13]. JADE is a software framework designed to make the development of agent applications easier and in compliance with the FIPA specifications for interoperable intelligent multi-agent systems. The goal of JADE is to simplify development while ensuring standard compliance through a comprehensive set of system services and agents.

To achieve such a goal, JADE offers the following list of features to the agent developer:

• FIPA-compliant Agent Platform, which includes the AMS (Agent Management System), the DF (Directory Facilitator), and the ACC (Agent Communication Channel). All these three agents are automatically activated at the agent platform start-up.

• Distributed agent platform. The agent platform can be distributed on several hosts (provided that the required ports are opened in the firewalls). Only one Java Virtual Machine is executed on each host. Agents are implemented as one Java thread and Java events are used for effective and light-weight communication between agents on the same host. Parallel tasks can be executed by one agent, and JADE schedules these tasks in a more efficient way than the Java Virtual Machine does for threads.

• A number of FIPA-compliant DFs (Directory Facilitator) can be started at run time in order to implement multi-domain applications. The concept of domain is a logical one as described in FIPA97 Part 1.

• API provided to allow registration of agent services with one or more domains (i.e. DF).

• JADE provides an integrated transport mechanism and interface (API) to send/receive messages to/from other agents

• FIPA97-compliant IIOP protocol is used to connect different agent platforms.

• Light-weight transport of ACL messages inside the same agent platform, as messages are transferred encoded as Java objects, rather than strings, in order to avoid marshalling and un-marshalling issues. When the sender or the receiver does not belong to the same platform, the message is automatically converted to/from the FIPA compliant string format. In this way, this conversion is hidden to the agent developer.

• JADE provides a library of FIPA interaction protocols.

• Automatic registration of agents with the Agent Monitoring Service (AMS).

• FIPA naming service, build in the platform, provides agents with a GUID (Globally Unique Identifier).

• Graphical user interface to manage several agents and agent platforms from the same agent. The activity of each platform can be monitored and logged.

The above characteristics make JADE a suitable platform for implementing a distributed data collection framework as it combines the flexibility and portability of Java with a very low memory footprint and CPU profile. This helps reducing the overhead of running JADE agents inside virtualized workloads in the cloud. The JADE container uses typically 32 Mb of memory for the Java heap, which is enough for most types of data collection.

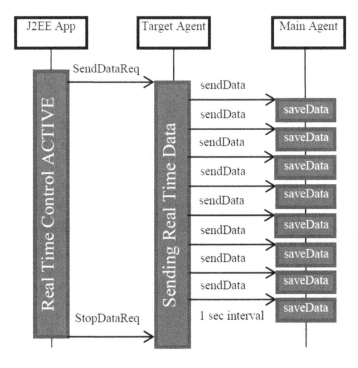

Fig. 7. Interaction diagram

The agent interaction in terms of real time data dialogue is presented in Fig. 7 and is similar for all data collection agents.

From Fig. 7 it can be seen that the normal operation of a data collection agent (target agent) is to send aggregate data to the main agent. When the end user accesses the web application and displays a page that requires real time data from a specific resource, the web application sends a message to the data collection agent. The data collection agent starts sending real time data to the main agent, which in turn stores it in the database. At this point, the database sends the change notification to the web application and the real time data is displayed to the user. Similarly when the real time data display control (graph, table, etc.) becomes inactive, the agent is notified to stop sending real time data. This approach reduces the network overhead of sending high granularity data, especially in large deployments, involving many agents.

In this agent architecture, scalability is implemented at the monitoring controller agent (or main agent) layer. This agent can aggregate data from other agents of the same type in a tree like model. This concept is illustrated in Fig. 8.

Each agent has a start-up switch that determines if the agent is the primary agent or if it is an intermediary agent used for data aggregation. Currently only the primary agent is storing data in the database. However, this agent is having an active/active setup in separate JADE containers, so that it assures high availability.

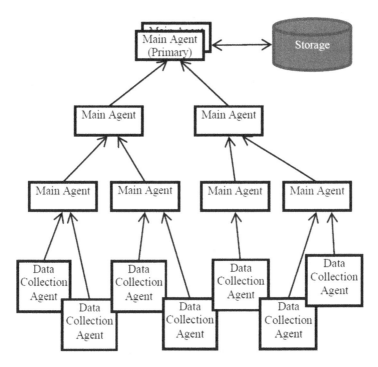

Fig. 8. Agent Model Scalability

The implementation contains an additional monitor agent that sends a FIPA inform message to each agent registered with JADE DF service. This assures that any agent failure can be detected by the monitor and handled accordingly.

4   Implementation Details and Experimental Results

The data collection agents are implemented using a JADE platform, in Java programming language. The generic code structure of a data collection agent is consisting in two behaviours: a ticker behaviour and a receiver behaviour. The first is triggering the polling mechanism that reads the target (resource/product/service) metrics. The second behaviour is listening for FIPA messages and processes them as required. The behaviours are defined by overriding the agentSpecificSetup() method:

```
protected void agentSpecificSetup() throws AgentInitializationException {
            super.agentSpecificSetup();
    //registering with JADE DF Service
    ServiceDescription sd = new ServiceDescription();
    sd.setType("RESOURCE_MONITORING_AGENT");
    DFAgentDescription dfTemplate = new DFAgentDescription();
    dfTemplate.addServices(sd);
```

```
SearchConstraints sc = new SearchConstraints();
sc.setMaxResults(new Long(10));
ACLMessage subscribe = DFService.createSubscriptionMessage(
                        this, getDefaultDF(), dfTemplate, sc);
//sending a subscription message to the JADE DF service
//this will allow monitor agent to find us
send(subscribe);
//Defining ticker behaviour
Behaviour ticker = new TickerBehaviour(this, 10000) {
  @Override
  protected void onTick() {
     try{
        collectDataAndSendMessage();
     }catch(Exception e){
        handleException();
     }
  }
};
//adding the behaviour to the agent
addBehaviour(ticker);
//defining message processor behaviour
Behaviour messageProcessor = new CyclicBehaviour(this) {
  @Override
  public void action() {
    ACLMessage msg= receive();
    if (msg!=null){
       processMsg(msg);
    }else{
       //wait until a new message arrives in the queue
       block();
    }
  }
};
//adding the behaviour to the agent
addBehaviour(messageProcessor);
}
```

The methods collectDataAndSendMessage() and processMsg() are implemented differently for each agent, depending on what target is the agent connecting to. However, these methods are defined in the DataCollectionAgent interface, which is implemented by every data collection agent. The UML class diagram showing the two interfaces implemented by the data collection agents is shown in Fig. 9.

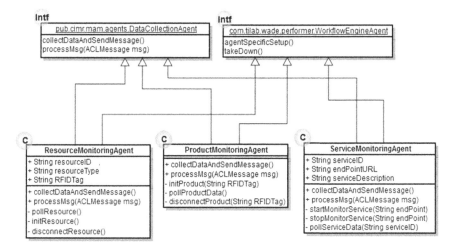

Fig. 9. UML class diagram

Each data collection agent is implementing both interfaces and implements the defined methods. Also, the agents provide implementation of specific methods and attributes used to connect and manage the lifecycle of the monitored target: resource, product or service.

UNIVERSITATEA POLITEHNICA DIN BUCURESTI

| Resource Monitoring | **Product Monitoring** | Service Monitoring | Reporting | Configuration |

Batch ID	Product ID	RFID Tag	Product Type	% Complete	# Operations	Entry Time	Exit Time	Makespan (s)
3	1	011000000FF0001FFFFF0001	H		5	9/7/2013	9/7/2013	173
3	2	011000000FF0001FFFFF0002	H		5	9/7/2013	9/7/2013	161
3	3	011000000FF0001FFFFF0003	H		5	9/7/2013	9/7/2013	149
3	4	011000000FF0001FFFFF0004	H		5	9/7/2013	9/7/2013	188
3	5	011000000FF0001FFFFF0005	H		5	9/7/2013	9/7/2013	120
3	6	011000000FF0001FFFFF0006	H		5	9/7/2013	9/7/2013	243
3	7	011000000FF0001FFFFF0007	H	60%	4	9/7/2013		88
3	8	011000000FF0001FFFFF0008	T	60%	4	9/7/2013		57
3	9	011000000FF0001FFFFF0009	T	60%	4	9/7/2013		76
3	10	011000000FF0001FFFFF000A	T	40%	2	9/7/2013		32
3	11	011000000FF0001FFFFF000B	T	40%	2	9/7/2013		43
3	12	011000000FF0001FFFFF000C	H	40%	2	9/7/2013		21
3	13	011000000FF0001FFFFF000D	H		0	9/7/2013		0
3	14	011000000FF0001FFFFF000E	H		0	9/7/2013		0
3	15	011000000FF0001FFFFF000F	H		0	9/7/2013		0
3	16	011000000FF0001FFFFF0010	H		0	9/7/2013		0

Fig. 10. Production tracking in monitoring portal

Fig. 10 presents a screenshot from the monitoring portal showing the production tacking page. The page contains a table presenting a real time view of the products in the product batch. The information presented in this table contains the ID of the product, the product type (in our case T shaped products and H shaped products), the % complete, the number of operations performed, the entry time and exit time and the make span.

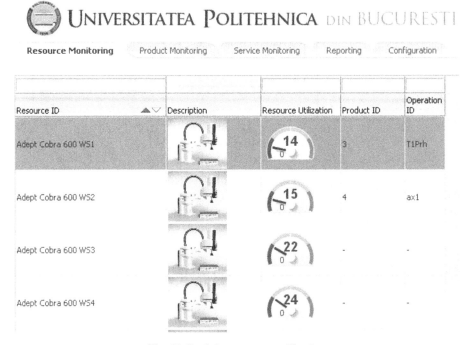

Fig. 11. Real time resource utilization

Fig. 11 presents a screenshot from the resource monitoring portal. This page shows real time information about each shop floor resource, including: resource identifier, the picture of the resource (descriptive purpose), the resource utilization computed for the current product batch, the current operation executed by the resource, the current product in the corresponding workstation and the operation being performed. The last three columns are updated in real time based on data sent by the product monitoring agents.

The reporting engine is based on Pentaho [14] reporting solution and is offering a set of predefined reports, including: Resource Operation History, Product Operation History, Resource Utilization Trend and Product MakeSpan Trend. At the same time, the system allows plugin like creation of new reports, using the Pentaho report designer tool.

Fig. 12. Real time resource utilization

Finally, Fig. 12 presents the service monitoring module. In the pilot implementation the service monitoring agent is monitoring three web services: create customer order (CreateCO), track customer order (TrackCO) and request offer. The table shows real time data including the invocations high and low for the time unit (default is 1 day, specifically the last 24 hours), average response time in ms and the success rate.

5   Conclusions

Real time monitoring in a holonic manufacturing system provides valuable information that can be used for tuning the system and for long term reporting. Service monitoring is also an important aspect specifically targeted to illustrate the information flow in the SOA oriented manufacturing system. It helps identifying the number of invocations and the success rates together with the service performance. Real time monitoring of services provides useful clues when diagnosing information flow bottlenecks in the system. Product monitoring provides real time information about material flows and together with resource stock monitoring represent direct integration points for supply chain applications in regards to the manufacturing system.

The real time data collection provides the opportunity to create a trigger mechanism for various conditions, like resource breakdown or stock depletion. At the same time, due to the fact that all monitoring data is concentrated centrally in the Monitoring Data Stream, complex triggers can be defined that can consider multiple input signals, like energy consumption for a specific operation, in conjunction with the actual target product where the operations are performed. Currently the prototype of the monitoring application supports only basic resource monitoring and service monitoring, but the trigger mechanism is considered for future developments.

References

1. Borangiu, T.: A service-orientated arhitecture for holonic manufacturing control. In: Rudas, I.J., Fodor, J., Kacprzyk, J. (eds.) Towards Intelligent Engineering and Information Technology. SCI, vol. 243, pp. 489–503. Springer, Heidelberg (2009)
2. Borangiu, T., Gilbert, P., Ivanescu, N., Rosu, A.: An Implementing framework for holonic manufacturing control with multiple robot-vision stations. Journal of Engineering Applications of Artificial Intelligence (2009) ISSN 0952-1976
3. Borangiu, T., Raileanu, S., Anton, F., Parlea, M., Tahon, C., Berger, T., Trentesaux, D.: Product-driven automation in a service oriented manufacturing cell. In: Proceedings of the Int. Conf. on Industrial Engineering and Systems Management, IESM 2011, Metz, May 25-27, pp. 978–972 (2011) ISBN 978-2-9600532-3-4
4. Morariu, C., Morariu, O., Borangiu, T.: Manufacturing Service Bus Integration Model for Implementing Highly Flexible and Scalable Manufacturing Systems. Information Control Problems in Manufacturing, IFAC Papers Online 14(1), 1850–1855 (2012)
5. Morariu, C., Borangiu, T.: Manufacturing Integration Framework: A SOA Perspective on Manufacturing. Information Control Problems in Manufacturing, IFAC Papers Online 14(1), 31–38 (2012)
6. Cappelli, W.: Magic Quadrant for Application Performance Monitoring. Gartner Research (2012)
7. McFarlane, D., Sarma, S., Chirn, J.L., Wong, C.Y., Ashton, K.: The intelligent product in manufacturing control and management. In: 15th Triennial World Congress, Barcelona, Spain (2002)
8. Meyer, G.G., Främling, K., Holmström, J.: Intelligent products: A survey. Computers in Industry 60(3), 137–148 (2009)
9. Ahmad, M., Nasreddin, D.: Establishing and improving manufacturing performance measures. Robotics and Computer-Integrated Manufacturing 18(3), 171–176 (2002)
10. Cai, J., Xiangdong, L., Zhihui, X., Jin, L.: Improving supply chain performance management: A systematic approach to analyzing iterative KPI accomplishment. Decision Support Systems 46(2), 512–521 (2009)
11. Zheng, L., Jing, X., Hou, F., Feng, W., Na, L.: Cycle time reduction in assembly and test manufacturing factories: A KPI driven methodology. In: IEEE International Conference on Industrial Engineering and Engineering Management, IEEM 2008, pp. 1234–1238 (2008)
12. Monostori, L., Váncza, J., Kumara, S.R.: Agent-based systems for manufacturing. CIRP Annals-Manufacturing Technology 55(2), 697–720 (2006)
13. Bellifemine, F., Poggi, A., Rimassa, G.: JADE: a FIPA2000 compliant agent development environment. In: Proceedings of the 5th International Conference on Autonomous Agents, pp. 216–217. ACM (2001)
14. Bouman, R., Dongen, J.V.: Pentaho Solutions: Business Intelligence and Data Warehousing with Pentaho and MySQL. Wiley Publishing (2009)
15. Shen, W., Wang, L., Hao, Q.: Agent-based distributed manufacturing process planning and scheduling: a state-of-the-art survey. IEEE Transactions on Systems, Man, and Cybernetics, Part C: Applications and Reviews 36(4), 563–577 (2006)

The Role of Distributed Intelligence in Warehouse Management Systems

Wenrong Lu[1], Vaggelis Giannikas[1], Duncan McFarlane[1], and James Hyde[2]

[1] Institute for Manufacturing, University of Cambridge,
17 Charles Babbage Road, Cambridge CB3 0FS, UK
{wl296,eg366,dcm}@eng.cam.ac.uk
[2] Six Works Ltd., 1Station Road, Foxton, CB22 6SA
www.ecommercefulfilment.co.uk

Abstract. Third-party-logistics warehouses have been gaining popularity in recent years mainly due to their ability to buffer the material flow along a supply chain and consolidate the products from various suppliers, which has a major impact on supply chain efficiency. The issues and challenges in warehouse management are similar in many ways to those faced in manufacturing control, resilience and adaptability being two such issues. Since distributed intelligence (DI) approaches have been extensively studied to address such issues in manufacturing control, this paper examines the possible adoption of DI approaches in warehouse management systems (WMS). Further, the paper discusses the challenges in warehouse management and compares these challenges with the characteristics of manufacturing problems that DI approach is suited to.

Keywords: distributed intelligence, warehouse management systems, order-picking scheduling.

1   Introduction

An increasing number of companies are outsourcing their logistics, especially warehouse functions, to third-party-logistics providers (3PLs) [1]. In comparison to traditional transport and warehousing services, 3PLs are more complex and encompass a broader number of functions [2] due to the need to adapt to different customer-specific requirements [3]. Therefore, as 3PLs continue to expand, higher performance is required from their warehouses, including tighter inventory control, shorter response time and managing a greater product variety [4]. In particular, these improvements are subject to different customers whose requirements vary. i.e. 3PLs are required to be more customer-oriented in order to be more responsive to orders with different requirements in an efficient manner.

Similar demands have occurred in manufacturing control where flexible and lean manufacturing become the mainstream requirements for many production plants [5]. Because of the inability to cope with a high degree of complexity and changes corresponding to the requirements, conventional centralised approaches applied to decision making in manufacturing control can be inadequate [6]. Therefore, a new class of

T. Borangiu et al. (eds.), *Service Orientation in Holonic and Multi-Agent Manufacturing and Robotics*, Studies in Computational Intelligence 544,
DOI: 10.1007/978-3-319-04735-5_5, © Springer International Publishing Switzerland 2014

approaches which adopts the concept of distributed intelligence (DI) have been extensively studied in recent years [7].

Considering this motivation, this paper will discuss the potential role of a DI approach in warehouse management through the following structure: section 2 will review the challenges of current warehouse management and section 3 will review the characteristics of manufacturing problems that DI is suited to. Section 4 will then discuss why DI approach would be a suitable solution for future warehouse management system and a vision of such adoption is described in section 5.

2   Current Issues in Warehouse Management

In this section, we will review the business trends and requirements for warehouse management, and the consequent limitation and challenges occurring in the current warehouse management systems.

2.1   Business Trends and Associated Requirements for Warehouse Management

Despite the rapidly changing landscape in warehouse operation, the current 3PL warehouse business objectives of providing desired service level at low cost will still remain the basic requirement in the future. However, additional requirements will arise to enhance the competitiveness of a warehouse firm. For modern warehouses, such trends lead to *"the rapidly changing preferences of customers, customer orders increasingly exhibit characteristics of higher product variety, smaller order size and request of reliably shorter response time"* [8]. In other words, the cost and space pressures are due to rising customer demands for faster and more tailored fulfilment[1]. i.e. the requirements for warehouse operations are improving the robustness (flexibility[2] and adaptivity[3]) for overall efficiency, which imply the needs of warehouse operations to provide agile solutions (subject to cost) in response to changes or disruptions.

2.2   Limitation and Challenges of Current Warehouse Management Systems

As customers become more demanding and competition becomes more intense, companies need to have information-technology tools to help their warehouse operation to be more reliable, flexible and efficient [9]. Therefore, Warehouse Management Systems (WMS), a group of software programs, have been developed for handling warehouse resources and monitoring warehouse operations efficiently and effectively, in order to enhance the productivity and reduce the operation costs [10].

---

[1] Aberdeen Group observed: *"the best performing companies are focused on winning in both these dimensions: by creating faster throughput and more workflow agility in their warehouses, they are able to satisfy customer demands while lowering logistics costs"* [38].

[2] *Flexible: being responsive to short-term changes of customer demands or in a timely manner*

[3] *Adaptive: being able to maintain the service level when mid-term changes/requirements are demanded by customers.*

In general, WMS is a software application that integrates bar-code system, radio frequency communications equipment, and peripherals [11], p393. WMS offer different functionalities which include managing storage locations, directing, receiving, put-away, replenishing, picking and packing activities, and planning outbound transportation [12]. Fig. 1 shows how these functions are connected and implemented in a standard WMS.

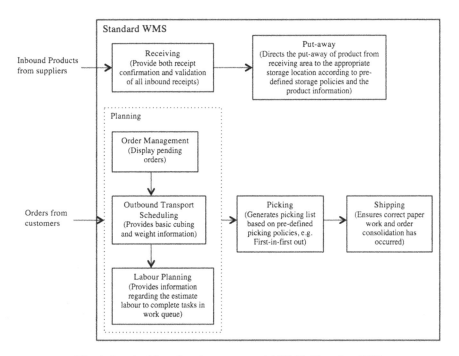

Fig. 1. Standard functions in a commercial WMS (Based on [11])

Although paper- or spreadsheet-based WMS can fulfil certain needs and manage stock accurately if well managed, a company needs to introduce a real-time WMS in order to compete effectively in today's fast moving technological world [9]. However, the current WMSs are incapable of providing accurate and timely warehouse operation information since they rely heavily on warehouse staff members to enter information manually or through a barcode system [10]. Hence, incorrect information is unavoidably input from time to time as human error is inevitable [13].

Currently, researchers are trying to solve these problems of accuracy and timeliness by focusing on providing real-time information of the warehouse operations using automatic ID system (such as RFID) [14][15][10] and wireless sensor network (WSN) [16][17][18]. Although such systems can capture the real-time information regarding operations and disruptions via the aforementioned information technologies, it is still a challenge for warehouse managers to make decisions using this information in a short response time when disruptions and changes occur [14].

A significant amount of research has focused on developing decision support models (including heuristics and algorithms) which aim to optimally manage the warehouse operations [4]. These models are designed using a hierarchical framework which assumes there is a master/slave relationship between higher and lower levels of control [19], which is based on deterministic and simplified assumptions. Due to their rigid assumptions and constraints, successful implementations of these models in current commercial WMS are rare [20].

To sum up, the service levels of warehouses could be significantly improved if the WMS is able to effectively deal with the changes of orders and products, since it will allow a more flexible order management offering for the customers. Moreover, if the system can manage better the disturbances such as shortage of resources or delay of arriving outbound trucks, better operation performance and lower costs would be obtained. Similar requirements in manufacturing control have been found suitable to be addressed by the concept of distributed intelligence, which is discussed in the next section.

3   Distributed Intelligence for Manufacturing Control: A Review

Manufacturing control is concerned with managing, controlling and monitoring the physical activities in the factory [21]. Such systems aim to decide what and when to produce, how and when to use the resources, and to release jobs optimally [22].

The development of distributed intelligence management systems was originally motivated by the evolution of industrial requirements for manufacturing control which has been focused on the move from static optimisation to dynamic optimisation and agility in order for control systems to be more reactive to environmental changes. i.e. a growing number of industrialists now want control systems that provide satisfactory, adaptable and robust solutions rather than optimal solutions that require meeting several hard assumptions [23]. Distributed intelligence control has been widely studied to fulfil this requirement. The main idea is that even though the solution obtained might not be optimal, by distributing the calculations over a number of controllers, real-time response can be achieved and complex problems can become tractable [24].

Van Dyke Parunak [25] concludes that challenges to manufacturing problems include desirability (some costs are more important than others), stochasticity (the real world changes unexpectedly), tractability (complex algorithms take too long to compute), chaos (small uncertainties lead to widely divergent outcomes), and decidability (no algorithm exists to predict the operational behaviour). In general, these complex problems, which DI approach is suited to, are those that conventional centralised solutions aren't appropriate. A review paper [6] summarised three characteristics of the complex problems that manufacturing and supply chain operations are facing:

- *Partial information availability* – Due to the uncertainty and distinguished process of the problems, at any time, each possible decision-making node has only part of the information required to make the decision.

- *Impracticality* – even if all information is available to each decision-making node, practical constraints such as time, cost and quality, inhibit a centrally based solution. Moreover, the variety of disruptions and changes are impractical to be modelled and solved using a centralised logic [26].
- *Inadvisability* - even if centralised decision making is practical, it might still be inadvisable due to the susceptibility of a single decision-making node to disruptions and changes, and the complexity of making long term changes under a centralised regime. i.e. since these scheduling functions of higher level controllers usually assume deterministic and simplified behaviour of their lower level components, unforeseen disturbances may invalidate the schedule at an operational level [26].

A number of distributed intelligent systems have been developed in recent years, see, for example, the review papers on Multi-agent system [23], Holonic manufacturing system [24] and Product Intelligence [27][28] are three suitable examples in this paradigm which introduced artificial intelligence technique in practice. Based on these developments, [7] proposed a definition for the distributed intelligence industrial system, of which should apply the following:

- There is a degree of autonomy associated with the operations of the elements of the system;
- The elements of the distributed system display the ability to reason solely or jointly;
- The elements have the ability to interpret the state of their environment and detect the intentions of other elements.

Beyond a significant focus on manufacturing control, we note that DI has also been examined in other industrial domains such as logistics [29][30][31] and maintenance and service [32][33]. Many of the comments made in this section regarding manufacturing could equally apply to these other domains. In the next section, we focus on the application of DI in warehouse management.

4   A Distributed Intelligence Approach to Warehousing

4.1   Overview

In theory, managing a warehouse has many similarities with managing production plans. i.e. the material flows in various stages in manufacturing is similar to products flows in various operational stages/areas in warehousing. In fact, from the interviews conducted with warehouse manages, we noted that in some instances warehouse operations are increasingly seeking to add "manufacturing" operations such as assembly and packaging to their offering. Hence the problems occurring when managing warehouse operations might have the characteristics mentioned in the previous section.

In a typical warehouse, an arriving item will be palletised, stored, replenished, picked, packed, accumulated and shipped in the corresponding areas, as shown in Fig. 2. Since order-picking operations has long been identified as the most labour-intensive and costly activity for almost every warehouse [34], and as the main operation that links different operational stages and areas as shown in Fig. 2, we will discuss the necessity of DI approach through the problem occurring in order-picking operation.

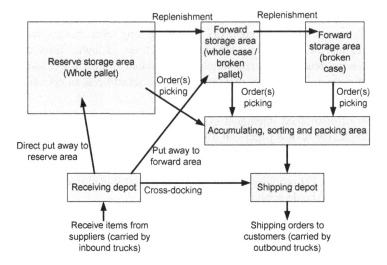

Fig. 2. Typical warehouse process flow (based on [39])

4.2   Order-Picking Operation

Order-picking operations can be described as the following: multiple pickers are assigned the task of transferring products in a warehouse from stationary storage locations to a common loading shed or depot, where the pickers begin and end the trips in the depot [35]. As shown in Table 1, there are four main decisions in scheduling the order-picking tasks. Firstly, the orders that need to be fulfilled should be selected and batched. Once the overall requested SKUs have been determined from step 1, number of picks per tour and corresponding pickers can be scheduled next. Finally, the storage locations to be visited (since one type of SKU might be available at multiple storage locations) and the associated routes for each of the picker can be generated. As in manufacturing control, characteristics of problems occurring when making the order-picking decisions are as the following:

- Partial information availability

Not all the information is available for decision-making. For instance, the information of the overall number of orders will only be partially available since new orders might arrive at any time during working shifts; aisle congestion can only be predicted using sophisticated picker travelling models; real-time status of the pickers such as their positions cannot be tracked by most of the WMS.

Table 1. Decision making criteria for order-picking operation

Decision making criteria	Order-picking decisions			
	1. Order selection and batching	2.Number of picks per tour	3.Assigning orders to pickers	4. Routing of picking tour per picker
Orders status	*			
SKUs[4] properties	*	*		*
Schedule of the following process	*			
Routing of order picking	*	*		
Volume of picking device		*		
Aisle congestion		*		*
MHE[5] availability			*	
Status of pickers			*	*
Storage location				*
Outbound position(s)				*

- Impracticality

Even though with all this information available, unforeseeable events such as change of orders, storage errors or delay of outbound transportation may invalidate the optimal decisions/schedules since they are based on static or deterministic information, as well as hard assumptions. It could be time and computation consuming to regenerate the schedule every time an unforeseeable event occurs. For instance, although several studies have been conducted to provide pickers' real-time position using Wireless Sensor Networks [17] or tracking model [36], the constraints of cost (amount of sensors or RFID systems required and associated deployment strategy) and computation-time (if the number of pickers and their associated transverse areas are large) could become impractical.

- Inadvisability

As shown in Table 1, several criteria are involved in decision-making in different steps, which could potentially lead to conflicting interest. For instance, when generating the optimal route for a picker, conflict interest might occur between optimal routing and order priority (or SKU's entry date, storage locations which contain different quantity of requested SKU). Therefore it would be particularly complex to cope with these conflicts under centralised regime since the weighted variables for different

---

[4] SKUs: Stock Keeping Units.
[5] MHE: Material Handling Equipment.

criteria may vary dynamically in different situations. For example, during peak periods when maximising warehouse throughput is the primary objective, optimal routing will generally have higher priority than other criteria. Whereas during off-season, SKU's entry dates might have higher priority than optimal routing since the primary objective is to gain better inventory management.

As discussed in the section 3, these three possible characteristics make a centralised (conventional) approach inappropriate and hence make the DI approach attractive. A vision for a DI approach to warehouse management is described in the next section.

5   A Vision for DI in Warehouse Management: Example of Order Picking Scheduling

In this section, we present, through an example, the potential application of the distributed intelligence paradigm in warehouse management and compare it with conventional approaches.

5.1   Problem Description

We focus our attention on the order-picking rescheduling problem. In this problem, new orders are arriving in the warehouse management system while static orders – orders that are known – are in the process of being picked. The rescheduling process refers to both the re-assignment of orders to pickers and the re-identification of the best route for the picking tour of each picker. The objective here is the minimisation of the total picking time across all pickers.

Apart from the introduction of new orders in the system, there are other reasons that might lead to the rescheduling of the order-picking plan, such as:

• Certain resources might become unavailable and additional resources might be introduced, e.g. a certain picker has an accident
• Disruptions might occur, e.g. MHE failures, changes in delivery dates, changes in rush orders status.

As discussed in the previous section, due to the stochastic nature of these unexpected events, only partial information is available regarding the total input, and a global solution to the dynamic problem is not possible [37]. Hence, an alternative objective is the minimisation of the picking time each time an unexpected event takes place. Although this does not guarantee the calculation of a global optimum, it can lead to schedules that do take into account the arrival of new orders, disruptions and other unexpected events.

We will present the potential application of a distributed intelligence approach in the problem discussed above using a simplified example regarding the arrival of new orders into a warehouse management system. Fig. 3 shows a part of a warehouse in which two pickers P1 and P2 are in the middle of their order picking process given a

predefined list of orders and a route to be followed. Let L_{ij} denote the different locations in the warehouse where products can be stored and let the circles and the squares
in Fig. 3 represent the locations for P1 and P2 to visit respectively, e.g. the picking list
for P1 is L51, L52, L53, L54, L33, L34. Also, let each picker be able to pick a product from a location if he is in the middle of an aisle in front of this location; that is P1
can pick the product in L51 and/or L41 from his current position. Let the pickers travel on the dotted lines and let the distance between two consecutive nodes on the line
be 1. Assuming that each picker has a capacity of 6 items, P2 is currently having two
spare spaces in his trolley.

Now, let a new order, which requires a particular SKU type with the quantity of
two, arrive in the system. The available locations that contain the required SKU are
marked with doted triangles. In the next section we will show how a DI approach
can be used in order to minimise the total picking time so as to fulfil all the current
orders.

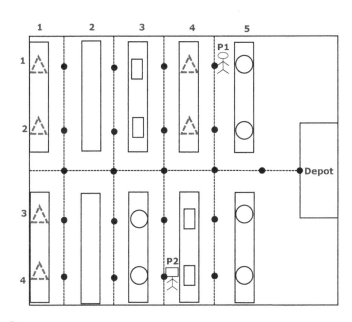

<p>⃝ the initial products on the picking list of Picker P1</p>
<p>▢ the initial products on the picking list of picker P2</p>
<p>△ the available storages that contain the items from the newly arriving order.</p>
<p>---- the aisle for the pickers to be travelled on</p>
<p>● represents points where storage locations can be assessed, intersections between aisles,
or the enter/exist of the warehouse</p>

Fig. 3. Order picking example

5.2   Conventional VS DI Enabled Rescheduling Approach

This section will compare two typical conventional systems with the DI system we envision. The scheduling mechanisms of these three approaches are shown in Fig. 4 and their associated results are compared in table 2.

Considering the most conventional approach that can be used for this problem, we can see that most optimisation algorithms for order picking scheduling are traditionally calculated off-line and consider the problem as being a static and deterministic one [37]. In such a conventional WMS, the new order will be scheduled for the next picking round i.e. after P1 and P2 pick all products from the current picking list.

Moving forward, we can identify other conventional approaches that incorporate the usage of real-time information and some knowledge-based method. In this example the new order will be assigned to P2 since he is currently having two spare spaces in his trolley. Moreover, among all the available locations that P2 can find the right product in, L41 and L42 will be chosen as they can minimise the picker's picking time compared to L11, L12, L13 and L14. The differences in the picking list for these approaches are presented in Table 2, where 8 units of travel distance have been saved.

In a WMS that uses a distributed intelligence approach, artificial intelligence could be applied to pickers or/and storage locations. Through these distributed intelligence, dynamic rescheduling could be enabled. As depicted in Fig. 5, we envision three potential frameworks for their negotiations:

a)   *Direct negotiation between pickers*, in which only picker agents will carry out the negotiations between each other, and cooperatively make decisions corresponding to the changes of environment.

b)   *Picker Storage-locations negotiation*, in which picking operations will be rescheduled by storage location agents based on the bids from picker agents.

c)   *Cluster negotiation*, in which picking tasks are (re-)allocated by the negotiation protocol, such as auction or bargaining, between two leader agents that represents the communities of storage-locations and pickers. For instance, newly arrival orders will be represented by order agents, which will update the request of products and quantities. The leader agents of storage-locations and pickers will then collect the updated information and inform the individual agents in their cluster. After the biding of individual agents, two leader agents will communicate the proposals submitted by their individual agents and negotiate a final schedule.

Assuming framework (a) is applied to the example aforementioned, pickers will now have the opportunity to communicate with each other and perhaps make a different decision for the final picking lists. First of all, the pickers will be scheduled based on the current orders using an optimisation algorithm to start their operation. When a new order arrives, the locations which contain the requested SKUs will flag themselves, indicating that they contain the products of the newly arriving order and the associated quantity. Pickers can then decide whether they should modify their own schedule based on the above information and the status of the other pickers. For the problem described above, P1 knows that he could pick the requested SKUs in the locations L41 and L42 as his is currently in the proper aisle. However, due to capacity limitations of his trolley, P1 needs to negotiate with P2 to see whether P2 can pick up the items in L33 and L43 (which were originally scheduled for P1). Since P2 is currently next to these locations and having two spare spaces in his picking trolley, P2 would accept the offer from P1 in order to minimise the total picking time. The rescheduled picking lists for P1 and P2 are shown in table 2, in which the overall travel time (distance) to fulfil all the orders is decreased compared to the previous two types of conventional approaches.

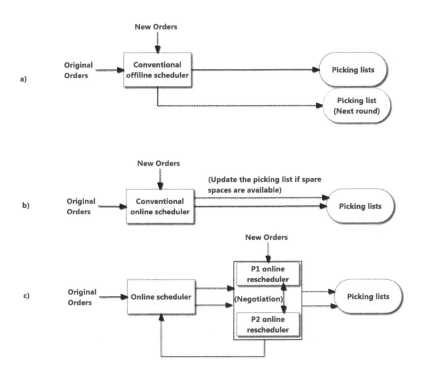

> ➢ Conventional approach(optimal algorithm)
> ➢ Conventional approach (knowledge-based method)
> ➢ Dynamic Rescheduling Approach (Enabled by DI)

Fig. 4. Different approaches to order-picking scheduling

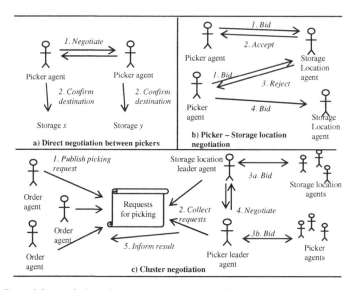

Fig. 5. Potential negotiations frameworks between intelligent pickers and storage locations

Table 2. Picking lists for different types of WMS

Type of WMS	Picker	Picking lists						Travel distance	
Conventional approach	P1	L51	L52	L53	L54	L33	L34	18	
(optimal algorithm)	P2	L44	L43	L32	L31			14	= *40*
	Next round	L41	L42					8	
Conventional approach	P1	L51	L52	L53	L54	L33	L34	18	= *32*
(knowledge-based method)	P2	L44	L43	L32	L31	L41	L42	14	
DI approach	P1	L51	L52	L41	L42	L53	L54	12	= *26*
	P2	L44	L43	L34	L33	L32	L31	14	

The results in Table 2 are simply intended to indicate differences in picking strategies that might evolve using conventional and DI approach. The travel distance results are not intended to highlight performance improvement but simply to show that different picking lists and hence travel time/distance might arise from different scheduling approaches.

We also note that were the scenarios in Fig. 3 and 4 extended to include a greater number of pickers then achieving optimality via the DI solution would become computationally more challenging. Realistically, it is more likely that a (sub-optimal) heuristic would be used in these circumstances.

6   Conclusion

As the increasing number of value added services, types of SKUs and customer demands, warehouse management has been gaining popularity in recent years aiming to improve efficiency and robustness especially for 3rd party warehouses. Since distributed intelligence (DI) approaches have been developed extensively to address similar problems and requirements in manufacturing control, this paper has discussed the potential role of the DI and provided a vision of such approach in warehousing.

A centralised WMS is always better than a distributed approach if all the information is available and deterministic, and no constraints (e.g. cost or process time) are associated with it. However, due to the uncertainty, complexity and dynamicity of the constraints and requirements in real life, the debate of whether DI approach is superior to a centralised one has been raised. Therefore it is necessary to study in some detail the trade-off between these two approach in warehousing so as to determine when, where, under what conditions and how a warehouse should adopt a DI approach rather than a centralised one.

References

1. Jung, D., Semeijn, J., Ghijsen, P.: Evaluating Third Party Logistics Relationships: When provider size matters (2008)
2. Africk, J.M., Calkins, C.: Does asset ownership mean better service? Transportation & Distribution (1994)
3. Large, R.O.: The influence of customer-specific adaptations on the performance of third-party-logistics relationships—document studies and propositions. Int. J. Logist.-Res. App. 10, 123 (2007)
4. Gu, J., Goetschalckx, M., McGinnis, L.F.: Research on warehouse operation: A comprehensive review. European Journal of Operational Research 177, 1–21 (2007)
5. Gupta, Y.P., Goyal, S.: Flexibility of manufacturing systems: concepts and measurements. European Journal of Operational Research 43, 119–135 (1989)
6. Marik, V., McFarlane, D.: Industrial adoption of agent-based technologies. IEEE Intelligent Systems 20, 27–35 (2005)
7. McFarlane, D., Parlikad, A., Neely, A., Thorne, A.: A framework for distributed intelligent automation systems developments. In: IFAC Proceedings Volumes (IFAC-Papers Online), pp. 758–763 (2012)
8. Li, L.: Supply chain management: concepts, techniques and practices enhancing the value through collaboration (2007)
9. Richards, G.: Warehouse management: a complete guide to improving efficiency and minimizing costs in the modern warehouse. Kogan Page, London (2011)
10. Poon, T.C., Choy, K.L., Chow, H.K.H., Lau, H.C.W., Chan, F.T.S., Ho, K.C.: A RFID case-based logistics resource management system for managing order-picking operations in warehouses. Expert Systems with Applications 36, 8277–8301 (2009)
11. Gattorna, J.: Strategic Supply Chain Alignment: Best Practice in Supply Chain Management. Gower Publishing, Ltd. (1998)
12. Faber, N., van de Velde, S.L.: Linking warehouse complexity to warehouse planning and control structure: an exploratory study of the use of warehouse management information systems. International Journal of Physical Distribution & Logistics Management 32, 381–395 (2002)
13. Sexton, J.B., Thomas, E.J., Helmreich, R.L.: Error, stress, and teamwork in medicine and aviation: Cross sectional surveys. British Medical Journal 320, 745–749 (2000)

14. Chow, H., Choy, K., Lee, W., Lau, K.: Design of a RFID case-based resource management system for warehouse operations. Expert Systems with Applications 30, 561–576 (2006)
15. Li, M., Gu, S., Chen, G., Zhu, Z.: A RFID-based Intelligent Warehouse Management System Design and Implementation, pp. 178–184. IEEE (2011)
16. Liu, T., Liu, J., Liu, B.: Design of intelligent warehouse measure and control system based on Zigbee WSN. In: 2010 IEEE International Conference on Mechatronics and Automation (ICMA 2010), pp. 888–893 (2010)
17. Rohrig, C., Spieker, S.: Tracking of transport vehicles for warehouse management using a wireless sensor network. In: IEEE/RSJ International Conference on Intelligent Robots and Systems (IROS 2008), pp. 3260–3265 (2008)
18. Shen, S., Wang, D.: Research on Warehouses Management Based on RFID and WSN Technology. In: 2010 2nd International Workshop on Database Technology and Applications (DBTA), pp. 1–4 (2010)
19. Kim, B.-I., Graves, R.J., Heragu, S.S., St. Onge, A.: Intelligent agent modelling of an industrial warehousing problem. IIE Transactions (Institute of Industrial Engineers) 34, 601–612 (2002)
20. Gu, J., Goetschalckx, M., McGinnis, L.F.: Research on warehouse design and performance evaluation: A comprehensive review. European Journal of Operational Research 203, 539–549 (2010)
21. Leitão, P.: Agent-based distributed manufacturing control: A state-of-the-art survey. Engineering Applications of Artificial Intelligence 22, 979–991 (2009)
22. Baker, A.D.: A Survey of Factory Control Algorithms That Can Be Implemented in a Multi-Agent Heterarchy: Dispatching, Scheduling, and Pull. Journal of Manufacturing Systems 17, 297–320 (1998)
23. Trentesaux, D.: Distributed control of production systems. Engineering Applications of Artificial Intelligence 22, 971–978 (2009)
24. Babiceanu, R.F., Chen, F.F.: Development and applications of holonic manufacturing systems: a survey. Journal of Intelligent Manufacturing 17, 111–131 (2006)
25. Van Dyke Parunak, H.: Characterizing the manufacturing scheduling problem. Journal of Manufacturing Systems 10, 241–259 (1991)
26. Van Brussel, H., Wyns, J., Valckenaers, P., Bongaerts, L., Peeters, P.: Reference architecture for holonic manufacturing systems: PROSA. Computers in Industry 37, 255–274 (1998)
27. McFarlane, D., Giannikas, V., Wong, A.C.Y., Harrison, M.: Product Intelligence in industrial control: Theory And Practice. Annual Reviews in Control 37 (2013)
28. Meyer, G.: Production monitoring and control with intelligent products. International Journal of Production Research 49(5), 1303–1317 (2011)
29. Kola, D., Giannikas, V., McFarlane, D.: Travel behaviour applied in freight transportation using intelligent products. In: 2012 2nd International Conference on Communications, Computing and Control Applications (CCCA), pp. 1–6 (2012)
30. Van Belle, J., Saint Germain, B., Valckenaers, P., Van Brussel, H., Bahtiar, R., Cattrysse, D.: Intelligent products in the supply chain are merging logistic and manufacturing operations. In: Preprints of the 18th IFAC World Congress, pp. 1596–1601 (2011)
31. Law Kozlak, J., Créput, J.-C., Hilaire, V., Koukam, A.: Multi-Agent Environment for Modelling and Solving Dynamic Transport Problems. Computing and Informatics 28, 277–298 (2009)
32. Brintrup, A., Ieee, M., McFarlane, D., Ranasinghe, D., Sanchez, T.: Will intelligent assets take off? Towards self-serving aircrafts, pp. 1–13 (2002)

33. Yang, X., Moore, P., Chong, S.K.: Intelligent products: From lifecycle data acquisition to enabling product-related services. Computers in Industry 60, 184–194 (2009)
34. De Koster, R., Le-Duc, T., Roodbergen, K.J.: Design and control of warehouse order picking: A literature review. European Journal of Operational Research 182, 481–501 (2007)
35. Rubrico, J.I.U., Ota, J., Higashi, T., Tamura, H.: Scheduling multiple agents for picking products in a warehouse. In: Proceedings 2006 IEEE International Conference on Robotics and Automation (ICRA 2006), pp. 1438–1443 (2006)
36. Kelepouris, T., McFarlane, D., Giannikas, V.: A supply chain tracking model using auto-ID observations. International Journal of Information Systems and Supply Chain Management 5, 1–22 (2012)
37. Rubrico, J.I.U., Higashi, T., Tamura, H., Ota, J.: Online rescheduling of multiple picking agents for warehouse management. Robotics and Computer-Integrated Manufacturing 27, 62–71 (2011)
38. AberdeenGroup: The warehouse productivity benchmark report (2006)
39. Tompkins, J.A.: Facilities Planning. John Wiley & Sons, NJ (2003)
40. Chiang, D.M.-H., Lin, C.-P., Chen, M.-C.: The adaptive approach for storage assignment by mining data of warehouse management system for distribution centres. Enterprise Information Systems 5, 219–234 (2011)

Part II
Holonic and Multi-Agent Technologies for Manufacturing Planning and Control

An Extended Contract Net Protocol
with Direct Negotiation of Managers

Doru Panescu and Carlos Pascal

"Gheorghe Asachi" Technical University of Iasi,
Department of Automatic Control and Applied Informatics, Iasi, Romania
{dorup,cpascal}@ac.tuiasi.ro

Abstract. This paper introduces a new coordination scheme for multi-agent systems. This is obtained by combining the contract net protocol with a direct negotiation between managers. The results regarding the resource allocation in a distributed system are favourable, as proved by the state space analysis obtained with a Petri net model. The envisaged application is for holonic manufacturing execution systems.

Keywords: Coordination, contract net protocol, multi-agent systems, coloured Petri nets, holonic systems.

1 Introduction

Coordination is a central problem for distributed control architectures. A widely used mechanism is the Contract Net Protocol (CNP), [1]. Besides the advantages CNP offers – a simple and general approach, it also determines some drawbacks. Thus, a distributed system that operates according to the CNP can fail in finding a solution though this exists and it can conduct to a result that is not the optimal one [2 - 4].

Different extensions were proposed for the CNP in order to diminish or eliminate its weak points. In [5], a classification of the coordination protocols according to the type of environment is done. A cooperative environment regards the case when agents either share a common goal or they have different goals, but their operation is directed towards the maximization of the agents' collective welfare. The second type of environment is composed of competitor agents, meaning they have distinct goals and they try to maximize their individual interests, no matter what is the global state. The respective work introduces an extension of the CNP for competitive agents, while the present paper is an attempt to enhance the result obtained when applying CNP in the case of a cooperative environment.

A distinct improvement of CNP is presented in [6]; here, in the same way as in the proposed extension, it is possible for a contractor to make bids for more goals, even when it would not have enough capabilities to satisfy all the goals. To solve potential conflicts, a special committing phase is introduced in [6]. In our approach, the procedure for resource allocation is detailed, including the needed message exchange; moreover, in our solution, a distinct feature is the way the decision is taken by managers.

Another research direction was to introduce into the coordination scheme an additional component – a centralized one. In different methodologies this supplementary agent has its operation tuned according to the information it possesses. The simplest role is that of a dispatcher, as in [2], when the central agent is a staff holon within a holonic structure. In this case, the centralized component knowledge on the other agents and environment is reduced. The application of different optimization criteria is possible too, when the central agent has certain a priori competency in the domain of problems to be solved and also information acquired from the other agents; such examples are presented in [7], when the centralized part appears as a planning and coordination agent consortium, and in [8] when it is named knowledge manager agent.

This paper considers a further possibility to enhance the CNP. In the normal CNP, the exchange of messages appears between one manager and one or more contractors, namely those agents receiving a goal from the manager. The basic procedure of CNP does not discuss about the case when more managers are operating and it happens that their goals regard common resources (contractors), though this is a common case. For an enhanced performance in such circumstances, an additional decisional mechanism is needed, besides the choice of the best bid done by each manager alone. While the previously mentioned involvement of a centralized component can be a solution [2] (namely, the central agent allows, delays, or denies the access of managers to resources), another coordination scheme, the one considered in this paper, can use a direct communication and negotiation between the managers that try to access common resources. First, the paper describes the new, extended form of the CNP. Then, the model-prototype system that allowed us to verify the performance of the proposed coordination scheme is briefly introduced. This analysis is presented in the following section, which deals with two illustrative scenarios. Then, an evaluation of the proposed scheme optimality is done, conducting to an additional enhancement of the coordination mechanism. The paper ends with certain conclusions, focused on explaining the applicability of the proposed coordination scheme for holonic manufacturing execution systems (HMESs).

2   The Extended CNP with Direct Negotiation of Managers

The proposed coordination scheme uses managers and contractors as common for CNP. Managers issue goals towards potential contractors when they are not able to solve a task by themselves. Contractors should provide bids and a problem appears when more managers need common resources for solving their goals. In such a case, the new solution is to initiate a message exchange between the managers that necessitate the same resource, allowing them to find a proper allocation. In comparison with the common CNP, certain differences appear with respect to the behaviour of managers and contractors. The operation of a manager should be guided by the following steps:

1. The manager announces the goal towards all potential contractors. In the proposed approach a manager is aware of all the agents that are able to solve its goal according to the information provided by the staff holonic agent existing in the considered architecture (some further details about this will be given in Section 3).

2. The manager waits to receive all the bids or until the deadline is reached (the standardization of FIPA is considered [9]). The manager can receive two types of bids. First, there are simple bids (SBs) from contractors that are not involved in another relation with a manager (these are the normal bids of the common CNP); this category also includes the refusals, which will be named negative bids (NBs). The second type refers the conditional bids (CBs). This new type appears when a contractor already made a bid towards another manager and still has not received the answer from that manager. In such a case, the CB will contain, besides its cost, the information about which is the other agent (manager) that previously requested the contractor.

3. For each CB being received, the manager sends a message requesting the contractor's discharge, towards the agent mentioned in the CB. Such a message will be named discharge request (DR), meaning that a manager asks another manager to free (disengage) a contractor (a resource). The DR contains a cost (a quantitative parameter) reflecting the benefit obtained by the manager (the sender of the message) if it gets the resource. The recipient of a DR should reply with a discharge answer (DA). This can be a negative answer, meaning the other agent decided to keep the resource, or a favourable answer, when the resource is freed.

4. When the manager has received all the information regarding a goal (both SBs and the DAs for CBs) or when the deadline is reached, it analyses the set of bids. This means a comparison between the elements of a set (see Fig. 1) containing both SBs and those CBs that are accompanied by favourable DAs (when a CB is followed by a negative DA it should not be taken into account, being equivalent with a NB). If the set of bids contains only NBs, then the manager fails in solving the goal. Otherwise the manager enters an analysing stage. This regards not only the choice of the best bid, but also the decision on the DRs that were possibly received with respect to certain bids of the analysed set of bids. The manager finds the best bid. If this is not accompanied by a DR, then the corresponding contractor is chosen and a contract is sent towards it. If that bid has been followed by a DR, then the manager has to compare the benefit obtained for the whole system when the resource is kept for itself with the benefit that could be obtained by the requesting agent (according to the information provided in the DR). If the decision is to free the resource, then a message containing a favourable DA is sent, and the corresponding resource is no longer considered in the evaluation process (a refusal message is sent towards that resource). In this case, the process of finding the best bid is reiterated. In the specific case when there is no bid to choose due to the fact that all received bids were followed by DRs and all these requests got favourable answers, the manager does not succeed to solve the goal. After a bid was selected, regarding the remaining, unselected bids there are two cases. For such a bid that was followed by a DR, the manager issues a favourable DA (as it decided not to use that resource) towards the demanding agent, as well as the refusal message towards the contractor. For an unselected bid that regards a contractor not being requested by any other manager, only the announcement of rejection is issued.

An important point to comprehend the above protocol refers the significance of bids and requests, as well as the way they are succeeding; Fig. 1 gives an overview regarding the classification of messages. Bids represent responses of contractors to goals. A goal can conduct to an SB or a CB; the SBs can be classified in favourable (marked with PB in Fig. 1) and NBs. A CB is to be coupled with a DR, which should conduct to a received DA (the various relations are marked with arrows in Fig. 1). The set of received DAs is divided into favourable ones (P_DA in Fig. 1) and nega- tive DAs (marked with N_DA in Fig. 1). Each P_DA determines the corresponding CB to become a PB, while an N_DA is coupled with an NB. As about the set of DRs received by a manager, Fig. 1 suggests that this set conducts to the set of issued DAs. Such an answer will be a negative one in two cases. If the manager has received an unfavourable answer from another manager in the decisional chain (the corresponding CB became an NB), then it cannot free a resource that was not able to gain and thus it issues an N_DA (examples of this type will be commented in Section 4). A PB can also conduct to an N_DA when the manager decides to keep the resource as being more profitable for the system (in Fig. 1 the arrow linking a PB with an N_DA is marked with S_B, as it regards a bid that was selected). A PB is followed by a P_DA that is issued if the manager decides to liberate the resource for a requesting agent (the bid was not selected, the label US_B being placed on the corresponding arrow in Fig. 1).

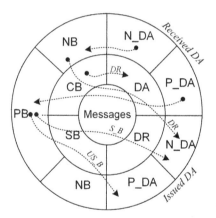

Fig. 1. Classification of messages in the extended CNP

As about the operation scheme for a contractor, this is quite simple to understand. It contains the following steps:

1. A contractor receives a goal.
2. It replies with an NB, if it is already committed, according to a received contract.
3. If the agent is free (it has not issued another bid or it has not exhausted its execu- tion capacity), it replies with an SB.

4. If the agent already made one or more bids towards other managers, exhausting its carrying out capacity, and without receiving a contract (an acceptance for its proposals), then it replies with a CB indicating both the cost of the bid and which is the last manager that previously received a bid.
5. If the agent receives a contract, then it enters the execution stage and according to its ending it will provide the proper answer.

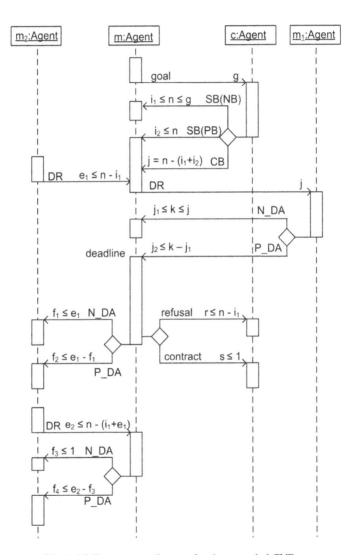

Fig. 2. UML sequence diagram for the extended CNP

To summarize the interaction between managers and contractors (all of them are instances of the Agent class), a UML sequence diagram is presented in Fig. 2. It is supposed that the initiator manager (m in Fig. 2) issues the goal towards g potential contractors. From them, only n agents are providing bids during the established deadline. There are $i_1 + i_2$ SBs (in the two categories, NBs and PBs) and j CBs. After sending the DRs (there are j DRs, one for each CB), the manager m receives k DAs. It can also receive some DRs, from other managers, related to the previously got bids (e_1 such requests appear before the deadline, and e_2 after it). As already explained, when the deadline is reached the manager m issues at most a contract (to simplify the explanation, it is supposed that the announced goal can be solved by assigning a single contract) and a number of refusals that can equal the number of received bids. In Fig. 2, $r = n - i_1$ if all the available resources are released for other managers; for those contractors that provided NBs (the number of NBs is i_1), there is no need for refusal. It is to mention that the new coordination scheme considers an errorless operation only for the initiator manager; that is why the number of DAs sent by the agent m equals the number of received DRs, while some of the messages to be obtained by the agent m may not be received ($n \leq g$ and $k \leq j$). The proposed extension of CNP targeted only the planning phase. According to the result of the execution stage (not represented in Fig. 2), which takes place according to the normal CNP, in the case of an execution failure the corresponding goal will be re-issued.

With respect to the operation of a contractor, a possible case is when an agent already made (in succession) a SB and a CB, and then it receives another goal while it still does not get answers for the previous bids. In this situation, the agent makes a further CB. About the information within the CB on the managers that already received a proposal from the same contractor, this can indicate the last manager to which a bid was sent (the previously presented operation cycles for managers and contractors considered this case), or the set of all managers that received bids. In this way, two variants of the coordination protocol result. First, the approach when the CB contains information on only the last manager to which the contractor interacted is presented; then, in Section 5, the second variant of the extended CNP is discussed.

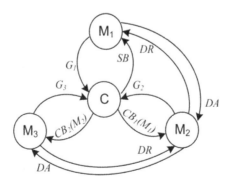

Fig. 3. The extended CNP operation when more managers compete for the same resource

A scenario with a chain of negotiations is explained in Fig. 3. The contractor C sends an SB towards the manager M_1 from which it has received the first goal. Then, it receives another goal from the manager M_2. For it, a CB (CB_1 in Fig. 3) is issued indicating M_1 as the manager towards which a previous bid was sent. Finally, it happens that another goal is got by the agent C from the manager M_3. Because it is yet not engaged with a contract, the agent C issues an additional CB (CB_2 in Fig. 3) indicating that the last previous bid was made for the manager M_2. Thus, the sequence issued by a contractor containing one SB and one or more CBs determines a negotiation chain between competitors for the same resource. The decision regarding which of them will use the resource is got according to the exchange of information between them (according to the DRs and DAs being sent and received). In the first variant, the communication is established between pairs of managers that successively requested the same resource; another possibility would be to calculate the maximal benefit of the entire set of solicitors, not only between sets of two managers, as explained in Section 5.

3 The Model Used for the Coordination Protocol Analysis

The proposed coordination mechanism was analysed by using a previously constructed model for a distributed system of the holonic type [10], [11]. HMESs regard a control architecture where the decisional entities are holons, mainly developed according to the principles of multi-agent systems. In fact, a holon has two components: a decisional entity that is an agent (holonic agent), and a structural component being either a physical device (a manufacturing equipment) or another holon/group of holons that has the role to execute the commands of the decisional part [10]. The dynamically formed groups of holons constituting structural components make up holarchies. Though holonic operation comprises some specific aspects, the issue addressed by this paper refers the interaction between holonic agents, meaning the discussion is about the agents' coordination.

For studying the behaviour of HMESs, a Coloured Petri Net (CPN) model was developed [10 - 12]. It is built as a hierarchical structure, containing successive layers of CPN models, for the entire HMES, holons, plans and the Belief Desire Intention (BDI) deliberative component of every holonic agent. Thus, the obtained model is a prototype as it includes the real system workflow and the procedural part of CPNs contains functions that can be used to develop holonic agents.

In the present research, for each holon only its holonic agent is considered (the structural component was not the object of this study), each plan contains a single action (because the only issue of interest was to find a resource allocation scheme), and the BDI mechanism was only involved to switch from sending a favourable answer when a plan can be completed to issuing an NB when the plan fails.

The analysing tool is the reachability graph (RG). In order to reduce its dimension, the set of contractor agents was replaced by a simpler model, which contains a simulation function (Contractor SIM in Fig. 4). The highest model layer for one of the experiments described in the next section (with three managers and a contractor) is displayed in Fig. 4, with the usual notations for places and transitions of Petri nets. The communication network is modelled according to an ideal operation, namely a message always arrives at its address and the order in which messages are issued is respected at receivers. The HMES structure contains the staff holonic agent (see Fig. 4) as a centralized component, which could be included in the coordination process, too [13]. In the present approach the staff holonic agent does not have an active coordination role, but it only provides the list of potential contractors for managers. Concerning this, in an initialization phase each agent informs the staff holonic agent about the types of goals it is able to solve.

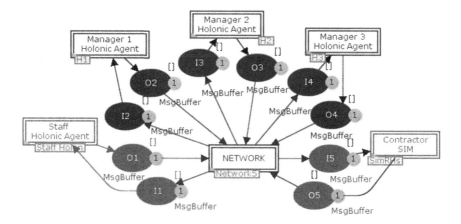

Fig. 4. The highest model layer for an experiment with three managers and a contractor

4   The Coordination Protocol Analysis

Two case studies were analysed. The first one is with two managers (named M_1 and M_2) and two contractors (resources). The managers issue goals that can be solved by both contractors, meaning a solution exists for the entire system (there are enough resources). In this case the RG contains eight dead markings; these are the final states, which give information on the result of the coordination process according to the token contents, [10 - 12]. The brief description of the eight possible results is:

1. The manager M_1 receives two SBs from the two contractors and two DRs from M_2, which received two CBs. The manager M_1 decides to free a resource so that it issues a favourable DA.
2. The manager M_1 receives two SBs and the DRs arrive at it after the end of its decisional process. Thus, because a resource is enough for it, two DAs are issued: a favourable one for the resource that was not chosen, and a negative DA for the resource being already engaged.
3. The manager M_1 receives two SBs and from the two DRs of M_2 one arrives before its decisional process end and the other after it. The final state reflects the case

when M_2 receives a favourable DA for that resource that is not chosen by M_1 (this can be related with either the first or the second DR).

4. The manager M_1 receives two SBs and decides about using one resource; its decisions are sent to contractors, and thus one contractor issues a PB towards the manager M_2, because the goal from M_2 arrives at that contractor after it got the answer from M_1.

The other four dead markings regard the cases when the manager M_2 is the one that receives SBs from contractors (the role of M_1 and M_2 is switched). The results of this experiment are displayed in Table 1; for comparison, the case of involving the common CNP and the case of using the staff holon (the centralized component) are presented together with the proposed extended CNP. These results show that the new coordination protocol always assures a solution for the considered combination of managers and contractors. When the common CNP is used the analysis highlights the possibility for a manager to fail in fulfilling its goal (two of the four final states obtained with the common CNP represent cases when a manager does not achieve its goal). A staff holon based solution is also safe, as presented in detail in [2]. The other information appearing in Table 1 refers the state space dimension (number of nodes and arcs), the time needed to obtain the RG and the number of messages exchanged between agents until obtaining a final state. About this last aspect, the two different numbers of messages for the extended CNP can be explained by the distinct communication processes (with more or less CBs and DAs) that are run until reaching the final states.

Table 1. RG for the experiment with two managers and two contractors

	Common CNP	Staff Holon	Extended CNP
Dead markings	4	2	8
Nodes	1126	719	2984
Arcs	1480	934	4776
Time to obtain the RG (seconds)	271	151	695
Msg. No.	17	19	20 or 22

The second experiment regards a system with three managers and a single contractor. This scenario is important because it reveals the way the decisional chain appears between more managers (see Fig. 3, too). The obtained RG showed the existence of three classes of dead markings, as follows:

1. The manager that receives the SB provides a negative DA to the requesting manager, which thus replies with a negative DA to the last manager.
2. The manager receiving the SB decides to free the resource for the requesting manager; this second manager decides to keep the resource, providing a negative DA to the last manager.

3. The manager that receives the SB and the first requesting manager decide to free the resource (they successively provide favourable DAs), so that the last manager is the one that gains the resource.

The results of this experiment are presented in Table 2, again in comparison with the other two coordination schemes. For the extended CNP, nine final states appear, but as already told, they can be grouped into three classes; the three dead markings of the same class conduct to the same resource allocation and they differ only with respect to the message exchange process. The difference of the RG dimension between the three methods presented in Tables 1 and 2 is commented in the final section.

Table 2. RG for the experiment with three managers and a contractor

	Common CNP	Staff Holon	Extended CNP
Dead markings	3	3	9
Nodes	12950	9241	53436
Arcs	19347	12634	84753
Time to obtain the RG (seconds)	2813	1505	10777
Msg. No.	18	21	24

5   On the Proposed Coordination Protocol Optimality

For a scenario in which more managers try to use the same resources, the optimality of the coordination protocol can be discussed. In the proposed scheme, when more than two managers are involved the optimum may not be reached. The example presented in Fig. 3 and analysed in the previous section is illustrative for this. Suppose that the greatest benefit is accomplished by manager M_3, while the lowest one is obtained when the contractor is used by M_2. In this case, the agent M_1 (which happens to gain a benefit greater than M_2, but lower than M_3) provides an N_DA to M_2, which in turn replies with an N_DA to M_3 because it was not able to get the resource. Thus the optimum is not obtained, the explanation being that the negotiation process (when the benefits of different managers are compared) is only performed pairwise between agents, and not across the entire set of managers. To improve our protocol, a new version was tested in which an attempt to use the information on benefits of all managers that compete for the same resources was considered.

The weak point of the previously presented negotiation scheme stems from the way in which the contractor issues a CB containing only the information about the last manager to which it transmitted a bid. Instead of this, the contractor can include into the CB the information on all the managers to which it already sent a bid. It means the fourth step of the contractor's operation cycle (see Section 2) should be correspondingly modified. Accordingly, the manager's working cycle is to be slightly adapted: during the third step (see Section 2), the manager has to send a DR towards each agent being mentioned in the received CB.

The way the coordination protocol is further improved can be understood by considering the previous example with three managers and a contractor. A scenario is presented in Fig. 5, when it happens that the last manager in the negotiation chain obtains the greatest benefit (B in Fig. 5). The benefit obtained by each agent if it gains the contractor is considered a positive number. Fig. 5 displays the messages exchanged by agents, their order being shown by the number attached to each arc (the goal G_1 is the first message, followed by the SB and so on). The bid CB_2 informs the agent M_3 about the other two managers that try to gain the contractor. Thus, M_3 issues two DRs (messages labelled with 8 and 9 in Fig. 5) towards M_1 and M_2 and it may obtain the resource, for example when it receives a favourable DA from M_1. This is the optimal solution for the considered scenario, which could not be obtained in the previous form of the coordination protocol.

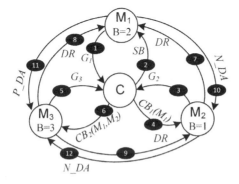

Fig. 5. The extended CNP operation with the decision based on the entire set of managers

With this improvement, the optimality of the proposed coordination scheme can be further analysed. Referring the already introduced example (see Figs. 3 and 5), six cases can appear according to the relation between the chronology of events – the order of bids received by managers, and the benefit they obtain; these six possibilities are presented in Table 3, where CB_1 is the first conditional bid made by the contractor and CB_2 is the second one.

Table 3. Possible cases according to the relation between bids and managers' benefits

Benefit / Bids	Maximum	Medium	Minimum
1	CB_2	CB_1	SB
2	CB_1	CB_2	SB
3	CB_2	SB	CB_1
4	CB_1	SB	CB_2
5	SB	CB_1	CB_2
6	SB	CB_2	CB_1

Each case appearing in Table 3 was analysed using the CPN model introduced in Section 3, by constructing the RG. This approach allowed us to determine all the final states possible to appear for the considered system. For example, the scenario of Fig. 5, which corresponds to the third row in Table 3, produces four dead markings depending on the events' order. Besides the case presented in Fig. 5, which corresponds to a negotiation between all three managers, three other happenings are possible. One is when the manager M_1 decides to keep the contractor because the DR messages arrive at it after its decision about awarding the contract to the agent C. In this case M_1 sends N_DAs both to M_2 and M_3, and certainly M_2 transmits an N_DA to M_3. Another final state corresponds to a negotiation carried out only between M_1 and M_2. M_1 sends an N_DA to M_2 (because its benefit is greater) and the DR from M_3 arrives at M_1 after it took the decision of awarding the contract. The fourth case corresponds to a negotiation carried out only between M_1 and M_3, happening when the DR from M_2 arrives at M_1 after the DR sent by M_3. This final state determines the same result as the one in Fig. 5, namely the agent M_3 gains the contractor. These four possible final states can be related to the way negotiation takes place: there is no negotiation, a negotiation between two agents (with two cases – between M_1 and M_2 or between M_1 and M_3) and a negotiation between all three managers.

Thus the RG-based analysis revealed that the obtained result is dependent on the time relation between the decisional instant of each manager (the moment when it decides to send a contract or a refusal to the contractor) and the arrival of DR messages at the manager agents. An interesting case is the one corresponding to the first row in Table 3. It regards the scenario when the agent obtaining the greatest benefit receives the last CB, while the SB is got by the agent with the minimum benefit (the difference from the case in Fig. 5 refers the managers M_1 and M_2: now M_2 gets the SB and M_1 the CB_1). The five dead markings obtained in the RG for this case are:

1. The agent M_1 takes the decision (awards the contract to agent C) before receiving the DRs; a non-optimal solution is obtained.
2. The agent M_1 receives the DR from M_2 and replies with a P_DA; M_2 takes the decision of awarding the contract to agent C before receiving the DR from M_3; a non-optimal solution is got.
3. Similar with the previous case, but now M_2 receives the DR from M_3 before its decisional instant, so that it provides a P_DA to M_3 (M_3 receives an N_DA from M_1 because M_1 already released the contractor for M_2); the optimal solution is reached.
4. The agent M_1 sends a P_DA to M_3 (it happens that the DR from M_3 arrives before the DR from M_2); in this succession of events M_2 receives an N_DA from M_1 and M_3 gets an N_DA from M_2, but again the optimal solution is obtained.
5. The agent M_1 receives the two DRs and replies with a P_DA for M_3 and an N_DA for M_2 (the decision of M_1 is obtained by comparing the benefits of all three managers); again the optimal solution is obtained.

It is to underline how the final states depend on the way the negotiation process takes place. The above five possibilities are: no negotiation; negotiation between M_1 and M_2; negotiation between M_1 and M_2, then between M_2 and M_3; negotiation between M_1 and M_3; negotiation between M_1, M_2 and M_3.

To conclude, regarding the optimality of the proposed coordination scheme, it is clear that this last version of the protocol, when a manager is informed by its contractor about all the other managers that try to gain the contractor can determine the optimal result more often than any of the other considered approaches. Even so, a non-optimal result is still possible. As stated in the above presented analysis, this appears only when a manager takes a decision on awarding a contract before the instant when all messages from other agents requesting that contractor arrive at it. It means that the optimal operation can be reached when the deadline for the manager's decision (see Fig. 2) is postponed as much as possible. It is clear that the right balance has to be found: a quick decisional process may conduct to a non-optimal result, a postponed decision can affect the system real time operation; the trade-off has to be judged depending on each application.

6   Conclusions

This paper deals with a new coordination protocol for agent-based systems, which is an augmentation of the CNP. The novelty of this approach regards the combination between the classical bidding mechanism and the direct negotiation of the entities (managers) that try to acquire resources. Its suitability was investigated and restricted to the case of cooperative entities, as holons of an HMES are. The proposed coordination scheme can determine an improved operation for an HMES because the managers involved in solving holonic goals can be the ones to better decide on the optimal resource utilization. For example, in the PROSA based holonic architecture that we used [14], which is named HAPBA [2], [10], [11], managers can be order or product holons. Such holons possess the information on which of the manufacturing commands is to be treated first in the case of limited resources in order to obtain the maximum benefit for the whole system. When the common CNP is used, it can happen that a resource is kept by a holon whose goal is less beneficial for the HMES in comparison with the goals of other holons. In the developed scheme, the order/product holons can directly negotiate and thus the optimal solution may be obtained.

The negotiation phase that was added to the CNP has as negative side effect an increasing of the message exchange process. As the results presented in Tables 1 and 2 show, the growth of message number is not significant for the proposed scheme, when compared with the common CNP. This can be explained as the additional message exchange appears only between those managers interested in acquiring the same resource, so that the cardinality of the set of negotiating agents should not be a great one. Nevertheless, the obtained results (see Tables 1 and 2) reveal an important extension of the RG. This is explained by the fact that the negotiation between managers determines a great number of new states. On one hand this means that the coordination process duration is increased, on the other it signifies that the decisions are taken after considering a larger set of possibilities, meaning the developed protocol can be closer to the optimal solution.

Though the new coordination scheme has been tested for only two cases, it is to notice that state space analysis supported by the used CPN models is much more than a simple simulation. It allows us to derive conclusions on the absence of deadlocks and the correctness of the provided solutions according to the token contents in the final states. The two experiments were attentively chosen, so that they can show how more contractors are treated by the same set of managers (the first considered scenario) and the way the negotiation is sequenced when more managers are involved (the second case).

Regarding optimality, the proposed coordination scheme is close to producing the optimal resource allocation in a distributed system through its second variant. Namely, when the negotiation is carried out between all managers that plan to use the same resource, the result conducting to the greatest benefit for the whole system can be obtained. To meet this case, the time interval that agents wait for messages from other managers before deciding about acquiring and locking a resource has to be tuned according to the problem to be solved. In this paper agents used simple plans, with only one action; about this, a further research would be to investigate the proposed coordination scheme behaviour for complex plans. Another future work regards testing the coordination scheme within an experimental manufacturing system, by developing the holonic agents with the Erlang programming environment.

References

1. Smith, R.G.: The contract net protocol: high level communication and control in a distributed problem solver. IEE Transactions on Computers 29, 1104–1113 (1980)
2. Panescu, D., Pascal, C.: HAPBA – A Holonic Adaptive Plan-Based Architecture. In: Borangiu, T., Thomas, A., Trentesaux, D. (eds.) SOHOMA 2011. SCI, vol. 402, pp. 61–74. Springer, Heidelberg (2012)
3. Hsieh, F.-S.: Analysis of contract net in multi-agent systems. Automatica 42(5), 733–740 (2006)
4. Borangiu, T., Raileanu, S., Trentesaux, D., Berger, T., Iacob, I.: Distributed Manufacturing Control with Extended CNP Interaction of Intelligent Products. In: Proc. of INCOM 2012, vol. 14, part 1, pp. 734–739 (2012)
5. Vokřínek, J., Bíba, J., Hodík, J., Vybíhal, J., Pěchouček, M.: Competitive Contract Net Protocol. In: van Leeuwen, J., Italiano, G.F., van der Hoek, W., Meinel, C., Sack, H., Plášil, F. (eds.) SOFSEM 2007. LNCS, vol. 4362, pp. 656–668. Springer, Heidelberg (2007)
6. Kadera, P., Tichy, P.: Plan, Commit, Execute Protocol in Multi-agent Systems. In: Mařík, V., Strasser, T., Zoitl, A. (eds.) HoloMAS 2009. LNCS, vol. 5696, pp. 155–164. Springer, Heidelberg (2009)
7. Fung, R., Chen, T.: A multiagent supply chain planning and coordination architecture. Int. J. Adv. Manuf. Technol. 25, 811–819 (2005)
8. Wang, X., Wong, T.N., Wang, G.: Service-oriented architecture for ontologies supporting multi-agent system negotiations in virtual enterprise. J. Intell. Manuf. 23, 1331–1349 (2012)
9. FIPA: The Foundation for Intelligent Physical Agents, Geneva, Switzerland (2002)

10. Panescu, D., Pascal, C.: On a holonic adaptive plan-based architecture: planning scheme and holons' life periods. Int. J. Adv. Manuf. Technol. 63(5-8), 757–769 (2012)
11. Pascal, C., Panescu, D.: Modeling a Holonic Agent based Solution by Petri Nets. Computer Science and Information Systems 9(3), 1287–1305 (2012)
12. Jensen, K., Kristensen, L.: Coloured Petri nets: modelling and validation of concurrent systems. Springer, New York (2009)
13. Panescu, D., Pascal, C.: On the Staff Holon Operation in a Holonic Manufacturing System Architecture. In: Proc. of ICSTCC 2012, pp. 427–432 (2012)
14. Valckenaers, P., Van Brussel, H., Wyns, J., Bongaerts, L., Peeters, P.: Designing holonic manufacturing systems. Robotics and CIM 14, 455–464 (1998)

Towards an Ontology for Distributed Manufacturing Control

Silviu Raileanu, Theodor Borangiu, and Stefan Radulescu

University Politehnica of Bucharest,
Dept. of Automation and Industrial Informatics, Romania
{theodor.borangiu,silviu.raileanu,
stefan.radulescu}@cimr.pub.ro

Abstract. This paper describes a solution for standardizing the structure and communication acting in a decentralized control architecture implemented through a multi-agent development framework. The standardization is done by defining an ontology that deals with syntactic and semantic validation of the content of exchanged messages.

Keywords: distributed manufacturing control, multi-agent system, ontology.

1 Introduction

During the last years the research associated to discrete manufacturing control had been focused on dealing with growing requirements on flexibility, reconfigurability and fault tolerance especially through the proposal of new decentralized architectures [1][2]. At this pace the control system's complexity grew [3] to such a level that it became difficult to be managed, leading thus to problems in interoperability both at machine-to-machine level (lack of standardization) and at human-to-human level (no common vocabulary) [4]. Some solutions have been proposed at the architectural level consisting of changing from a many-to-many implementation with several different interfaces to an implementation based on a common communication infrastructure and standard interfaces [5]. Practically this means implementing SOA [6] in manufacturing control. This is done using a software architecture with a unique entity in charge of event handling, protocol conversion, queuing and buffering a.o. [5]. Based on the Enterprise Service Bus concept the Manufacturing Service Bus (MSB) concept arose. The main disadvantage of this approach is that the MSB becomes a single point of failure. Also, the fact that the MSB is in charge with protocol conversion, message queuing and buffering might result in lag in communication and difficulties in developing new standard interfaces for the new added control entities.

A solution would be the usage of a multi-agent framework as follows:

1. *agentification* of control entities: each control entity has an associated software agent in charge with the decision making process and with the communication through a standard medium. These agents will be seen as automation objects – abstraction of mechanical devices with encapsulated intelligence [7]) allowing thus component reusability;

2. development of a *common vocabulary* which defines the elements of the system, their attributes and the relation between them. This corresponds to what is called an ontology [8].

The term ontology refers to a formal and common vocabulary for a shared domain [9]. With its origins in the Artificial Intelligence the concept has extended to the systems engineering domain where it is used to share common understanding of information structures (by both human and software agents), to analyse domain knowledge and then use it in control systems design [7]. Most of the proposed control architectures tackle the agentification of entities, but consider that the vocabulary is generic only to a certain level (basic entities), the rest (properties, methods, a.o.) being associated to the method of implementation [10]. Nevertheless, there are propositions of ontologies used for the manufacturing domain but they tend to categorize the elements of a manufacturing system for learning purposes [11], cost estimation [12], description of manufacturing domain applications [13], a.o. From current research [4] there is a lack of standardization both at machine-to-machine level and at human-to-human level.

In this above mentioned context (types of ontologies) the paper presents an ontology draft for the manufacturing domain. The ontology standardizes the processes of resource scheduling, operation execution, traceability and system monitoring. This ontology will be modelled in Protégé (protege.stanford.edu) and its application will be shown using JADE (jade.tilab.com). The remaining of the paper is structured as follows: presentation of the requirements for the control system, analysis of the processes associated to the control framework ontology realization using Protégé, and implementation using JADE.

2   The Manufacturing Control Framework

The current work results as a formalization of the structure (ontology concepts) and communication acts (ontology agent actions) of the architecture presented in [14]. This is based on the PROSA architecture [15] from which it takes the basic elements modelled as autonomous and communicative entities composed of an informational and, most of the times, a physical counterpart. These elements are called holons [15]. The composing holons are of the following types: resource holons, order holons (containing both the offline information needed for product manufacturing – product recipe – and online information needed for operation receipt) and staff/supervisor holon. This structure is extended with: (i) a strategy switching process performed by the multi-functional *Staff Holon* entity, (ii) a *service-orientation* of the planning, scheduling, monitoring and execution processes through the usage of a multi-agent platform, (iii) the aggregation of the information related to accessing the resources into an entity called *Resource Service Access Model (RSAM)* which is distributed and replicated on several machines to avoid single point of failure and (iv) entities in charge of mediating conflicts between order holons in case of conflicting interests – *mediator(s)*.

A functional view of this architecture is presented in Fig.1. It is characterized by:

- *A layered structure* which separates the business processes (business layer) from centralized planning and scheduling (centralized MES layer) and decentralized scheduling (decentralized MES layer) and execution (resource layer). This characteristic allows switching strategy from hierarchical to heterarchical to cope with both resource and demand changes. The main purpose of this switch is resource downtime and makespan minimization;
- *Attachment of agents to control entities* (agentification).

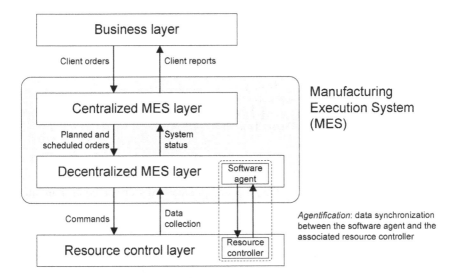

Fig. 1. Layered control architecture

Based on the layered architecture presented in Fig.1 the purpose of the article is to formalize the *structure* and the *communication acts* occurring only in the decentralized MES layer. It is here that the informational counterparts of the order and resource holons are found. These informational counterparts along with other elements (Mediator and RSAM) are implemented as software agents in a multi-agent framework. The process of associating a software agent to a physical entity, creating thus what is called in the manufacturing domain a holon [15], is called agentification as previously described. Therefore, the decentralized MES layer is materialized through a multi-agent platform (Fig.2) composed of:

Order Holon (OH) Agent. The decentralized (heterarchical) control architecture is characterized by the fact that its operation is product-driven. Thus, the Order Holon agent representing the decisional part of the order being manufactured is in charge of operation scheduling, monitoring and execution (routing and processing). It executes on an Intelligent Embedded Device (IED) attached to the item being produced. The whole assembly is located on a pallet in order to be transported through the manufacturing system (the shop floor).

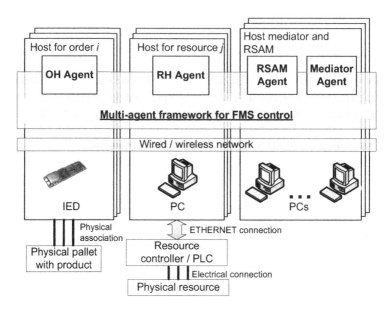

Fig. 2. Overview of the software system controlling the product-driven automation layer

Resource Holon (RH) Agent. In the proposed product/order driven control architecture the resources are seen as service providers to the products being manufactured. A general resource offers two types of operations: transportation for routing and physical transformation for processing. The Resource Holon agent is located on the PC attached to the resource controller and acts as an intermediary between heterogeneous equipment. Its main task is to offer a standard interface both at data and protocol level and to implement simple interaction behaviours for sustaining the OH's execution.

RSAM Agent. It is in charge of continuously collecting information about the RHs' behaviours (services offered, timelines, accuracy of performed operations, a.o.) and offering it in a concise manner upon request. This information is used by the OHs for the collaborative real-time decision on operation scheduling. In order to avoid failure it is designed as a replicated agent running on multiple PCs. The following types of information is contained by the RSAM:

- Information related to the *accessibility of services* (what operation is done on each resource and how resources are interconnected) used in decentralized production scheduling. Thus, the RSAM will guide the OH requiring services directly to the desired resource, eliminating the unnecessary inquiries about availability for jobs addressed by all OHs to all resources;
- Resource *service history* allowing to discard RHs from an early stage in operation scheduling due to poor service quality.

Mediator Agent. It provides the functionality of mediating between the schedules proposed by the OHs in order to avoid multiple usage of the same RH in the same interval of time. Just like in the case of the RSAM the mediator functionality is distributed among a set of PCs in order to provide fault tolerance.

3   Information Analysis for the Processes Associated to the Control Framework

As described before, the communication acts taking place inside of the decentralized MES layer concerning the fabrication of a product are materialized through the creation of an OH and the sequential insertion of all OHs in the manufacturing system. The series of events are as follows: (1) RHs register operations (in the form of services) to RSAM, (2) OHs search services on the RSAM, (3) OHs compute their personal schedule (all operations composing an OH are allocated on resources), (4) OHs synchronize schedules between them (the newly inserted OH with the existing OHs), (5) OHs execute schedules, (6) OHs monitor the state of RHs.

The main processes are: *resource scheduling*, *operation execution* and *system monitoring*. The interactions between the entities involved in these processes will be detailed in the following part of the article and from this description the conceptualization of the manufacturing control domain in the form of an ontology will result. The description of the ontology will be detailed in chapter 4.

3.1   Resource Scheduling Process

In order for the execution of a product to start, the needed resources have to be added to the manufacturing system and their operations registered as services to the RSAM (message 2, Fig.3). Together with the resources services the mediators services are registered to the RSAM (message 1). This is called the **registration process** and represents the first part of resource scheduling.

The **resource scheduling process** (Fig.3) deals with the decentralized allocation of all the operations of all orders on the available resources. Information is exchanged through standardized messages in conformity with the FIPA specification (www.fipa.org).

Each OH Agent updates its information about the existence of resources and operations through RSAM inquiry (messages 3&4) and about the resources availability through RHs inquiry (messages 5&6). After this update a local schedule is computed which contains all allocated operations, the resources on which they are allocated and the time interval in which they should be executed. This schedule proposition is sent to the Mediator Agent in order to eliminate possible conflicts between schedules from different OH Agents (message 7). If the response (message 8) is positive then the schedule is validated on the needed resources (messages 9&10). Otherwise (message 8 is negative) the schedule is recomputed based on the restrictions received from the Mediator Agent and the validation phase is redone. Only after the validation phase is passed the reservation can continue.

Since there is a limited number of OHs in production and these are introduced sequentially some priority levels can be established (for example, the oldest OHs have higher priority) in order to avoid and endless loop for this decentralized scheduling process.

A detailed description of the messages contents is given in Fig.3.

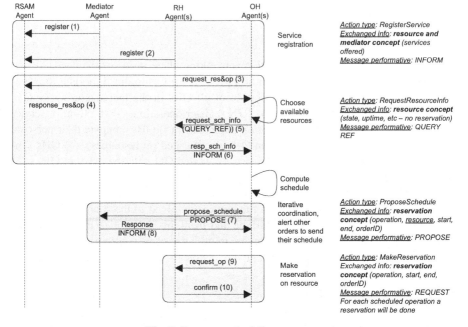

Fig. 3. Resource scheduling process

3.2   Operation Execution Process

The process consists of a simple communication between the OH Agent (initiator) and the RH Agent (participant). The scope of the process is the execution of a given operation. The exchange of messages concerning this process along with the message content is described in Fig.4.

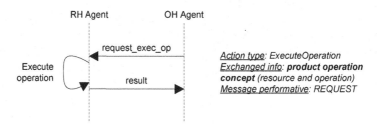

Fig. 4. Operation execution process

3.3   Operation Monitoring Process

The operation monitoring process (Fig.5) has 4 facets:

- OH Agents requiring information from RH Agents concerning reservations previously made (messages 1&2). This is the normal operating mode, where

once the schedule established, it is executed by the associated OH which also periodically monitors the RHs;
- RH Agents updating the RSAM (message 3);
- OH Agents cancelling a reservation on a RH Agents (message 4&5).

A detailed description of the messages contents is given in Fig.4.

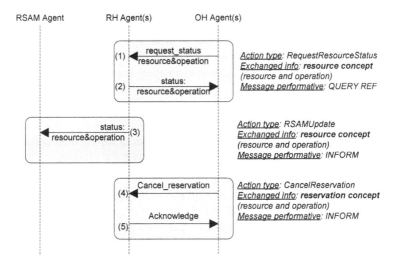

Fig. 5. Operation monitoring process

4   Ontology Definition

In order to support the requirements of the manufacturing decentralized control system presented before (*resource scheduling*, *operation execution* and *system monitoring*), a corresponding ontology has been developed. Defined initially as a set of terms along with their properties and relations between them, the concept of ontology has suffered minor modifications when being applied to the multi-agent programming [7] in terms of the specialisation of the composing elements according to the process they are involved in. The operation of a multi-agent system (implementing the decentralized control system presented before) is based on message exchange and decision making. Therefore, the ontology deals with the standardization of the *structure* of the communicative language (syntax/grammar) through concepts and with the *meaning* of the messages through agent actions [16]. Concepts are complex structures characterized by a set of properties. They are associated to entities/sub-entities in the control system. Agent actions are a special kind of concepts that represent the actions performed by the agents. When two agents interact it is useful to characterize the messages associated to this interaction with a combination of concepts (structure) and agent actions (meaning). Through an analysis of the information associated to the control framework and given the previously mentioned information (concept VS agent action), the content of the ontology (Fig.6) consists of:

   i. Concepts of two types:

- Main concepts used for modelling the elements of the control system:

  o Order - concept associated to an agent in charge with order scheduling, execution and monitoring;

  o Resource - concept associated to an agent in charge with a resource generic decisional process: monitor state and execute operations;

  o RSAM - concept associated to an agent responsible with uniform access to information;

  o Staff - concept associated to an agent in charge of the system's strategy;

  o Mediator - concept associated to an agent in charge with the mediation of all orders schedules;

- Support concepts used by the main concepts to sustain their definition:

  o Resource operation - concept that models the operations that each *resource* can execute;

  o Product operation - concept modelling the operations that have to be executed to assemble the product associated to the current *order*;

  o Reservation - concept that models a reservation made by an order (OrderID) on a resource (ResourceID)

   ii. Agent actions are associated to the message exchanges presented in Fig.3-5 and are as follows:

- Register service (RegisterResourceService, RegisterMediatorService): the action taken to *register the services* of the resources and mediators to RSAM

- Request resource information/schedule (RequestResourceInfo): the action taken by an OH Agent in order to *find out the schedule of a resource*

- Propose schedule (ProposeSchedule): action taken by the OH Agent in order to *synchronize its schedule* with the schedules of other OH Agents through the use of a Mediator Agent

- Make reservation (MakeReservation): action taken by the OH Agent in order to *make a reservation* on a given resource

- Execute operation (Execute Operation): action taken by the OH Agent in order to *execute a reservation* on a given resource

- RSAM update (RSAMUpdate): action taken by the RH Agent in order to periodically *update the information from the RSAM*

- Request resource status (RequestResourceStatus): action taken periodically by the OH Agent in order to be assured that the *scheduled resources are online*/still usable

- Cancel reservation (CancelReservation): action taken by the OH Agent in order to *cancel a reservation* on a given resource

Each element (concept and agent action) is characterized by properties whose meanings are according to their names (e.g.: *element*ID – the identification of the given element). The majority of the properties are simple descriptions as stated by their names with some exceptions, the ones whose names start with *has*:

- *hasProductOperation* (Order) is a vector of ProductOperation type containing the operations that must be executed in order to obtain the final product;
- *hasReservation* (Resource) is a vector of Reservation type (reservation ID, order ID, start time, end time, a.o., see Fig.6) that contains all reservations made on the current resource;
- *hasResourceOperation* (Resource) is a vector of ResourceOperation type that contains all operations that can be performed by the current resource;
- *hasResources* (RSAM) is a vector of Resource type that contains all information concerning the resource accessibility and their capabilities.

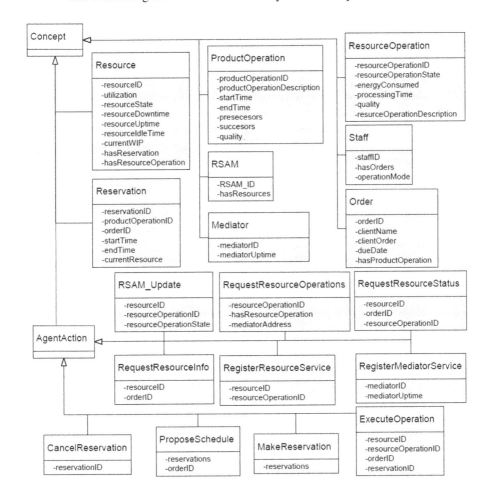

Fig. 6. Manufacturing ontology for scheduling, monitoring and execution

5   Conclusions

The validation of the proposed ontology was done using Protégé for the ontology generation and JADE as the multi-agent development framework which implements the FIPA standards. An experiment concerning a part of the resource scheduling process had been done in order to test the correctness of the exchanged messages and to measure the execution times. The experiment contains the following agents: one RSAM (whose functionality is realized by the Directory Facilitator existing in JADE), four resources launched together and n orders launched sequentially. All agents have been executed locally, on a single machine, using three run configurations in the following order:

- 1st *configuration* – a RSAM agent which waited for resource services to register;
- 2nd *configuration* – all (four) resource agents with 16 operations each. These agents connected to the RSAM and registered their services;
- 3rd *configuration* – a batch of n order agents lunched sequentially. Each order has 5 operations and a single order agent was launched at a given time; it connected to the RSAM in order to find where possible operations could be done and based on the resources' previous utilization (the resource that was the most underused was chosen) made reservations. By repeating this phase n times (n being the number of orders in a batch of products) the scheduling of all orders results taking into account the resources utilization. The three phases are depicted in Fig.7 using the JADE Sniffer Agent functionality.

The number of messages exchanged is as follows:

- At *initialization* (associated to the 1st and 2nd configurations) of the system 8 messages are exchanged between the resource agents and the RSAM agent: one message from each resource containing all the available operations in the form of services and 4 messages from the RSAM, one to each resource, as acknowledges;
- At *utilization* phase (3rd configuration) a set of maximum 28 messages are exchanged: 5 inquiries to the RSAM for each operation associated to the order, 5 responses from the RSAM, 4 inquiries (one for each resource) in order to discover their utilization, 4 responses from the resources, 5 requests for reservation (one for each operation) to the chosen resources and 5 acknowledges from the resources. In this case due to the flexibility of the system a single resource was chosen.
- The process can continue for all the n orders in the batch, order number 2 being also seen in Fig.7.

The rapid (less than 1 second) and uninterrupted flow of messages proves correctness of the protocol. Besides the implementation, the advantages of the proposed ontology and its associated implementation technique are:

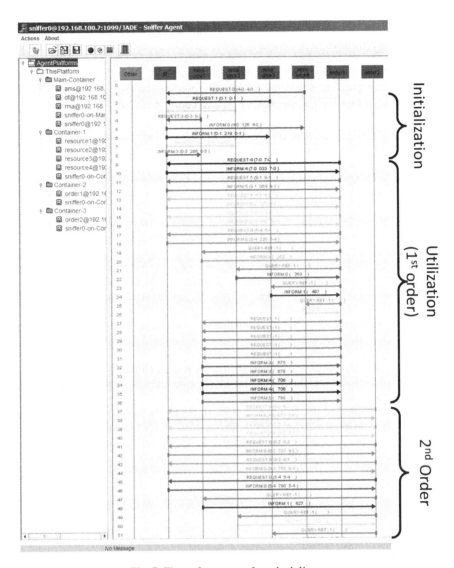

Fig. 7. Flow of messages for scheduling process

- easy generation and regeneration of the ontology through the usage of Protégé application and its plugin (Bean Generator, [17]) which generates directly the Java classes further used in JADE: it takes the developer less programming time by avoiding the manual writing of standard classes (concepts and actions);
- simple code reuse and information sharing between agents: after being generated, the standard files are directly imported by different agents;

- separation between domain knowledge (ontology in Protégé) and decision behaviour (agents in JADE) which results in easy development of multi agent applications.

For future work three directions are taken into account: running the composing agents on different physical platforms, add order coordination through the use of the mediator agents, and testing the monitoring and execution processes.

References

1. Borangiu, T., Gilbert, P., Ivanescu, N.A., Rosu, A.: An implementing framework for holonic manufacturing control with multiple robot-vision stations. EAAI 22(4-5), 505–521 (2009)
2. Sallez, Y., Berger, T., Raileanu, S., Chaabane, S., Trentesaux, D.: Semi-heterarchical control of FMS: from theory to application. Engineering Applications of Artificial Intelligence 23(8), 1314–1326 (2010)
3. McFarlane, D., Giannikas, V., Wong, A.C.Y., Harrison, M.: Intelligent products in the supply chain - 10 years on. In: Borangiu, T., Thomas, A., Trentesaux, D. (eds.) Service Orientation in Holonic and Multi agent, SCI, vol. 472, pp. 103–117. Springer, Heidelberg (2013)
4. 1st Cambridge Workshop on Product Intelligence, Cambridge, UK (September 24-25, 2012), http://www.ifm.eng.cam.ac.uk/research/dial/events/cambridge-pi-workshop/
5. Boyd, A., Noller, D., Peters, P., Salkeld, D., Thomasma, T., Gifford, C., Pike, S., Smith, A.: SOA in Manufacturing Guidebook. In: MESA International, IBM Corporation and Capgemini Co-Branded White Paper (2008)
6. Demirkan, H., Kaufmann, R., Vayghan, J., Fill, H.-G., Karagiannis, D., Maglio, P.: Service-oriented technology and management: Perspectives on research and practice for the coming decade. Int. J. on Electronic Commerce Research and Applications 7, 356–376 (2008), doi:10.1016/j.elerap.2008.07.002
7. Obitko, M., Mařík, V.: Ontologies for Multi-Agent Systems in Manufacturing Domain. In: DEXA Workshop, pp. 597–602 (2002)
8. Uschold, M., Gruninge, M.: Ontologies: Principles Methods and Applications. Knowledge Engineering Review 11(2), 93–155 (1996)
9. Gruber, T.R.: A Translation Approach to Portable Ontology Specification. Knowledge Acquisition 5, 199–220 (1993)
10. Raileanu, S., Berger, T., Sallez, Y., Borangiu, T., Trentesaux, D.: The Open-Control Concept in a Holonic Manufacturing System. In: Proc. of the 18th International Workshop on Robotics in Alpe-Adria-Danube Region - RAAD, Brasov, Romania, May 25-27, p. 86 (2009) ISBN 978-606-521-315-9
11. Lagos, N., Setchi, R.: A manufacturing ontology for e-learning. In: Proceedings of the I*PROMS* 2007, Cardiff, UK, pp. 552–557 (2007) ISBN 978-1-904445-52-4
12. Lemaignan, S., Siadat, A., Dantan, J.Y., Semenenko, A.: MASON: A Proposal for an Ontology of Manufacturing Domain. In: Proceedings of the IEEE Workshop on Distributed Intelligent Systems: Collective Intelligence and its Applications, DIS 2006, pp. 195–200 (2006)

13. Borgo, S., Leitão, P.: The role of foundational ontologies in manufacturing domain applications. In: Meersman, R. (ed.) OTM 2004. LNCS, vol. 3290, pp. 670–688. Springer, Heidelberg (2004)

14. Borangiu, T., Raileanu, S., Anton, F., Parlea, M., Tahon, C., Berger, T., Trentesaux, D.: Product-driven automation in a service oriented manufacturing cell. In: International Conference on Industrial Engineering and Systems Management, Metz, France, May 25-27, pp. 1392–1400 (2011) ISBN 978-2-9600532-3-4

15. Van Brussel, H., Wyns, J., Valckenaers, P., Bongaerts, L., Peeters, L.: Reference architecture for holonic manufacturing systems: PROSA. Computers in Industry 37(3), 255–274 (1998)

16. Fabrici, D.: Defining Ontologies in a Multi Agent Scenario Using the JADE Framework. In: 4th Slovakian-Hungarian Joint Symposium on Applied Machine Intelligence (2006)

17. http://protegewiki.stanford.edu/wiki/OntologyBeanGenerator_4.0

On the Team-Based Goal-Oriented Development for Holonic Manufacturing Systems

Gabriela Varvara

"Gheorghe Asachi" Technical University of Iasi
Department of Automatic Control and Applied Informatics
Blvd. Dimitrie Mangeron 27, Iasi, 700050, Romania
gvarvara@ac.tuiasi.ro

Abstract. This contribution makes a reconsideration of the holonic organizational concept based on analogies with team formation and execution. It presents the team modelling concepts that support holonic structure and behaviour and the development of multi-agent framework based on the aggregation of agents in teams. Some drawbacks of the classical BDI reasoning mechanism and enhancement solutions are identified. In order to offer a practical argument of this perspective, a basic holonic control scenario was implemented using the GORITE team-based framework. It was adopted a goal-based top down development strategy for the execution model. The agent BDI reasoning and execution engine of framework was used to offer an improved solution to sensorial integration for holonic manufacturing systems.

Keywords: BDI reasoning, multi-agent systems, team programming, process modelling, goal oriented development, Holonic Manufacturing Execution Systems.

1   Introduction

During the last years the control and coordination solutions for complex, large scale manufacturing systems had been intensely studied. Notable and correlated efforts were made in the direction of distributed artificial intelligence and multi-agent design and implementation [5], [11]. Despite the abundant research results, there are still current problems solved by ad-hoc methods. This is a major drawback in the context of the global resources access, flexibility and scalability requirements for services and goods production [2], [10]. The manufacturing ongoing dynamic imposes rapid responses to both market demand and production conditions. Consequently, substantial research effort was made in the directions of interaction modelling among entities in complex systems, group activities coordination, development of autonomous reasoning mechanisms and multi-agent framework implementation based on intelligent agents and robust communication mechanisms [4], [11]. Although there are a large number of research reports on this topic, the multi-agent frameworks still have a limited use in supporting real manufacturing systems [6], [11]. This is the result of the organization of a classical production process into planning, programming and control activities [6] sustained by different well established methodologies and implementation technologies where integration problem is slow and remains an open problem.

An integrating solution could come from the enhancement of the reasoning mechanisms at individual and collective level and from a reconsideration of the goal achievement representation at all the levels of the manufacturing process.

A multi-agent implementation based on the direct execution of goals, recursive organizational model and   BDI (Belief Desire Intention) reasoning mechanism at all the levels would offer better solutions to disruption events or concurrent access to resources [2].

This paper focuses on BDI reasoning aspects concerning short and long time commitment during intentions achievement, team formation and explicit representation and execution of team hierarchies of goals as they are implemented in the Goal Oriented Teams (GORITE) framework. It also offers an implemented solution to model the behaviour of a manufacturing system relative to the integration of its environment observations. The paper is organized in five sections and a paragraph of conclusions. Section 2 reviews the main aspects concerning the BDI intelligent agent reasoning. Section 3 refers to the correlation between the team organization and the holonic model. Section 4 is devoted to the presentation of the implemented reasoning mechanism of the GORITE framework both at agent and team level in order to execute goal oriented scenarios. Section 5 presents an implemented example that integrates sensorial information through cognitive reasoning mechanisms and reveals the enhancements compared to a classical partially specified plan solution. Some conclusions and future work directions are formulated in the last paragraph.

2   Belief-Desire-Intention (BDI) Reasoning Mechanism – Principles and Operation

The BDI reasoning model is intensively used to solve different types of problems that concern the artificial intelligence. Multi-agent systems implement this paradigm at agent level. Consequently, an autonomous reasoning mechanism is propagated at every level of a multi-agent system. It is used for selecting the route to achieve a certain goal and for making valuable reconsiderations of goal achievement when this route fails [7], [14]. This way of thinking brings design advantage like simplicity, maintainability and performance because similar constructs are repeated during system integration. There is also a major disadvantage – an increased complexity during system development.

From a pragmatic perspective, a more sophisticated mechanism is needed to solve both the environment and the individual actions monitoring, as it is the case of the embedded agents. For these situations where simple reconsideration of plans could block the agent actions, only ad-hoc solutions are reported [6]. This blocking problem could be solved by the means of a procedural embedding of the plan hierarchy specification [7], [8] at the reasoning level.

Moreover, at the BDI agent level [14] the beliefs do not change for a long time period, their adjustment being imposed by processes from outside the model scope, and there is no distinction between the way the agents treat their own beliefs and those shared with other agents. The BDI reasoning model assumes that intentions are selected from the possible desires. The agent makes a commitment to achieve its intentions by adopting them one by one, as goals, and developing actions toward their fulfilment. This commitment mechanism assumes further that the agent persists on an intention as long as it deliberates on the desires that will become intentions and makes plans to achieve them.

When no major time restrictions are imposed to the deliberating process, intentions reconsideration will bring benefits as an agent could adapt better to its internal / external environment modifications.

According to Wooldridge [14], three agent commitment strategies could be employed, ranging from a *blind commitment* where an intention is maintained until it is believed to been achieved, through *single-minded commitment* where an intention could be maintained on blind commitment basis or until it is believed to be impossible, *to open-minded commitment* where both already mentioned conditions are simultaneously considered.

3   A Holonic Perspective for Team Oriented Programming

The current multi-agent programming frameworks are facilitating mainly one of the two major agent characteristics namely, the autonomy and the mobility. However, the integration of the individual behaviour at the collective level in compliance with some organizational principles remains an open research and experimental problem.

An interesting aspect for modelling the BDI rational agent concerns the collective intention formation mechanism. Some authors [6] sustain the idea of using the same model for individual and group intentions, some are against it. In this respect there are different theories for the development of the collective mental state [9], namely:

1. The **SharedPlans** theory – the commitment of an agent to act in group activities is based on two fundamental intentions – to do an action and accepting that a proposition holds.
2. The **Joint Intention** theory – the team is responsible to hold all the joint intentions and the individual agents commit to inform the other agents if a goal is achieved, it is impossible or it is no longer relevant as the circumstances have been changed.
3. The **Cooperation Domain** theory [1] – the interactions required by the achievement of a group (also named team) goal are encapsulated into a cooperation domain that offers the interaction protocol for agent interactions during the execution of a particular task. This protocol does no longer apply for the behaviour of the agents outside the cooperation domain.

The Jack Teams environment [1] is the first implementation where a team entity is capable of reasoning and coordinates its members' behaviours towards the achievement of a common goal. The team is modelled explicitly and separate from its agents, as an entity with its own beliefs and desires, capable of reasoning about its intentions and of making commitments to achieve its goals.

Beginning with 2006, Jarvis et al. [7], [8] explored the relationship between the team and the holon concept [3]. Similarities were observed concerning the fundamental properties of the holon - both the team and the holon are identifiable, self-contained parts of an organizational structure. The role software concept and reasoning intentional attitude mechanism could represent the duality part-whole from the holon definition. This idea gives opportunities to apply multi-agent team-based implementation for the control and coordination of Holonic Manufacturing Systems (HMS).

The GORITE environment [1], [6], [12], [13] shares with Jack Teams the team concept as a BDI reasoning entity, modelled explicitly in terms of roles and goals. It brings notable improvements referring to:

— An explicit representation of goals that replace the Jack team/agent procedural representation based on plans
— The behaviour description in terms of goal-based process models.

According to GORITE perspective, a system is a collection of goals achieved by teams of agents working together on a common data context and complying with a well-defined business process. This metaphor allows the passage from individual autonomy with ad-hoc constraints necessary to obtain the organization coherence to an autonomy related to the conformation with organizational rules and procedures. Consequently, the design will focus on the group behaviour required by a specific application context. It is to notice that GORITE environment, a Java platform for implementing Team Oriented designs, can model better the holonic organizational paradigm.

4   GORITE Teams and HMS Design

The GORITE team is made of other teams and/or individual agents in the same way that the holon is recursively organized. It is a *Performer* for a large team/multiple teams that contain it, as illustrated in Fig. 1.

The team behaviour is encapsulated in the process model separately defined from the sub-teams behaviour as an orchestration of the services provided by sub-teams following a top-down design methodology. It defines all the team behaviours required by a specific application covering design details related to the complexity of organization, goals and associated activities. Consequently, the team behaviour results from the coordination of the embedded Performers behaviour.

The team goals are achieved through the capabilities of the role entities. Roles are fulfilled for a specific scenario with autonomous agent members engaged in team activities.

The team goal execution is handled by an *Executor* entity, as unique representative of the team members. Its associated object traverses and executes explicitly all the goal hierarchies. Thus a procedural representation of plans/sub-plans hierarchy becomes no longer necessary. Furthermore, the Executor makes the shared data context accessible to all the participants to the execution of goals.

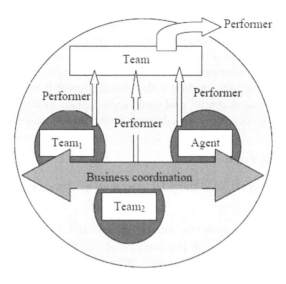

Fig. 1. GORITE team – holon analogy

4.1   GORITE and Goal-Oriented Design

According to the BDI reasoning paradigm implemented by GORITE, the goal is a desire that a Performer wants to achieve. On the other hand, intentions are goals that the Performer is committed to fulfil. The reasoning process begins with a phase of intention selection followed by planning of the achievement strategy. For no time constraints imposed to the system behaviour, GORITE allows the resumption of this process in order to exploit the environment changes. This is made dynamically, during the goal execution.

The goal specification contains its achievement strategy. Goals can be decomposed in sub-goals and represented as hierarchies in the form of process models that GORITE is able to execute. The root of all the goal hierarchies is the class *Goal* that encapsulates an idea about goal achievement. The current intentions are represented by *Goal.Instance* type objects.

From the behaviour perspective, the intentions could be achieved immediately or as a result of a goal decomposition process. As a consequence, the GORITE goal execution engine implements two possible execution routes:

— "Task semantics" – the goal achievement is made through the direct invocation of the method execute() from the goal class. It applies to all the leaf nodes of the goal hierarchy excepting the TeamGoal, BDIGoal and directly executable Plan instances.

— "BDI semantics" – the goal achievement is made through sequential execution of an applicable set of goal hierarchies. The goal fails if all its associated hierarchies fail. The sub-goals with the same priority are, by default, executed as they appear in the set. An explicit selection of goal execution could also be defined.

Classes applying task semantics specify the Performer behaviour and, consequently, they are named behaviour classes, and added to the process model by the control overlay classes that organize their execution in a procedural manner. For the achievement of control overlay goals, the execution engine replaces the direct execution of their classes execute() method with the execution of their sub-goals according to a control mechanism. A control goal could apply equally to leaf and non-leaf nodes of the goal hierarchy. Its achievement depends on the achievement of all its sub-goals.

4.2   The Situated Team

Some shortcomings in the classical BDI reasoning mechanism appear. They regard the following aspects:

— The lack of a cognitive reasoning model with an explicit representation of the perception, deliberation and action processes along with their relationships;
— The environment interaction made only through the beliefs, as internal processes and the missing of a clear separation between short time and long-time goals.

These were overcome by GORITE framework. It brings enhancement related to the explicit use of the cognitive reasoning mechanisms, allowing short and long time deliberation followed by short time, long time or event-driven actions.

The Perception Representation

The GORITE perception, named percept, is modelled by a *Perceptor* object. This specifies the Performer responsible for handling the percept, the goal that will be initiated to handle the percept and the ToDo group that will be managed by the goal execution. The ToDo group is the stack of active goals that are currently pursuing or will be in the near future.

The information about the external environment can be accessed from a specific data context, named "perceipt". This is updated by the percept processing during the invocation of the perceive() method of the Perceptor object.

The goal that will be initiated to handle the percept must be defined and added to the Performer object. This goal will access the "perceipt" data context by the execute() method. When the perception result has to be a new order of intention execution inside the ToDo group, a meta-goal has to be defined and added to the Performer. GORITE framework allows also internal reflection, as perception about the beliefs of a Performer. It is modelled by a Reflector instance that triggers a percept process every time the data context is changed.

The Deliberation Process

The GORITE deliberation process is related to the concurrent goals execution mechanism and is based on the group concept. The default execution mechanism gives access to the execution engine, for a small continuous time, for the intention being in the top of the stack.

The order of the intentions inside the ToDo Group can be changed by the current Performer through the execution of a meta-goal. The execute() method of the meta-goal will be invoked automatically at the beginning of each access cycle to the execution engine.

An element from the ToDo Group is manipulated by a meta-goal through:

— The value returned by the last invocation of the method execute(), representing the intention state (one of the values STOPED, BLOCKED, CANCELLED).
— The return of isRunning() method that represents an indicator of the intention willingness to progress. If the return value is 'true' the intention wants to progress, if it is 'false' it does not.

When a new intention is added to the ToDo Group, its state has the STOPPED value and it wants to progress.

Any meta-goal can access the data context information about the intentions added/removed from the last execution cycle along with the last executed intention.

Usually, a meta-goal is used to direct the progress of all the active intentions from ToDo Group in fair manner through the focus concept. Each sub-goal has allocated a maximum number of execution cycles. While it has focus its execution state is monitored. When it has completed or it is blocked, the focus will move to the next active sub-goal.

As a result, the GORITE framework allows the reasoning processes development during the intention executions.

The Action of the Situated Team

The GORITE execution mechanism is reflected in the Executor actions that handle the achievement of the team intentions. It allows concurrent execution by exploiting the state of the active intentions from a ToDo Group. The intention execution will always follow the Performer reasoning about the execution order of the current active intentions and about the creation of new intentions. During this reasoning process the Executor is free to achieve other active intentions.

The GORITE action means the way an execute() method progresses. The execution engine has the same semantics for actions as for goals:

— The task action is associated to a behaviour task goal. A single invocation of the goal execute() will be made, and during this invocation the Executor will not be able to process other goals.
— The goal action is associated with a BDI goal or a Team goal. It takes a longer time and it is initiated by the Performer that monitors its progress and notifies the Executor of its completion. The Executor will be free to process other goals during action execution as it will be notified about its completion.

Both task actions and goal actions are implemented by the Action class. For a goal action the monitoring of the execute() method progress is facilitated by the argument 're-entry', a flag set by the Executor on 'false' at the first invocation of the execute() and on 'true' for the subsequent invocations.

The Team Modelling

In order to facilitate team modelling, GORITE promotes the following analogies: the team and the agent are similar from the BDI reasoning perspective; the role and the team behaviour are synonymous concepts.

The GORITE team is modelled by the Team class, a sub-class of type Performer. A particular behaviour of a team is performed by a TaskTeam instance, a virtual structure within a team. This particular behaviour results from a set of roles added to the Task-Team instance. The GORITE role is defined as a set of related goals. Once defined a TaskTeam, its roles must be allocated to the team members (agents/teams) through an establishment process (the invocation of establish() method). As a result the role filler from the populated TaskTeam will be added to the data context.

The TaskTeam behaviour will be further specified as a process model. As this specification should be independent of the Performers allocated to the TaskTeam, it will be made through the use of the team goals. A team goal appears as a leaf node in the process model. It will be delegated to be executed by the filler of the role able to achieve the goal. This one will provide one or more process models with the same name as the received goal. The associated process models are goals hierarchies that form the applicable set for the team goal and will be executed in a normal manner. When there is more than one filler for a role, the team goal will be performed in a concurrent manner.

5   An Experimental HMS Implementation Using GORITE Situated Team

5.1   The Manufacturing Setup and Scenario

In order to test the GORITE enhancements in a holonic scenario using the team-holon analogy and the goal-oriented execution, an experimental manufacturing system was considered. It consists of a robotized cell, named Cell1, selected from a manufacturing system. It was created around one 6 d.o.f. industrial robot, see Fig. 2, able to operate with a machine tool, a conveyor and several storage devices. The manufacturing system also contains a second robotized cell that was not used for the current experiments, but will be used in the future extension of the manufacturing scenario. Despite its simplicity, the Cell1 setup can solve palletizing and assembling goals, including also sub-goals for part processing on the machine tool, which is a CNC mill. The holonic model for Cell1 comprises an order holon responsible for managing the goal at the cell level and its time constraints, two product holons: one for the sub process of pallet stack construction and the other for obtaining the necessary pallet layout, and resource holons allocated to the physical devices from the cell, namely, the robot R1, the machine tool and the storage device within the robot R1 area (item no. 7 in Fig .2), named M1. The palletizing scenario, as illustrated in [10] consists of an order to obtain a stack of two pallets with the layout displayed in Fig. 3.

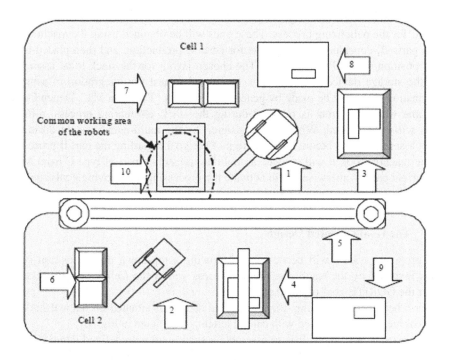

Fig. 2. The experimental manufacturing system:
1. Industrial robot (R1), 2. Industrial robot (R2),
3. Machine tool, 4. Computer vision system,
5. Conveyor, 6, 7. Storage devices,
8, 9. Robot controllers, 10. Storage/Assembly table

The pallet 1 will be placed at the bottom of the stack. The final position of the assembled stack will be on the assembly table, as depicted in Fig. 2. For the proposed assembling scenario it was made the assumption that the storage M1 is able to provide enough parts of types B and D in order to achieve any palletizing order.

Pallet 1

B		D
	C	
B		C

Pallet 2

B	C	B
	D	C
D	C	B

M1

B	D	

Fig. 3. The pallets and the storage device S1 layout

As can be noticed (see Fig. 3), the storage device M1 is not able to provide the parts C for the palletizing process. These ones will be obtained using the machine tool from parts B, considered as raw parts for parts C production, and then placed to the right position on the Pallet1 by R1. The chosen layout for the stack to be assembled and the storage device M1 is simple, but useful to test the integration of sensorial information. This will be made by generating an event "Part C in M1". Depending on the time when the event is released during the stack assembling process, different goals will be activated. When it is registered by the multi-agent system during the pallet assembling and before the execution of the goal regarding the part B processing on the machine tool, it will activate a goal that transfers a part of type C from M1 to the current pallet; otherwise it will activate the processing and moving goals specified in the cell1 normal process model.

5.2   The Goal Oriented Design

The proposed system will be designed following a top-down decomposition of the goals involved by the stack assembling process within the Cell1. This process will foster the GORITE goal-oriented execution.

For a better understanding, the assembling cell goals situated on the leaf nodes of the goal hierarchy are named with capital letters, as depicted in Fig. 4.

The main goal of the cell is to assemble a stack with a predefined number of pallets. This will be achieved by executing a number of times the palletizing goal (Make_Pallet), the loop being controlled by an EndGoal achieved when the assembled number of pallets is equal with the required number of pallets.

Moreover, to make a pallet means to move the pallet to the final position and then to fill the pallet with parts according to the required layout. As no predefined precedence of MOVE_PALLET and Assembling_Pallet goals is imposed, they will be achieved according to a SequenceGoal control overlay mechanism. It means that the sub-goals are executed in a default order and Make_Pallet goal fails if one sub-goal fails.

In order to achieve the goal Assembling_Pallet, MOVE_A, MOVE_B and Assembling_C sub-goals have to be repeated a number of times, different from one sub-goal to another. For this situation a RepeatGoal control overlay has to be applied to obtain all the replicas needed to fill the pallet according to its specified layout (see Fig. 3.). The goal Assembling_Pallet succeeds if all the replicas, processed as parallel branches, succeed. The execution control parameters, indicating the number of repetitions for each sub-goal are accessible from a data context multi-valued element.

The goal Assembling_C will be achieved according to a SequenceGoal control overlay mechanism with predefined goal order. It will succeed if all its sub-goals, as specified in Processing_Assembling_C goal, succeed. The goal MOVE_C is out of the scope of cell1 process model. It will be activated by perception of the event that a part of type C was delivered to M1. As a result, by using GORITE design strategy to integrate sensorial information at the agent decision level, the process model that has to be executed in normal situation is simple and coherent. The exception situations are captured and treated by separate working contexts.

The assembling cell goals hierarchy represents the process model of the proposed manufacturing cell. The goals will be activated by traversing the hierarchy from top to down, goal by goal, and executed according to their associated control goals and the GORITE goal execution engine. The goal decomposition, as depicted in Fig. 4., represents an explicit representation of the goals execution that foster the understanding of goals correlations and the future extensions.

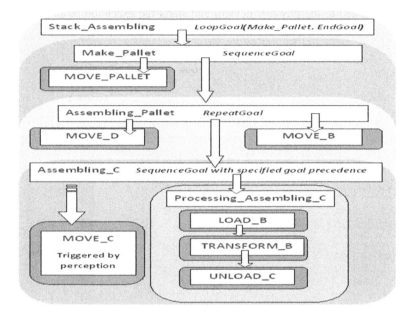

Fig. 4. The stack assembling goal decomposition

5.3   Situated Team-Based Implementation

The GORITE application for the proposed system is implemented using the team concept. Referring to the above mentioned holonic organization, the following analogies were made:

— The order holon, the product holon that controls the sub-process of pallet stack construction and the product holon for obtaining the necessary pallet layout are all represented by the TaskTeam associated with the Team that models the manufacturing Cell1.
— The resource holons allocated to the physical devices from the Cell1 are modelled by the Performer associated to these resources.

The application is entirely developed in Java, standard edition v1.7, using the GORITE packages from the release v8RC04. It illustrates the holarchy formation through team agent association, the BDI reasoning mechanism at organization level and the flow of execution based on goal hierarchy. The planning, scheduling and synchronization aspects are out of the scope of the presented solution.

The class structure, as depicted in Fig. 5., was developed by considering the following aspects:

The Main Class

The Assembling main class represents the entry point of the application. It starts the GORITE engine, creates an instance of the Team associated to Cell1, adds the Performers associated to the physical devices and starts the execution of an assembling order.

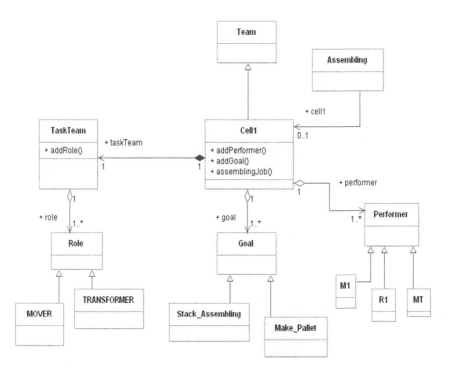

Fig. 5. The multi-agent system class diagram

The Team Class

The Team class is the central point of the implementation. It has the following main objectives:

— The definition of the TaskTeam that will perform the current manufacturing order and add role capabilities to do the job. Two roles, MOVER and TRANSFORMER are necessary to carry out the stack assembling. As specified by addRole() method, the Performer that will fill the role TRANSFORMER will assume the achievement of a single goal, TRANSFORM_B, while the role MOVER requires abilities to accomplish five goals (MOVE_B, MOVE_D, MOVE_PALLET, LOAD_B, UNLOAD_C.

— The definition and the instantiation of all the behaviour and control goals as represented in process model. This will be made through the return values of goal associated methods.
— The definition of the method that executes the establishment of the TaskTeam object. At the end of the establishment process the team members will fulfil the roles defined by the Team and an appropriate data context will be created.
— The definition of the Team constructor that associates a Team instance with its own goals.

The rest of the application consists in the definition of the Performers robot R1, machine tool MT and the storage device M1 associated to the physical devices assigned to Cell1.

In the next paragraph, the Performer for the robot R1 is sketched, with emphasis on the sensorial information integration programming aspects.

Sensorial Information Integration at the Robot R1 Level
During a typical manufacturing scenario, a part of type C is obtained only by processing a part of type B on the machine tool. If the sensorial event "part C in M1" could be treated during the manufacturing process it will bring opportunistic benefit regarding the palletisation time.

The purpose of the R1 performer implementation was to define the goals associated with the assumed MOVER role and, further, to obtain a rapid rational reaction to the provision of the storage device M1 with parts of type C.

A solution for the same scenario was proposed in [10]. It applies the classical BDI reasoning model as it is implemented in Jack® environment, and has some drawbacks, namely:

— The deliberation on the environment changes is possible only before the start of plans execution;
— The perception processing is procedurally embedded in the set of applicable plans.

This paper offers a new solution based on the explicit representation of the perception processing at the Performer intention level. It enhances the dynamic reaction of the performers as their reasoning processes are carried in a permanent manner, based on the active and near future intentions. Moreover, GORITE framework offers support for the concurrent execution of goals.

For the proposed manufacturing system, the robot R1 monitors the environment for the event "part C in M1" and then searches for the goal MOVE_C in the default ToDo group to replace the goal LOAD_B that becomes no longer necessary. In order to implement this sensorial integration, as illustrated in Fig. 6., the following steps are necessary:

I. To create an object of type Perceptor in the Performer R1 class. It opens a perception channel by specifying that the current Performer handles the perception referring to "part C in M1" through the goal named MOVE_C.

To define inside the constructor method of the Performer R1 class:

II. The goal MOVE_C that will be further added to the R1 goal hierarchy as a top level goal that imposes the Executor to leave the current executed goal and change the focus to allow the intervention of a meta-goal. This will be done if all the goals achieved by the R1 Performer will be executed as goal actions allowing resumption of the execute() method. When this method is stopped or blocked, the execution engine or a meta-goal can promote for execution another active goal in the top of the ToDo group.

III. The meta-goal named "top promotion" that promotes the goal MOVE_C in the top of the R1 ToDo Group if no machine tool processing is started. It traverses the R1 ToDo group looking for a Goal.Instance whose name is "LOAD_B" placed in the top that will be executed immediately. Then its state is forced to FAILED and the MOVE_C goal promoted in the top of the ToDo group.

IV. The meta-goal ToDo group named "percepts" will be added to the R1 default ToDo group.

V. A Thread class that initiates the perception processing through perceive() method invocation. Then the resulting Runnable object will be started in order to detect the event "part C in M1".

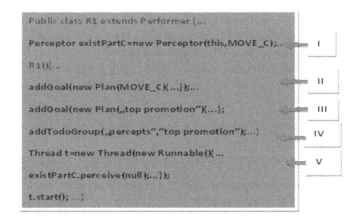

Fig. 6. The solution for "part C in M1" sensorial event integration

When running the application, similar output will be produced:

```
[0] Goal LOAD_B STOPPED running
Analyze perception
[0] Goal LOAD_B BLOCKED waiting
Analyze perception
[0] Goal LOAD_B BLOCKED waiting
[1] Goal MOVE_C STOPPED running
Promote MOVE_C
Analyze perception
```

```
[0] Goal MOVE_C BLOCKED waiting
[1] Goal LOAD_B BLOCKED waiting
Eliminate LOAD_B
Analyze perception
[0] Goal MOVE_C BLOCKED waiting
...
```

As it can be seen, the goal MOVE_C is promoted for immediate execution as soon as the perception reviewing process detects a new part C in M1 and the manufacturing context is favourable (the machine tool does not start the processing of a type B part).

This experiment tests the sensorial information integration during goal execution and is promising for future more complex scenarios.

6   Conclusions

This paper presents a new solution for the integration of sensorial information for a holonic experimental system. It is based on GORITE framework using team-holon analogy and explicit goal execution mechanisms. As proved by the developed application, the team orientation of this framework allows clear definition of complex system behaviours using top-down design methodologies. As a consequence, the control and coordination for manufacturing systems with holonic organization can be completely and intuitively modelled with GORITE framework. The goals have an explicit representation and can be handled directly in a BDI reasoning context at the system level. The team acts according to the classical BDI reasoning mechanisms selecting different routes for goal achievement, but in an enhanced manner, as it can reconsider the way to succeed when the current context has changed. These were the favourable premises of the presented implementation that offers better results in comparison with the classical planning methods. The system reacts rapidly and dynamically to its environment changes, thus obtaining visible personal benefits.

The presented work is supposed to be extended in order to study more complex manufacturing scenarios that could exploit the Robot R2, its associated storage device M2 (item no. 6 in Fig .2.) and the conveyor from Cell2, for parts C provision and to implement the holon reactivity to its interior environment (reflection on its own beliefs), too.

References

1. Agent Oriented Software Pty Ltd, Carlton South, Australia: JACK™ Intelligent Agents – Teams Manual, release 5.3 (2008)
2. Breman, R., Norrie, D.: From FMS to HMS. In: Deen, M. (ed.) Agent Based Manufacturing Advances in the Holonic Approach. Springer, Heidelberg (2003)
3. Brussel, H.V., Wyns, J., Valckenaers, P., Bongaerts, L., Peeters, P.: Reference architecture for holonic manufacturing systems: PROSA. Computers in Industry 37, 255–274 (1998)

4. Evertsz, R., Fletcher, M., Frongillo, R., Jarvis, J., Brusey, J., Dance, S.: Implementing Industrial Multi-agent Systems Using JACK™. In: Dastani, M., Dix, J., El Fallah-Seghrouchni, A. (eds.) PROMAS 2003. LNCS (LNAI), vol. 3067, pp. 18–48. Springer, Heidelberg (2004)
5. Giret, A., Botti, V.: Engineering Holonic Manufacturing Systems. Computers in Industry 60, 428–440 (2009)
6. Jarvis, D., Jarvis, J., Rönnquist, R.: Multiagent Systems and Applications: Development using the GORITE BDI Framework, vol. 2. Springer, Heidelberg (2013)
7. Jarvis, J., Jarvis, D., Rönnquist, R., Jain, L.: Holonic execution: a BDI Approach. SCI, vol. 106. Springer, Heidelberg (2008)
8. Jarvis, J., Rönnquist, R., McFarlane, D., Jain, L.: A Team-Based Holonic Approach to Robotic Assembly Cell Control. Journal of Network and Computer Applications 29, 160–176 (2006)
9. Papadimitriou, C., Singh, M.P., Müller, J.P. (eds.): ATAL 1998. LNCS (LNAI), vol. 1555. Springer, Heidelberg (1999)
10. Panescu, D., Varvara, G.: On the sensorial Integration for a Holonic Manufacturing System. In: Al-Dabass, D. (ed.) UKSim, Third European Modelling Symposium on Computer Modelling and Simulation, Athens, Greece (2009)
11. Pěchouček, M., Mařík, V.: Industrial deployment of multi-agent technologies: review and selected case studies. Autonomous Agent Multi-Agent Systems 17, 397–431 (2008)
12. Rönnquist, R., Jarvis, D.: Interoperability with Goal Oriented Teams (GORITE). In: Fischer, K., Müller, J.P., Odell, J., Berre, A.J. (eds.) ATOP 2008. LNBIP, vol. 25, pp. 118–128. Springer, Heidelberg (2009)
13. Rönnquist, R.: GORITE, Intendico Pty, Ltd. (2012),
   http://www.intendico.com/gorite
14. Wooldridge, M.: Reasoning About Rational Agents. The MIT Press (2000)

Extraction of Priority Rules for Boolean Induction in Distributed Manufacturing Control

Nassima Aissani[1], Baghdad Atmani[1],
Damien Trentesaux[2], and Beldjilali Bouziane[1]

[1] Laboratoire d'informatique d'Oran "LIO"
Département d'informatique, Faculté des sciences, Université d'Oran
BP 1524, EL M'Naouer, Es Senia, 31000 Oran, Algérie
[2] UVHC, Le Mont Houy, F-59313 Valenciennes cedex 9, France
Research: "Production, Services, Information" - PSI Team, TEMPO Lab. Res. Centre EA4542
aissani.nassima@yahoo.com, atmani.baghdad@gmail.com,
damien.trantesaux@univ-valenciennes.fr,
bouzianebeldjilali@yahoo.fr

Abstract. In reactive manufacturing control, the allocation of resources for tasks is achieved in real time. When a resource becomes available it chooses one of the tasks in its queue. This choice is made according to priority rules which are designed to optimize costs, time, etc. In this paper, the aim is to exploit a Job Shop scheduling log and simulations in order to extract knowledge enabling one to create rules for the selection of priority rules. These rules are implemented in a CASI cellular automaton. Firstly, symbolic modelling of the scheduling process is exploited to generate a decision tree from the log and simulations. Secondly, decision rules are extracted to select priority rules for execution in a specific situation. Finally, the rules are integrated in CASI which implements the decisional module of agents in a distributed manufacturing control system.

Keywords: Dynamic scheduling, priority rules, inductive learning, distributed manufacturing control.

1 Introduction

Scheduling occupies a special place and plays a privileged role in workshop control within companies, at both tactical and operational level of the decision-making process [11]. What is most concerned is finding a compromise between *optimizing* in terms of cost, time, etc... and *responsiveness* for coping with increasingly growing competition. On the one hand, solving a scheduling problem means to schedule in time a set of tasks whilst respecting various constraints (deadlines, limited resources), in order to optimize a criterion such as total time, idle time, etc.... On the other hand, workshops are subject to hazards that are both internal (machine breakdown, lack of staff...) and external (supply delays, price variations, unexpected arrival of orders...). Therefore, distributed heterarchical and semi-heterarchical architectures have been

T. Borangiu et al. (eds.), *Service Orientation in Holonic and Multi-Agent Manufacturing and Robotics*, Studies in Computational Intelligence 544,
DOI: 10.1007/978-3-319-04735-5_9, © Springer International Publishing Switzerland 2014

developed in which intelligent entities agree to work together without following a specific plan and avoiding the master/slave relationships present in hierarchical architectures. Their cooperation strategies are based on full local autonomy, self-organization, minimum global information and enhanced communication capabilities [6]. In this paper, the aim is to develop a distributed manufacturing control system able to make suitable decisions for optimal behaviour. A decision-making process in manufacturing scheduling produces a lot of data. This data can be used in data mining processes to make a decisional engine which can be used for resolving scheduling problems. So, an approach will be developed allowing one to extract knowledge from the scheduling system log and from a simulated system. This knowledge will be used to develop a decisional engine which can be included in a decisional module of entities constituting a distributed manufacturing control system.

This paper is organized as follows: First, the problem is introduced with a review of the literature relating to distributed dynamic control. Then, the motivations behind our research are explained, showing the lack of research using past events in the future decision-making process and their inefficiency in reactive scheduling because of their centralized operation and slowness. Next, we introduce an accurate multi-agent model that has already been used in previous studies, but with some modifications. After the proposed model is described, we explain how Boolean modelling can be used to extract knowledge in order to generate rules which can be employed in the agent's decisional module. Then, the experimental results are analysed showing the reliability of the proposed approach. Finally, our conclusions are presented, summarizing our contribution and introducing our prospects for future research.

2   Distributed Dynamic Control and Knowledge Extraction: State of the Art

2.1   Dynamic Scheduling, Heterarchy and Multi-Agents

According to the classifications in [9], there are three main types of dynamic scheduling: *proactive* scheduling, *reactive* scheduling and *hybrid* scheduling. One point that dynamic control and dynamic distributed scheduling activities have in common is the fact that they both adopt a heterarchical model for agent-based manufacturing systems [24].

With heterarchical architectures, it is possible to use a Multi-Agent System (MAS) model because the agents are independent and are able to receive information from their environment, act on that information and generally behave in a rational manner. [1] experimented this approach to control job shop scheduling and scheduling production and maintenance tasks in the petroleum industry [2]. [10] proposed a multi-agent negotiation/bidding mechanism-based approach for automated guided vehicles and machines within a manufacturing system, showing how all that can provide feasible scheduling in a real-time environment in comparison with the frequently used dispatching rules.

Dynamic scheduling means providing a response to when a resource is released, which task among those in the queue is chosen to be executed? The work of researchers in this area is to define a selection policy. These policies are often based on AI techniques (Artificial Intelligence) such as Multi-agent systems [13], multi-agent systems coupled with machine learning [1, 2, 3] or holonic systems [3; 12]. However, it is possible to improve this type of model by creating a module capable of making scheduling decisions by choosing to execute one of the 15 known priority rules at a time, which is what we aim to do in this paper. [15] pointed out the importance of knowledge-based manufacturing, distributed manufacturing, and their integration, which is expected to balance the aspects of optimization, autonomy, and cooperation.

A knowledge base can be produced. [25] described a distributed knowledge base for manufacturing scheduling. This distributed knowledge base enables information on scheduling problems, and the corresponding methods of resolution, to be shared in a wider search space. This knowledge base can be easily and continuously updated by both the end users and the designers of the scheduling methods alike. The most important problem with this kind of approach is the quantity of data that the system has to provide to create knowledge, and because every user can update the knowledge base, the efficiency and the quality of this knowledge is not assured. Other researchers have tried to use data mining for knowledge extraction [14]. This class of models uses past events to help with the current decision-making process.

In the next section, a review of the literature is provided relating to the use of data mining in the field of manufacturing, and more particularly the use of data mining for extracting priority rules.

2.2   Rule Extraction Using Data-Mining

In modern manufacturing, the volume of data is growing at an unprecedented rate in digital manufacturing environments with the use of barcodes, sensors, vision systems, etc. These data may be related to design, products, planning and control, performance, etc., and may include patterns, associations and dependencies. However, the use of accumulated data has been limited, which has led to the "rich data, poor information" problem [26]. The manufacturing data collected contains valuable information and knowledge that could be integrated in the manufacturing system to improve decision-making and enhance productivity.

Generally, priority rules are based on criteria such as deadlines, operative times, arrival time, size of queues, etc. (FIFO (First In First Out), SPT (Shortest Processing Time)). Machine learning has most often been devoted to choosing the best priority rule to use by learning from simulated data. A hybrid model based on decision trees and neural net-works using back propagation was developed by [26] to determine new priority rules in the presence of noise and to predict the performance of these rules. Discovered rules are specific to the manufacturing system studied, unlike the known ones. In [19], a five-phase process was proposed for the construction of a knowledge-based model: the definition of control parameters, generation of training examples, and the definition of rule activation conditions. Another knowledge-based approach is presented in [7]; their study is based on the determination of the relevant parameters for the rule selection, using a neural approach. Using a simulation process in a job shop, [12] proposed an evolutionary optimization approach based on data mining to produce knowledge about the system's behaviour. [12] showed that the correct assignment of priority rules is very important to strengthen the performance measures in a flexible manufacturing system (FMS). This knowledge is centralized for the entire system. An attributes-oriented induction approach was used to characterize the relationship between the sequence of operations and their attributes. The knowledge gained can be used to analyse the effect of decisions made at any stage. [16] proposed a neural networks-based approach to select in real time, each time a resource becomes available in a flow-shop, the most suitable dispatching rule. [14] introduced a new methodology for generating scheduling rules using data mining. This approach includes pre-processing the scheduling history and its transformation into a set of appropriate learning bases. The knowledge extracted can be later used for scheduling new tasks.

In [17] they proposed a Decision Tree, learning from the scheduling data by applying a classification method. More recently, several researchers have used similar approaches to learning rules in scheduling in specific environments. For example, [21] investigated the use of an inductive learning mechanism to generate new heuristics to schedule a series of instructions. [22] proposed a data mining-based framework for job shop scheduling problems. The proposed methodology is based on the implicit assumption about the ability of Tabu search to move intelligently within the solution space while at the same time learning about the embedded knowledge related to the lines of thinking behind these intelligent moves. [18] proposed a data mining-based approach to discover previously unknown priority dispatching rules for a single-machine scheduling problem.

These studies show the interest of inductive learning in discovering priority rules. However, most of them developed a centralized decision-making process, which makes the system vulnerable. In addition, the inference engine is generally cumbersome, especially in the presence of large amounts of data.

3   Definition of the Approach and the Architecture

The approach developed is summarized in this section. As seen in Fig. 1, this approach is composed of three phases:

The simulation phase: the manufacturing system is designed and simulated using a simulation engine such as ARENA[1] ; several scenarios can be launched using different kinds of priority rules (FIFO, LIFO...) in order to gain a vast amount of experience and consequently a vast amount of data corresponding to a designed manufacturing system. The second phase is the learning phase: this phase is dedicated to the data-mining process. Indeed, in this phase the data collected from the simulations are coded to be integrated in the data-mining process using learning algorithms resulting in conjunctive rules.

---

[1] Rockwell ARENA 13.5. ARENA (*Modeling Corporation*).

The last phase is <u>Multi-agent system operation</u>: in this phase the rules discovered in the second phase are integrated in the agent's decisional module. The system is then launched to control the manufacturing system. When decisions are made a log file is saved in order to record the system's history. This data can be used again to improve rules learned in the second phase, constituting a loop.

4   Multi-Agent Framework for Manufacturing Control

In this section, the multi-agent system architecture developed is presented. This system is used to control a job-shop and to resolve a scheduling problem. We start by presenting the agents, then their interactions.

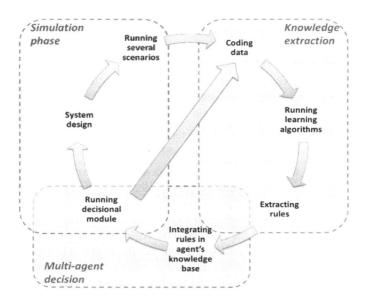

Fig. 1. Approach cycle

4.1   Agent's Generic Structure

High-level control and operational-level control are clearly differentiated in our agent model. The *high-level control* represents the intelligence in the agent, which means the decision-making, information processing and organizational interaction capabilities (e.g. decision-making based on a market-like approach functionality will reside in this layer). On the contrary, the *operational* layer, though it has a certain intelligence, is ruled by the decisions made at the level above (e.g., routing decisions to go from resource A to resource B, resource B being the destination decided by the high-level control).

Two additional characteristics are proposed: the *division* between "decision-making" and "data processing", and the *inclusion of specialized interfaces* at each level.

The first one is inspired by the different nature of both processes. While data calculation could be required on a high-frequency basis, decision-making processes are event-oriented. The data processing modules provide the decision-making entities with information and data, and keep the variables and parameters updated based on the decisions made and the current system conditions. Finally, the two-layer model counts on specialized interfaces at each level to have a clear understanding of the necessary protocols and technological requirements depending on the type of information they transmit. Inter-agent communication is also possible and is also handled by the same interface units.

4.2   Multi-Agent System Organization

Three types of agents were developed for the system presented in this paper:

Order Agent (OA) represents product requests in a manufacturing system. It uses its operational layer to create other sub-agents and assign to them parts of the problem so they can work on a solution in a distributed way.

Product Agents (PA) are the sub-agents of the order agent. Each product is assigned a specific task-sequence according to its type and a particular objective.

Resource Agents are service providers for product agents. To execute its services, the "decision-making" layer will contain the cellular decisional module developed, and which will be presented later.

The order agent (OA) sends production sequences to each product agent (PA) and PAs send execution requests to the resources concerned (RA). RAs cumulate a number of PA requests for execution and the RA launches its cellular decisional module to choose which priority rule to execute in order to sort the requests from PAs in its queue, and sends execution information to the PA (start date). The PA receives numerous propositions and chooses the best (the earliest date) and sends confirmation to the corresponding RA which changes the task status from reservation to execution. Each decision is recorded in the OA's memory in order to constitute its example base for learning. Fig. 2 presents a UML sequence diagram showing the interactions between these agents and their behaviour.

5   Boolean Modelling of the Priority Rule Induction

In this section, the learning phase of the approach (section 3) is detailed. The first step involves creating and coding data in preparation for the knowledge extraction process. The execution of the learning algorithm then uses and develops coded data to create conjunctive rules that, according to the job description, can predict which priority rule to choose. The ARENA simulation tool was used to generate a database related to the job-shop scheduling, which will be operated by automatic rule extraction. Arena is widely used for manufacturing simulations generating a considerable quantity of data as suggested in [23].

Firstly, the manufacturing system will be described.

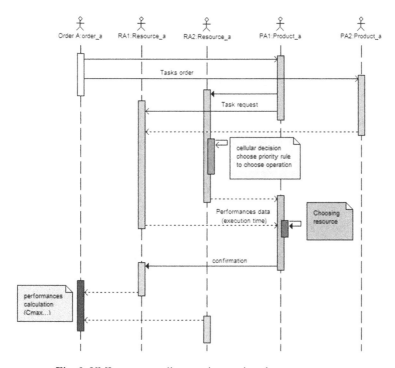

Fig. 2. UML sequence diagram: interactions between agents

This problem is known to be NP-hard in the sense that no solution in polynomial time is achievable [8]. To describe this system, parameters, indexes and variables are presented first, and then the constraints are detailed (inspired from [27]).

Parameters

P   set of jobs to be executed, P={1,2,...n}

R   set of resources, R={1,2,...m}

I_j set of operations of job j, I_j={1,2,...,$|I_j|$}, $j \in$ P

O_{ij} operation number i of job j,

S_e type e of operation O_{ij}.

p_{ij} processing time of operation i ($I \in$ I_j) for product j

R_{ij} set of possible machines for operation O_{ij}

Notation for indexes
$j \in \{1,2,...n\}$: for jobs; $r \in$ R: for resources and $i \in \{1,2,...,|Ij|\}$: for operations,

Notations for variables
t_{ij} completion time of operation O_{ij} ($i\in$ Ij), $tij \in$ N,
μ_{ijr} binary variable set at 1 if O_{ij} is performed on machine r, 0 otherwise,
b_{ijkl} a binary variable set at 1 if O_{ij} is performed before O_{kl}, 0 otherwise,

The objective of the MILP is to minimize the makespan C_{max}, where:

$$C_{max} = max_{\forall i \in I_j \forall j \in P} \ t_{ij} \tag{1}$$

The constraints imposed in the MILP are:

Disjunctive constraints: A machine can process one operation at a time and an operation is executed by only one machine. Machines have unlimited queuing capacity.

$$t_{ij} + p_{kl} * \mu_{klr} + BM * b_{ijkl} \le t_{kl} + BM \ \ \forall i,k \in I, \forall j,l \in P, \forall r \in R_{ij} \tag{2}$$

where BM is a large number

$$b_{ijkl} + b_{klij} \le 1 \ \forall i \in I_j, k \in I_l \ \forall j,l \in P, \tag{3}$$

$$\sum_{r \in R_{ij}} \mu_{ijr} = 1 \ \forall i \in I_j \ \forall j \in P, \tag{4}$$

Precedence constraints: ensure the task sequence of a product. The completion time of the next operation takes the completion time of the previous one and the processing time into consideration.

$$t_{(i+1)j} \ge t_{ij} + p_{(i+1)j} \ \forall i \in I_j, \forall j \in P \tag{5}$$

The remaining constraints of the MILP ensure the type of each variable.

$$t_{ij} \ge p_{ij} \ \forall i \in I_j, j \in P \tag{6}$$

$$b_{ijkl} \in \{0,1\} \forall i \in I_j, j \in P, \forall k \in I_l, l \in P \tag{7}$$

$$\mu_{ijr} \in \{0,1\} \forall i \in I_j, j \in J, \forall r \in R_{ij} \tag{8}$$

In the second step, this system was simulated in ARENA and the data created is used for learning. This will be explained in the next section.

5.1   Data Acquisition

In this paper, and as an initial experiment, a simple instantiation of the job shop model consisting of two machines and three products was used. The objective was to minimize the total scheduling duration (Cmax).

The job shop data are presented in Table 1. The model was then implemented (Fig. 3): "Create" nodes create and launch products (jobs) in the system; "Process" nodes represent the resources, and "Assign" are blocks that can parameterize the different nodes. The basic idea of the simulation in this case is to simulate the behaviour of the workshop over time and load machines with a free choice of available operations. This approach implies the notion of priority rules.

Table 1. Job shop (3×2)

jobs	$R(O_{i,j})$	$P_{i,j}$
J1	**1, 3**	**2,1**
J2	1,2	2,3
J3	2,3	1,1

The simulation was performed on a computerized machine equipped with an Intel Core (TM) 2 Duo 2 GHz processor with 1.99 GB memory and a Windows XP service pack 2 operating system.

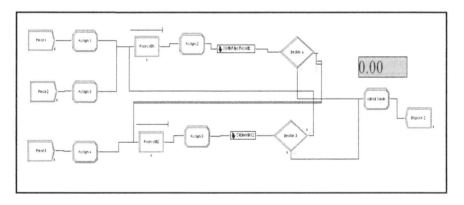

Fig. 3. Transcription of the job shop model in ARENA 13.5

An important number of experiments for each machine were tested, and for each experiment we considered:

— The execution time of operations on both machines;
— The arrival of the operations on the machines;
— The waiting time of operations;
— The priority rules, to investigate the influence of these rules on the calculation of the objective function (min C_{max}).

In order to collect as much information as needed for the knowledge extraction process, there were applied 4 different priority rules: FIFO "First In First out", LIFO "Last In First out", LAV "Lowest Attribute Value" and HAV "Highest Attribute Value", and different inputs: execution time and arrival time. The output considered was C_{max}.

An important number of experiments were tested. Table 2 shows a learning sample obtained by simulation (scenarios).

Let us consider $\Omega = \{w1, w2... w25\}$ - a set of observations and $X = \{X1, X2, X3\}$ - the set of attributes, called descriptors, for each observation:

X1: Operation time of job Ji, which is equal to:

- D1: if operation time (P) of job 1 is equal to P of job 2;
- D2: if P of job 1 is greater than P of job 2;
- D3: if P of job 1 is lower than P of job 2.

X2: Arrival: Arrival time of job Ji in queue, which is equal to:

- E1: If At of job 1 is equal to At of job 2;
- E2: If job 1 accesses resource before job 2;
- E3: If job 2 accesses resource before job 1.

X3: Wait: waiting time of job in queue, which is equal to:

- A1: If waiting time of job 1 is equal of job 2;
- A2: If waiting time of job 1 is greater than job 2;
- A3: If waiting time of job 1 is shorter than job 2.

Table 2. Learning sample obtained through simulation (parcel)

job1	Duration	start	wait	end	job2	duration	start	wait	end	Rule	Cmax
1	6	1	3	10	2	4	0	0	4	FIFO	10
1	6	1	3	10	3	5	0	10	15	FIFO	10
2	4	0	0	4	3	5	0	10	15	FIFO	10
3	5	0	0	5	1	6	1	9	16	FIFO	10
3	5	0	0	5	2	4	0	5	9	FIFO	10
1	6	1	9	16	2	4	0	5	9	FIFO	10
1	6	1	3	10	2	4	0	0	4	LIFO	10
1	6	1	3	10	3	5	1	9	15	LIFO	9

The variable to predict Y is a priority rule which takes values in the set of rules:

$C = \{$FIFO, LIFO, LAV, HAV$\}$, with:
$Y: \Omega \rightarrow C = \{$ FIFO, LIFO, LAV, HAV$\}$
    $Wi \rightarrow Y (Wi) = cj$

Table 3 shows the learning sample after codification. The experimental data are now filtered (removing gaps and duplication), coded and ready for learning and extracting knowledge. This step is detailed in the next section.

5.2   Data Mining Using Cellular Induction

The *Extraction of Classification Rules from Structured Data* (ECRSD) problem uses the principles of machine learning and inductive or deductive supervised learning methods. Among the inductive ECRSD methods, we focus on decision trees [20], in particular the cell induction graphs [5], because the classification function is expressed by a graph which can be converted into production rules.

For this extraction we used Weka^{CASI}[2] , which is an augmentation of standard WEKA integrating CASI (*Cellular Automata for Symbolic Induction*, which will be presented later) [5] for Boolean cellular representation of discovered rules.

Table 3. Coded learning sample Ω (parcel)

duration	start	wait	rule
D2	E3	A2	FIFO
D2	E3	A3	FIFO
D2	E3	A2	LIFO
D2	E3	A3	LIFO
D2	E3	A2	LAV
D3	E2	A3	LAV
D2	E1	A3	HAV

5.2.1   Construction of an Induction Graph and Generation of Conjunctive Rules

The objective of supervised learning is to find a prediction model φ, also known as an *assignment* or *classification*, allowing observations ω to be obtained from the learning population Ω, for which we do not know the class but we know the state of all the exogenous variables to predict this value using φ. The development of Ω requires two samples to be taken from the population Ω, denoted Ω_A and Ω_T. The first one says learning will be used to build the model φ and the second says testing will be used to test the validity of φ [28]. Thus, we assume for any individual that its description's value and its class are known. If the procedure φ is considered consistent, then we can generalize its use for all population observations in the system. Due to φ, we can calculate the class for each observation, knowing only its description.

The supervised machine learning by induction is therefore proposed to provide tools to extract the prediction model φ, from the information we have on the learning sample Ω,. This model can take the form of a neural network, a graph or an induction of a cellular automaton [5]. The general learning process that Weka^{CASI} applies to a population is organized in four steps:

1. <u>Data acquisition and preparation</u>, obtained by simulation, which consists in using the various data pre-processing techniques already integrated in the Weka environment (which is done in step 5.1);

---

[2] WEKA (Waikato Environment for Knowledge Analysis)
   http://www.cs.waikato.ac.nz/ml/weka

2. The development of an <u>Induction Graph Model</u> φ using Weka^{CASI} is summarized in three steps:

   a. Initiation of induction graph by cellular automaton;
   b. Generation of cellular production rules;
   c. Validation of cell rules;

3. <u>Model validation</u> by exploiting all the visualization and analysis methods already included in the Weka platform.

5.2.2.   Rule Generation

Provided that our sample is representative of the original population, we can deduce four rules: *If* condition *Then* Conclusion. The four classification rules from a decision tree model are presented in Fig. 4.

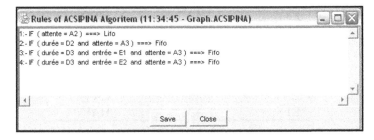

Fig. 4. Generation of conjunctive rules by Weka

The generation of these rules is made on the same machine described in (5.1). In the next section, discovered rules are analysed and a Boolean codification is proposed to use these rules in the cellular decisional module (Cellular Machine CASI [5]).

5.3   Analysis and Interpretation of Results, Boolean Codification and Validation

The interpretation of the discovered rules (Drules) is as follows:

R1: *IF (A2 = waiting) => Lifo*

If the waiting time of job1 is greater than that of job2 then the rule applied is LIFO.

R3: *IF (Operating = D2 and waiting = A3) => Fifo.*

If the operating time of job1 is greater than the operating time of job2 and if the waiting time of job1 is less than that of job2 then the rule applied is FIFO.

R2: *IF (Operating = D3 and start = E1 and waiting = A3) => Fifo*

If the operating time of job1 is greater than the operating time of job2 and if the arrival time of job1 is identical to that of job2 and if the waiting time of job1 is lower than job2 then the rule applied is Fifo.

R4: *IF (Operating = D3 and start = E2 and waiting = A3) => Fifo*

If the operating time of job1 is greater than the operating time of job2 and if the arrival time of job1 is greater than that of job2 and if the waiting time of job1 is lower than job2 then Fifo is applied.

In order to use these rules in the agent's decisional model, facts and rules must be Boolean encoded. The cellular machine CASI consists of two layers: "CELFACT" and "CELRULE" [5], which can represent a knowledge base. The first layer, called CELFACT, represents the fact base, and the second layer, called CELRULE, represents the rule base.

In each layer, the contents of a cell determine whether and how it participates in each inference step: at every step, a cell can be active or passive, can take part in the inference or not. It is assumed that there are 9 cells in the CELFACT layer, and 4 cells in the CELRULE layer according to the conjunctive rules extracted (tables 4 and 5). The states of cells are composed of three parts: EF, IF and SF, and ER, IR and SR which are the input, internal state and output parts of the CELFACT and CELRULE cells, respectively. IF and IR are not used in our case. Any cell i in the CELFACT layer with input $EF(i) = 1$ is regarded as representing an established fact. Any cell j of the CELRULE layer with input $ER(j) = 0$ is regarded as a candidate rule. When $ER(j) = 1$, the rule should not take part in the inference. The incidence matrices RE and RS represent the input/output relation of the facts and are used in forward chaining (Tables 6 and 7). One can also use RE as the output relation and RS as the input relation for backward chaining.

A goal fact, which is the basic cycle of an inference engine in forward chaining, traditionally operates as follows:

1. Search for applicable rules (evaluation and selection).
2. Choose one of these rules, for example Ri (filtering).
3. Apply and add the conclusion part of Ri to the fact base (execution).

The cycle is repeated until the goal fact is added to the fact base, or stops when no rule is applicable.

A cycle of an inference engine is made up of two local transition functions: δfact (equation 9) and δrule (equation 10), where δfact corresponds to the evaluation, selection, and filtering phases and δrule corresponds to the execution phase.

$$\delta\text{Fact}: (EF,IF,SF,ER,IR,SR) \xrightarrow{\delta\text{fact}} (EF,IF,\textbf{EF},\textbf{ER}+(\textbf{R}_E^T \cdot \textbf{EF}),IR,SR) \tag{9}$$

After applying this function we have:

$$\text{EF} = \text{EF} \quad \text{SF} = \text{EF} \quad \text{ER} = \text{ER} + R_E^T \times \text{EF} \quad \text{and} \quad \text{SR} = \text{SR}$$

$$\delta\text{Rule}: (EF,IF,SF,ER,IR,SR) \xrightarrow{\delta\text{rule}} (\textbf{EF}+(\textbf{R}_S \bullet \textbf{ER}),IF,SF,ER,IR,\overline{\textbf{ER}}) \tag{10}$$

After applying this function, we have:

EF = EF + (RS × ER); SF = SF; ER = ER and SR = $\overline{ER}$,

where R_E^T is the transposed matrix of R$_E$ and $\overline{ER}$ is the negation of ER.

Initially, all the cells in the input layer are passive Celfact (EF = 0), except those that are the initial evidence (EF = 1). EF and ER are input states of facts and rules, IF and IR are internal states and SF, SR are output states of cells. In this example, A3 and D2 are activated. Then, R2 is reached, and the LIFO fact is deduced using the δrule (Rule transition function). Then, the LIFO rule is executed in this case, at this moment (Tables 6 and 7).

To test the efficiency of these rules, they were used to control a job shop defined as follows (Table 8):

First, LIFO and FIFO rules were used. For the last experiments, discovered rules were used with different job arrival times. The Cmax obtained from each experiment were compared and the results are presented in Table 9.

Table 4. Celfact

	EF	IF	SF
Wait = A2		1	
Wait = A3	1	1	1
Duration = D2	1	1	1
Duration = D3		1	
Arrival = E1		1	
Arrival = E2		1	
FIFO		1	
LIFO	①	1	
HAV		1	

Table 6. Input Matrix R_E

	R1	R2	R3	R4
Wait= A2	1			
Wait = A3		1	1	1
Duration=D2		1		
Duration=D3			1	1
Arrival=E1		1		
Arrival =E2				1
FIFO				
LIFO				
HAV				

Table 5. CelRule

	EF	IF	SF
R1		1	1
R2	1	1	1
R3		1	1
R4		1	1

Table 7. Output Matrix R_S

	R1	R2	R3	R4
Wait= A2				
Wait = A3				
Duration=D2				
Duration=D3				
Arrival=E1				
Arrival =E2				
FIFO	1			
LIFO		1	1	1
HAV				

These limited experiments show that discovered rules, integrated in the resource agent, can provide good scheduling solutions compared to traditional priority rules (LIFO and FIFO). These discovered rules constitute a knowledge base distributed in a set of agents, thus making the control system reliable compared to previous work in this field. In addition, the use of cellular automaton for inference (CASI) makes the system very reactive because Boolean induction is trivial compared to traditional symbolic inference. These results are very encouraging to extend the sample base and discover more rules to create a relevant decisional cellular module.

Table 8. jobshop 4x2

jobs	R(Oi,j)	Pi,j
J1	1, 3	2,2
J2	1,2	2,3
J3	2,3	1,2
J4	2,2	1,3

Table 9. Cmax of each kind of rule

Arrival time	FIFO Cmax	LIFO Cmax	Drules Cmax
0000	10	13	10
5000	13	11	10
0500	13	10	9
0050	13	11	10
0005	13	13	10

6   Conclusion

The main objective of this paper was to present an approach based on a Boolean knowledge foundation for dynamic scheduling resolution. This approach results in decisional modules which contain rules that, according to a workshop state, can select which priority rule to execute. According to a data-mining process, we started by preparing a learning base. Simulation is often used to generate a learning base; then conjunctive rules were deduced. Next, Boolean codification was applied to prepare this base for cellular induction.

This engine was integrated in the decisional module of agents which constitute the distributed manufacturing control system. The experiments have shown that with the new Boolean induction the model provides good prediction/decision results. Using the CASI system reduces the storage size of the rule base and the validation time [4]. The validation step proved the efficiency of the discovered rules on a simple Job Shop. The objective now is to further pursue the experiments to ensure the reliability of the approach on more general, complex cases compared to a traditional inference engine.

References

1. Aissani, N.D., Trentesaux, D., Beldjilali, B.: Efficient and effective reactive scheduling of manufacturing system using SARSA-Multi-objective-agents. In: MOSIM 2008: 7th Int. Conf. on Modelling & Simulation, pp. 698–707. Lavoisier (2008)
2. Aissani, N.D., Trentesaux, D., Beldjilali, B.: Dynamic Scheduling of Maintenance Tasks in the Petroleum Industry: a Reinforcement Approach. EAAI: International Scientific Journal Engineering Applications of Artificial Intelligence 22, 1089–1103 (2009) ISSN: 0952-1976
3. Aissani, N.D., Trenteseaux, D., Beldjilali, B.: Dynamic scheduling for multi-site companies: a decisional approach based on reinforcement multi-agent learning. JIM: Journal of Intell. Manufacturing 22, 1089–1103 (2011), doi:10.1007/s10845-011-0580-y

4. Amrani, F., Bouamrane, K.: Towards a cellular indexing in a case based reasoning approach: Application to an urban transportation system regulation. In: Proceedings of the 2010 Conf. on Bridging the Socio-technical Gap in Decision Support Systems: Challenges for the Next Decade, pp. 321–332. IOS Press, Amsterdam (2010)
5. Atmani, B., Beldjilali, B.: Knowledge Discovery in Database: Induction Graph and Cellular Automaton. Computing and Informatics, Journal 26(2), 171–197 (2007)
6. Bousbia, S., Trentesaux, D.: Self-organization in distributed manufacturing control: State-of-the-art and future trends. In: IEEE International Conference on Systems, Man & Cybernetics, vol. 5, p. 6 (2002)
7. Chen, C.C., Yih, Y.: Identifying attributes for knowledge-based development in dynamic scheduling environments. International Journal of Production Research 34(6), 1739–1755 (1996)
8. Conway, R.W., Maxwell, W.L., Miller, L.W.: Theory of scheduling. Dover Publications (2003)
9. Davenport Andrew, J., Beck, J.C.: A survey of techniques for scheduling with uncertainty. Technical report, IBM and ILOG (2000)
10. Erol, R., Sahin, C., Baykasoglu, A., Kaplanoglu, V.: A multi-agent based approach to dynamic scheduling of machines and automated guided vehicles in manufacturing systems. Appl. Software Computing 12, 1720–1732 (2012)
11. Esquirol, P., Lopez, P.: L'ordonnancement. Economica, Paris (1999)
12. Huyet, A.L.: Optimization and analysis aid via data-mining for simulated production system». European Journal of Operational Research 173, 827–838 (2006)
13. Katalinic, B., Kordic, V.: Bionic assembly system: Concept, structure and function. In: Proceedings of the 5th IDMME, Bath, UK (2004)
14. Li, X., Olafsson, S.: Discovering dispatching rules using data mining. Journal of Scheduling 8, 515–527 (2005)
15. Monostori, L., Csáji, B.C., Kádár, B., Pfeiffer, A., Ilie-Zudor, E., Kemény, Z., Szathmári, M.: Towards adaptive and digital manufacturing. Annu. Rev. Control 34, 118–128 (2010)
16. Mouelhi-Chibani, W., Pierreval, H.: Training a neural network to select dispatching rules in real time. Comput. Ind. Eng. 58, 249–256 (2010)
17. Olafsson, S., Li, X.: Learning effective new single machine dispatching rules from optimal scheduling data. Int. J. Prod. Econ. 128, 118–126 (2010)
18. Premalatha, S., Baskar, N.: Implementation of supervised statistical data mining algorithm for single machine scheduling. J. Adv. Manag. Res. 9, 170–177 (2012)
19. Priore, P., Garcia, D.D., Quesada, I.F.: Manufacturing systems scheduling through machine learning. In: Neural Computation, NC 1998, Vienna, Austria, pp. 914–917 (1998)
20. Quinlan, R.: C4.5: Programs for Machine Learning. Morgan Kaufman Publishers (1993)
21. Russell, T., Malik, A.M., Chase, M., van Beek, P.: Learning heuristics for the superb-lock instruction scheduling problem. IEEE Transactions on Knowledge and Data Engineering 21(10), 1489–1502 (2009)
22. Shahzad, A., Mebarki, N.: Data mining based job dispatching using hybrid simulation-optimization approach for shop scheduling problem. Eng. Appl. Artif. Intell. 25, 1173–1181 (2012)
23. Thomas, P., Thomas, A., Suhner, M.-C.: A neural network for the reduction of a product-driven system emulation model. Prod. Plan. Control 22, 767–781 (2011)
24. Trentesaux, D., Pesin, P., Tahon, C.: Distributed artificial intelligence for FMS scheduling, control and design support. Journal of Intelligent Manufacturing 11, 573–589 (2000)

25. Varela, M.L.R., Aparício, J.N., Silva, S., do, C.: A Distributed Knowledge Base for Manufacturing Scheduling. In: Camarinha-Matos, L.M. (ed.) Emerging Solutions for Future Manufacturing Systems. IFIP, pp. 323–330. Springer, Boston (2005)
26. Wang, K., Tong, S., Eynard, B., Roucoules, L., Matta, N.: Review on application of data mining in product design and manufacturing. In: Proceedings of the 4th International Conference on Fuzzy Systems and Knowledge Discovery, vol. 4, pp. 613–618 (2007)
27. Zambrano, G.R., Pach, C., Aissani, N., Bekrar, A., Berger, T., Trentesaux, T.: Control of myopic behaviour in heterarchical production systems: a holonic based framework. EAAI: Journal of Intelligent Manufacturing 22, 1089–1103 (2012), doi:10.1007/s10845-011-0580-y
28. Zighed, D.A., Rakotomalala, R.: Graphes d'induction: apprentissage et data mining. Hermes Science Publications (2000)

Supply Chain Management Using Multi-Agent Systems in the Agri-Food Industry

Ait Si Larbi El Yasmine[1,2], Bekrar Abdel Ghani[2],
Damien Trentesaux[2], and Beldjilali Bouziane[1]

[1] Univ. Oran, LIO, Department of Computer Sciences, University of
Oran, ES-Sénia, BP 1524. El M'naour, Oran, Algeria
[2] Univ. Lille Nord de France, F-59000 Lille, France, UVHC, TEMPO Lab.,
"Production, Services, Information" team, F-59313 Valenciennes, France
`aitsilarbi@gmail.com`,
`{abdelghani.bekrar,damien.trentesaux}@univ-valenciennes.fr`,
`bouzianebeldjilali@yahoo.fr`

Abstract. This paper focuses on Multi-Agent Systems (MAS) applied to supply chain management (SCM) in the agri-food industry. The supply chain (SC) analysed includes the suppliers, manufacturers and distribution centres, considering that orders are sent by clients to the distribution centres. Many original constraints, such as the capacity of the suppliers and the manufacturers as well as balancing stocks, are supported in our contribution. Our proposal also considers some practical issues in agri-food SCM, such as expiry dates. The aim of our method is to find a near-optimal solution that minimizes costs and time throughout the SC process, whilst favouring reactivity. An AUML model shows the functioning of the MAS in the SC. The results obtained regarding duration and costs related to the execution of client orders were compared with those obtained using a heuristic that solves the optimization problem in the dynamic case, and a mathematical model in the static case.

Keywords: agri-food supply chain, multi-agent systems, optimization, reactivity.

1 Introduction

Companies extend beyond their organizational and physical borders; this is characterized by an externalization of activities and the development of partnerships. Externalization strategies lead them to join other companies seeking complementary of resources, for example [1].

Inventory management in a store or on a local site is currently a common task. However, when it is applied to an extended company, it requires other mechanisms to be studied, modelled and validated (such as communication a.o.) [2].

In recent years, due to competition and increasingly restrictive markets, SCM is encountered in different domains [3]. SCM must be optimized, but at the same time needs to remain reactive.

T. Borangiu et al. (eds.), *Service Orientation in Holonic and Multi-Agent Manufacturing and Robotics*, Studies in Computational Intelligence 544,
DOI: 10.1007/978-3-319-04735-5_10, © Springer International Publishing Switzerland 2014

SC optimization has been widely studied, particularly in terms of minimizing duration and costs in the global SC stages. [4] proposed an approach to solve the dynamic placement of stocks. [5] proposed the application of a fuzzy mathematical programming approach to solve production allocation and distribution. [6] presented a two-level planning approach for the design and optimization of production and distribution in the case of an agrochemical SC.

In Agri-food Supply Chains (ASC), certain effective constraints are rarely considered in the literature, such as deteriorating inventory and balancing stocks. In a previous study, we addressed such specific SC and focused on optimization by providing a mathematical formulation [7]. In this paper, we intend to address the reactivity of this kind of SC, taking into account the specific constraints which were not considered in [7]. A multi-agent model is proposed and tested. With the aim of addressing fault tolerance (which prevents a system from failing) and allowing communication, the MAS proposed consists of agents in a heterarchical architecture.

This paper is structured as follows: Section 2 presents a literature review of papers dealing with ASC. In section 3, we present a generic model proposed for ASC. In section 4, we discuss our implementation of the MAS in the SC case study and carry out some experiments. Finally, we present our conclusions and some future prospects.

2   Literature Review

2.1   Agri-Industry and Agri-Food Supply Chains

The term *agri-food supply chain* (ASC) is used to describe activities from the production to the distribution that bring agricultural or horticultural products from the farm to the table. Generally, an ASC is made up of the organizations responsible for the production (farmers), distribution, processing, and marketing of agricultural products to the end consumers [8]. Agri-food enterprises operate in a complex and dynamic environment. To meet the increasing demands of consumers and different partners, agri-food enterprises have to innovate and design new products, processes and solutions for cooperation in ASC networks.

The major importance of food quality, food safety and weather-related variability makes ASC different from other SCs. Other characteristics include the limited shelf life of products, the variability in demand and prices, making ASC more complex and harder to manage than other SCs [8]. Logistics operations are also critical in agribusiness management due to high consumer standards in terms of food quality and safety. The food industry is of course characterized by features that distinguish it from other sectors: current food quality and safety standards require the continuous monitoring of products and agricultural supplies throughout the SC. Food has a high perishability, and thus requires efficient supply chains in terms of time [9]. This fact creates a need for SCM automation and optimization.

There is also considerable variation in supply due to seasonality of agricultural production, weather conditions and the biological nature of agricultural products, resulting in input unpredictability[9].

Finally, a large portion of products from the food sector is sold to international markets. In addition, raw materials and imported products are also sourced from different countries. This is a challenge for the food industry in terms of distribution [9].

Thus, the agri-food sector faces global challenges: increased globalization and competition, highly differentiated and segmented food production, complex requirements regarding quality assurance, flexibility in the provision of food and efficiency in the sector's organization and processes [10]. Some of these problems cause product deterioration due to expiry dates being exceeded.

Many studies have been conducted to try and overcome these problems. We can cite [8], who drew up a list of studies related to ASC and focused on planning models. Mathematical optimization is the most common method of modelling the food production stage of a food supply chain (FSC) (see [7]). Existing agricultural optimization models are of course static and deterministic. However, very few of these are able to capture stochastic or dynamic elements of FSCs, and most of them only analyse a single stage of the FSC – food production.

Among these studies, we can cite [11], which addresses a blending and shipping problem in a company that manages a wheat SC and formulates it as a Mixed Integer Linear Model (MILP).

2.2   Inventory and Deteriorating Product Management in Agri-Food SC

Inventoried goods can be classified based on: obsolescence, deterioration and no obsolescence/deterioration. The inventory models available in the literature can be classified into:

a-   Models for a fixed shelf life inventory;
b-   Models for a random shelf life inventory;
c-   Models for an inventory which decays according to a proportional decrease in inventory in terms of its utility or physical quantity.

In our case, we have to consider the first category because of the nature of agrifood products characterized by expiry dates. If a product remains unused by the end of its shelf life, it is considered as out of date and must be disposed off.

According to [12], fixed shelf life products in inventory are usually depleted according to either a first-in-first-out (FIFO) or a last-in-first-out (LIFO) issuing policy.

Deteriorating inventory was originally studied by [13] and [14]. Concerning inventory models for fixed shelf life perishable products, many studies have been undertaken. We can cite [15], who derived near-myopic bounds on order quantities and used the bounds to evaluate the performance of the resulting heuristics.

Replenishment and inventory management represent a challenge in SC research. [16] considered various forms of replenishment of raw materials, components and end-products in each node. Moreover, [4] proposed an approach to solve the dynamic placement of inventory to meet challenges concerning compliance with the date of receipt in the machine industry. [5] proposed the application of a fuzzy mathematical programming approach to solve production allocation and distribution in a SC network. [17] proposed a policy for optimizing inventory in a multi-echelon supply chain and considered products with time-sensitive deterioration rates.

In addition, [18] presented a review of inventory systems taking into account inventory deterioration. Their study presents contributions relating to inventory control of perishable items since 2001. Various inventory models can also be found in the same paper.

2.3 Limits of Optimization and Reactivity in Agri-Food SC

The optimization techniques presented in the review have limitations regarding ASC. The models proposed do not take into account product categories at manufacturing level.

Concerning reactivity, [19] modelled a multi-echelon system in which disruptions can occur at any stage. [20] incorporated uncertainty and process dynamics into the enterprise. [21] examined demand-side reactive strategies for supply disruption. [22] developed a framework for responsive SC, and in [23] the authors used a Multi-Agent System (MAS) for an ASC. [24] described an intelligent food SC model that improves efficiency within the SC. [25] proposed a MAS to represent and integrate the decision-making processes of actors within a food grain SC. [26] adopted a system dynamics methodology as a modelling and analysis tool, and provided a review of the literature relating to the use of MAS in SCM.

In industry, there are software applications for planning and performance optimization, which include materials requirement planning (MRP), manufacturing resources planning (MRPII), enterprise resource planning (ERP), and advanced planning and scheduling (APS). These systems have been extended to inter-organizational processes but lack interface and communication abilities. Due to these flaws, collaboration and coordination, which are very important in SCM, become too costly and ineffective.

However, agent technology is well-suited to supporting collaboration in SCM. The reason for adopting agent technology is based on achieving three goals [24]:

(1) Data, control, expertise, or resources are inherently distributed.
(2) The system is naturally regarded as a society of autonomous cooperating components.
(3) The system contains legacy components, which must be made to interact with other, possibly new, software components.

The actors in the SC have their own resources, capabilities, tasks, and objectives (1st goal). They cooperate with each other autonomously to serve common goals, but also have their own interests (2nd goal).

Food supply networks are dynamic in nature, characterized by the flow of information, materials, and funds across numerous functional areas within and between the actors in the supply chain. In addition, supply flows are usually managed by information systems, but the effectiveness of SCs is impeded by the lack of coordination and integration between these systems (3rd goal) [24].

To capture the dynamic, stochastic, and multi-faceted elements of an ASC, they must be modelled as complex adaptive systems [23]. Knowing that an ASC is a system of autonomous interconnected entities [23], multi-agent simulation seems to be well-suited to modelling ASCs and capturing decision-making interactions and adaptations of autonomous actors in the food SC.

Throughout this section, we have deduced that the studies presented are insufficient for taking some constraints, such as product deterioration, into account. In [7], we presented a mathematical model to meet the needs of such systems without taking into account this constraint and reactivity. So, in this study we explored the MAS paradigm in order to integrate such constraints.

It is specified that an agent is a computer system component (software) located in an environment that is capable of autonomous actions within that environment in order to meet its design objectives [27]. So, in this paper, the term agent corresponds to the software problem-solving entities which have specific functions for processing the inputs received relating to the ASC studied.

The agents have the ability to control their internal states and their behaviour with the aim of achieving their objectives.

3   A Generic Model of an Agri-Food Industry SC

3.1   Case Study

The ASC model on which our studies are based was presented in [7]. It is inspired from a SC including seven enterprises of the Metidji Group, an Algerian company. One provides transportation services and the others produce flour, semolina, snacks and also process corn.

Enterprise integration requires the integration of work systems which concern each company within the Holding (Fig. 1).

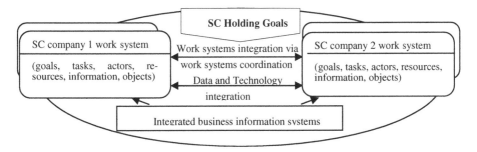

Fig. 1. Work systems and integrated business information systems in the ASC

Work systems integration means managing interdependencies between tasks, resources and actors. So, there is a need for solutions that ensure the implementation of these activities.

ADEPT (Advanced Decision Environment for Process Tasks) [24] solves process-related problems, including judgments and information from the company's marketing, sales, research, development, manufacturing and finance departments, by viewing it as a community of negotiating, service-providing agents where each agent represents a distinct role or department and is capable of providing services [24].

Inspired by this idea and the work in [15] and [28], we propose a multi-agent architecture to represent the control of the ASC studied. The following section describes our model according to AUML notations for agent-based modelling representations.

3.2   ASC Model with AUML

Regarding the work of [24], there are linkages between SC and MAS, especially relating to structure; a supply chain consists of multiple parties working on multi-stage tasks, and a MAS is a set of different types of agents with various roles and functions. In addition to what we said in section 2.3, the use of MAS can be justified by the fact that a supply chain is dynamic and the supply dynamics create inefficiencies; agents can be created or isolated from a MAS to avoid these problems. There are also many tasks in supply chains; agents can coordinate other agents' tasks with the aim of forming a higher-level system.

A methodology and a modelling notation are required for the development of MAS. AUML is a recent approach based on UML [29], proposed to represent various aspects of agents by introducing agent class diagrams and protocol diagrams [30].

Inspired by [31] and [32], who used the AUML approach to model SCM, we present three levels of representation of the SC studied. The first level represents the overall protocol which is considered as a package (we use the notation: agent name/ agent role: class), as shown in Fig. 2.

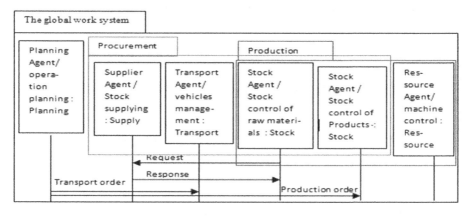

Fig. 2. The global protocol

In Fig. 2, the interactions between the packages (functions) have been represented. Reusability of these packages can be applied later to represent scenarios at a specific heterarchical level.

Level 2 concerns the interactions between agents. Fig. 3 shows the scenario for expiry dates. It is assumed that the stock agent detects a deteriorated raw material. After verification of the raw material in the laboratory, the supplier has two choices depending on the results: refuse the extension of the shelf life or accept and deliver a certificate of extension.

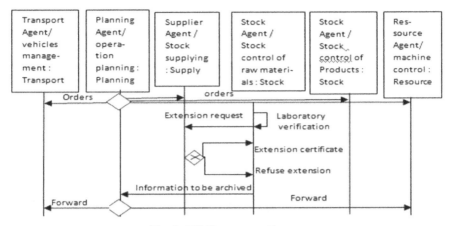

Fig. 3. AUML sequence diagram

Level three concerns the specification of the internal processing of agents. The processing of the resource agent is presented in Fig. 4.

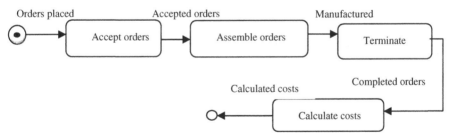

Fig. 4. AUML resource agent activities diagram

3.3 Implementation

This section provides some details of how the MAS model was implemented.

In order to make a MAS model ready for industrial applications, an association called FIPA[1] has proposed standards for MAS designers [3].

The prototype was implemented using JADE[2]: middleware that can be used to develop agent-based applications in compliance with FIPA specifications. JADE

[1] The Foundation for Intelligent Physical Agents (See http://www.fipa.org/).
[2] Java Agent DEvelopment Framework (See http://jade.tilab.com/).

facilitates the development of complete agent-based applications by means of a run-time environment implementing the life-cycle support features required by agents.

Fig. 5 shows our MAS applied to the heterarchical architecture of the ASC studied (five companies are represented here). For each manufacturing company, the resource agent, the raw material stock agent and the product stock agent can be found.

Each agent executes a set of behaviours. The most sensitive task was defining the moments when an agent switches from one behaviour to another.

There is communication between agents of the same heterarchical level; a planning agent is only used for supervision purposes and issuing the initial orders.

For example, a resource agent may communicate with the other resource agents in each company to solve a problem concerning a resource failure.

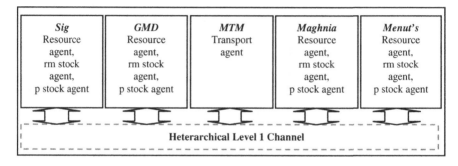

Fig. 5. Metidji Holding Multi-agent system

3.4   Experimentation

To illustrate the solution proposed, some examples were developed to validate our methodology using the case study introduced previously.

As presented in section 3.1, we have a global model which can produce a configurable element characterized by role settings, constraints and acts of communication.

We used data and configurations from our previous work in [7]. The results obtained with a heuristic simulating this behaviour and the results obtained from the application of the MAS model are compared, including the deterioration constraint (sequence diagram presented in Fig. 3). We worked on dynamic scenarios and the response time of the MAS was negligible in all samples.

Table 1 shows the global costs, when using parameters (T is the sampling period):

C: Customers,

P: Products,

D: distribution centres,

M: manufacturers,

S: suppliers,

E: raw materials,

Table 1. Results obtained by the MAS including the deterioration constraint and the heuristic

Configuration C,P,D,M,S,E,T	Heuristic application results	MAS application results
3,4,5,3,5,4,4	45223	45315
6,6,5,2,5,5,4	259940	259965
10,8,5,5,6,6,4	4911323	4912142

These results focus on the dynamic case, but it is still necessary to demonstrate whether this model achieves good performances in the static deterministic case. To do so, we compared the results obtained with bounds provided by the previously implemented MILP [7]. In this model, the deteriorating product constraint was relaxed. Table 2 shows the results obtained for the global costs.

Table 2. Results obtained with the MAS and the Mathematic model

Configuration C,P,D,M,S,E,T	MAS application results	Mathematic model results
2,2,3,3,3,3,4	41315	38940
4,3,4,2,3,3,5	159900	153070
10,12,5,5,6 ,10	5512142	5486871
10,5,5,5,5,5,15	5102566	5033481
15,5,5,5,5,5,15	6617883	6605339
15,5,5, 5,5,5,10	2894522	2874916

These results are encouraging because the MAS model could meet expectations in terms of responsiveness. From an optimization point of view in a deterministic context, the MAS was also efficient. However, it can be noted that when the values of the parameters C, P, D, M, S, E and T increase, the MAS becomes more efficient. This can be explained by the use of a heterarchical architecture.

4   Conclusion and Perspectives

In this paper, we proposed a multi-agent solution for SCM. The heterarchy principle was applied to a MAS. This is a promising solution considering the samples used in the experimentation and the system reactivity gained with a heterarchical architecture.

An original constraint for the agri-food industry was also considered: deteriorating goods and raw materials. Another highlight of our work is that an AUML model is presented. This representation was a good tool for checking our conceptual system.

Nonetheless, there seem to be many interesting prospects. First, the model proposed can be extended to include wholesalers and details relating to the acquisition of raw materials. The reactivity of the system can change relative to the response time. So, we intend to integrate the optimization mechanisms with the MAS to obtain a globally effective system that can optimize production over a long time period and still react effectively to perturbations. This will of course involve some new challenging issues such as the management of decision consistency when conflicts arise between a solution proposed by the optimization mechanisms and a solution proposed by the MAS system.

References

1. Benaouda, A., Zerhouni, N., Varnier, C.: Une approche MA coopératifs pour la gestion des ressources matérielles dans un contexte multi sites de e-manufacturing. In: Proceedings of MOSIM 2006 (2006)
2. Pinedo, M.L.: Planning and scheduling in manufacturing and services. Springer Series in Operation Research (2005)
3. Benyoucef, L., Jain, V.: Editorial note for the special issue on 'Artificial Intelligence techniques for Supply Chain Management'. Engineering Applications of Artificial Intelligence 22, 829–831 (2009)
4. Funaki, K.: Strategic safety stock placement in supply chain design with due-date based demand. Int.J. Production Economics 135, 4–13 (2010)
5. Bilgen, B.: Application of fuzzy mathematical programming approach to the production allocation and distribution supply chain network problem. Expert Systems with Applications 37, 4488–4495 (2010)
6. Sousa, R., Shah, N., Papageorgiou, L.G.: Supply chain design and multilevel planning: An industrial case. Computers & Chemical Engineering 32, 2643–2663 (2008)
7. Ait Si Larbi, E., Bekrar, A., Trentesaux, D., Beldjilali, B.: Multi -stage optimization in supply chain: an industrial case study. In: MOSIM 2012: 9th International Conference on Modeling, Optimization & SIMulation, Bordeaux, France (2012), oai: hal.archives-ouvertes.fr: hal-00728633
8. Ahumada, O., Villalobos, R.J.: Application of planning models in the agri-food supply chain: A review. European Journal of Operational Research 195, 1–20 (2009)
9. Mangina, E., Vlachos, I.P.: The changing role of information technology in food and beverage logistics management: beverage network optimization using intelligent agent technology. Journal of Food Engineering 70, 403–420 (2005)
10. Schiefer, G.: New technologies and their impact on the agri-food sector: an economists view. Computers and Electronics in Agriculture 43, 163–172 (2004)
11. Bilgen, B., Ozkarahan, I.: A mixed-integer linear programming model for bulk grain blending and shipping. Int. J. Production Economics 107, 555–571 (2007)
12. Goyal, S.K., Giri, B.C.: Recent trends in modelling of deteriorating inventory. European Journal of Operational Research 134, 1–16 (2001)
13. Ghare, P.M., Schrader, G.F.: A model for exponentially decaying inventory. Journal of Industrial Engineering 14, 238–243 (1963)
14. Ping-Hui, H., Hui Ming, W., Hui-Ming, T.: Optimal ordering decision for deteriorating items with expiration date and uncertain lead time. Computers & Industrial Engineering 52, 448–458 (2007)
15. Nandakumar, P., Morton, T.E.: Near myopic heuristic for the fixed life perishability problem. Management Science 39(12), 1490–1498 (1993)
16. Lim, S.J., Jeong, S.J., Kim, K.S., Park, M.W.: A simulation approach for production–distribution planning with consideration given to replenishment policies. International Journal of Advanced Manufacturing Technology 27, 593–603 (2006)

17. Wang, K.J., Lin, Y.S., Yu, J.C.P.: Optimizing inventory policy for products with time-sensitive deteriorating rates in a multi-echelon supply chain. International Journal of Production Economics 130(1), 66–76 (2011)
18. Bakker, M., Riezebos, J., Teunter, R.H.: Review of inventory systems with deterioration since 2001. European Journal of Operational Research 221, 275–284 (2012)
19. Schmitt, A.J.: Strategies for customer service level protection under multi-echelon supply chain disruption risk. Transportation Research Part B45, 1266–1283 (2011)
20. Puigjaner, L., Laınez, J.M.: Capturing dynamics in integrated supply chain management. Computers and Chemical Engineering 32, 2582–2605 (2008)
21. Shao, X.F.: Demand side reactive strategies for supply disruptions in a multiple-product system. Int. J. Production Economics 136, 241–252 (2012)
22. Gunasekaran, A., Kee-Hung Lai, T.C., Cheng, E.: Responsive supply chain: A competitive strategy in a networked economy. Omega 36, 549–564 (2008)
23. Meter, K.: Evaluating Farm and Food Systems in the U.S. In: Systems Concepts in Evaluation: An Expert Anthology, pp. 141–159. EdgePress, Inverness (2006)
24. Mangina, E., Vlachos, I.P.: The changing role of information technology in food and beverage logistics management: beverage network optimization using intelligent agent technology. Journal of Food Engineering 70, 403–420 (2005)
25. Goel, A., Zobel, C.W., Jones, E.C.: A multi-agent system for supporting the electronic contracting of food grains. Computers and Electronics in Agriculture 48, 123–137 (2005)
26. Georgiadis, P., Vlachos, D., Iakovou, E.: A system dynamics modelling framework for the strategic supply chain management of food chains. Journal of Food Engineering 70, 351–364 (2005)
27. Wooldridge, M., Jennings, N.R.: Intelligent agents: Theory and practice. Knowledge Engineering Review 10(2), 115–152 (1995)
28. Kishore, R., Zhang, H., Ramesh, R.: Enterprise integration using the agent paradigm: foundations of multi-agent-based integrative business information systems. Decision Support Systems 42, 48–78 (2006)
29. Huget, M.P., Odell, J., Bauer, B.: The AUML Approach, Methodologies and Software Engineering for Agent Systems. Multi agent Systems, Artificial Societies, and Simulated Organizations 11, 237–257 (2004)
30. Park, S., Sugumaran, V.: Designing multi agent systems: a framework and application. Expert Systems with Applications 28, 259–271 (2005)
31. Huget, M.-P.: Agent UML Class Diagrams Revisited. In: Kowalczyk, R., Müller, J.P., Tianfield, H., Unland, R. (eds.) NODe-WS 2002. LNCS (LNAI), vol. 2592, pp. 49–60. Springer, Heidelberg (2003)
32. Odell, J., Parunak, H.V., Bauer, B.: Extending UML for Agents. In: Proceedings of the Agent-Oriented Information Systems Workshop at the 17th National Conference on Artificial Intelligence, pp. 3–17 (2000)

Part III
Service Orientation in Manufacturing Management and Control

A Generic Service System Activity Model with Event-Driven Operation Reconfiguring Capability

Theodor Borangiu, Monica Drăgoicea, Virginia Ecaterina Oltean, and Iulia Iacob

University Politehnica of Bucharest, Faculty of Automatic Control and Computers
313, Spl. Independentei, 060042-Bucharest, Romania
theodor.borangiu@cimr.pub.ro, monica.dragoicea@acse.pub.ro

Abstract. This paper presents a proposal on developing an activity-based generic model of a service system (SSyst) that realises business oriented, IT-based intensive service processes. The approach proposed here to conceive and design the SSyst model takes into consideration the service systems' lifecycle that includes the interactions between the four stakeholders' categories: service provider (with suppliers and service outsourcing), service customer, competition, and compliance bodies (law, financial, environment, a.o.). The high level description of the lifecycle of a service process is used in order to further define, structure and analyze the SSyst model from a triple perspective: (1) the stages of the service's lifecycle; (2) core activities for services; (3) the activity type components of a composite service. The proposed SSyst model is generic, meaning that it can be applied to generate different types of services, ranging from already existent services (that only need to be configured according to the customer needs) to completely new services (that have to be integrated based on the provider's service repository and outsourced services). The services generated in a SSyst type service system must be planned and they must receive resources and capacities optimally allocated for service delivery.

Keywords: service systems, service lifecycle, value-co-creation, service design, service processes.

1 Introduction

Today a new generation of services is under development and new tools are used to design, develop, implement and monitor more complex services. While the new discipline of Service Science proposes a novel perspective on creating value in a continuous interaction between the service provider and service consumer [1], [2], more complex IT tools and methodologies are used to approach the process of developing new consumer-in-the-loop services. In this respect, service and service systems related literature reflects the fact that the service customer takes part in the process of service production and delivery [3], [4], [5], and this is the most important difference between manufacturing and service operations [6], [7], [8].

One of the first methodologies in service design was *service blueprinting*, a visual tool used in the Service Engineering process that was developed as a practical

T. Borangiu et al. (eds.), *Service Orientation in Holonic and Multi-Agent Manufacturing and Robotics*, Studies in Computational Intelligence 544,
DOI: 10.1007/978-3-319-04735-5_11, © Springer International Publishing Switzerland 2014

technique for service innovation and service improvement [9]. Service blueprinting is a customer-focused approach that helps visualize the working processes associated to service development, highlighting points of customer contact in service provision. Even though service blueprinting exploits the process nature of services [10], capturing service related activities and events occurring between service providers and service consumers, this service design methodology is not appropriate to dynamically express service customization in a process of value co-creation, being used mainly in the service design stage of the service lifecycle [11].

Also, recent references introduce the *Service Systems Engineering* as a methodology to approach service system design that addresses a service system from a lifecycle, cybernetic and customer perspective [12], [13], [14]. In this perspective, a service is seen as the outcome generated in a service system that has to fulfil customer expectations [15]. The Service System Engineering approach is intended to produce both good service outcome and a good process of service delivery to be well perceived by the customers [16].

The above mentioned perspective on service systems' engineering opens a new line of thought over service system design, pushing further a recent trend in service design and innovation that relates to improving operational capacity of companies for developing and managing new and complex services [17]. In this respect, *Business Process Modelling* techniques were proposed as a mean to foster understanding of the integration needs among different components and service system entities to define the activity, event, information and perception flows required for the governance, operations, administration, management and provisioning of new services [18], [8], [13]. Formalising process configurations of service delivery systems fostering collaboration between service actors (provider, customer, competition and compliance) aligned with organisational practice [19] lead to a new generation of IT-enabled service systems driven by customer requirements and collaboration between actors in order to co-create value through continuous interaction and service system reconfiguring [20].

Some current references present work in progress in this respect, but these attempts of modelling service systems from different perspectives are still in early stages of development. For example, in [21] different aspects of service modelling are presented, taking into consideration a component model, a resource model, and a process model. In [13] the Service Systems Engineering methodology is used to describe an intelligent transportation system and a dynamic Smart Grid service system. The modelling framework SEAM is investigated in [22] to be applied to the design and analysis of viability in service systems.

This work proposes a novel approach for the development of an activity-based model of a service system that extensively utilises methods and mechanisms for *iterative value co-creation*. It defines a continuous *interaction* between the service provider and the service customer, where value propositions are accepted (in terms of service activities and cost specifications) and it creates the Service Level Agreements (SLA) that determine service contracting, taxation and invoicing.

The paper is organised as follows. Section 2 describes the main components of the generic service system activity model (SSyst). Section 3 presents a detailed view on the Operation Management component of the SSyst activity model, while section 4 proposes the integration of service follow-up activities in the proposed model. Concluding remarks and further development directions are presented in section 5.

2   Development of a Generic SSyst Activity Model

This section considers the *activity model* of a generic Service System (SSyst), in an approach derived from the *service's lifecycle* [8]. It also indicates the interactions between the four stakeholders: Service provider (including his suppliers), Service customer, Competitors and Compliance bodies (government, legal national and EU service operating framework, authorities) [23]. The proposed SSyst *lifecycle model* includes four main stages (see also Fig. 1): (1) **Customer Order Management**; (2) **Service Management**; (3) **Service Operations Management**; and (4) **Service Taxation** and **Invoicing**.

These stages are mapped onto the four well-known *core service activities*: (a) Design and Development; (b) Delivery; (c) Operations Management; (d) Marketing. Fig.1 presents the three perspectives from which the SSyst activity model is further interpreted, namely: (a) Core service activities; (b) Stages of the service lifecycle; and (c) Components of the service system activity model.

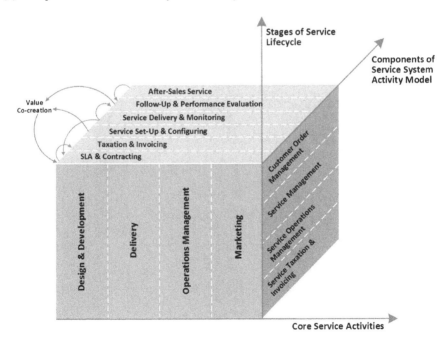

Fig. 1. The three perspectives for the Service System activity model interpretation

162   T. Borangiu et al.

Fig. 2 offers a global representation of the generic service system (SSyst) activity model developed in this framework and the interconnections of its components.

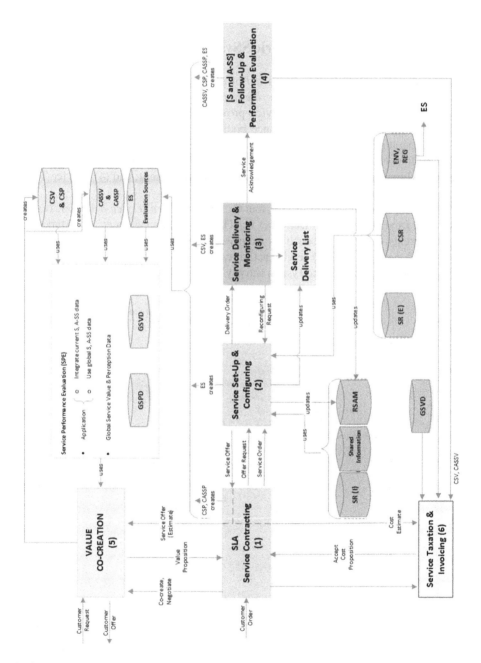

Fig. 2. Global view of the Service System activity model and component interconnections

As can be seen in Fig. 2, in the **Customer Order Management** (COM) stage a Service Level Agreement (SLA) is reached through close interactions between the service provider and the service customer. To reach a SLA, the requested service is set up and configured using:

- Planning, scheduling and allocation tasks;
- Service performance evaluation (SPE) data - relative to available, previously created value and perception data;
- Customer Relationship Management (CRM) strategy and Supply Chain Management (SCM) information;
- Comparative evaluation of similar services offered by the competitors;
- Compliance constraints imposed by national and EU legal and taxation specifications.

The *Service Set-Up and Configuring* component, which is called by the SLA component of the **Customer Order Management** (COM), is part of the **Service Management** (SM) stage.

Reaching the SLA is realized through an iterative and interactive process involving the two main stakeholders, service provider and service customer, the following two types of business items being generated:

- the *registered service contract*, generated for a pure service or an after (product) sales contract related to that product, and
- the *invoice*, usually accompanied by a *taxation form*.

The iterative SLA process generates value which is co-created by the service customer – provider interaction, with the primary goal of enhancing the service. The *Service Performance Evaluation* (SPE) component is extensively used during this process.

A more detailed representation of the iterative value co-creation process (VCo-C) shows in Fig. 3:

- the customer actions,
- the provider's back office activities - these activities are carried out mainly by the Information System customized for the organization's Service System activity model, and using its databases, and
- the customer – provider interactions.

The iterative SLA process features customer – provider interactions from a dual perspective: (a) Service specifications (options) agreement, and (b) Cost negotiation, as part of the Service Taxation and Invoicing (STI) component.

The two activity components (SLA and STI) realize the customer – provider interaction in a dual negotiation activity (service specification / cost) controlled in parallel by two strategies:

- Analysis of customer needs (service specification strategy update), and
- Cost strategy update for agreed service specifications.

These strategies are repeatedly run in the provider's back-office, where the two activity components (SLA and STI) receive the most recent updates from the interacting components: SLA receives service cost updates from the STI whereas STI receives service specification updates from the SLA.

During this iterative, combined negotiation process, the initial service specification confirmations and cost estimate (produced in the back-office by the ***Service Configuring and Set-Up*** component) are eventually updated, considering the global perception data (in the GSPD database) and global value data (in the GSVD database) for the requested service.

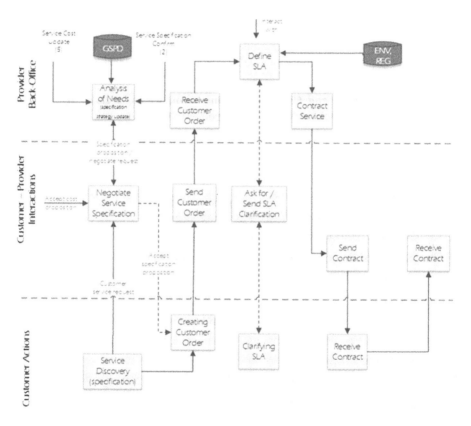

Fig. 3. Interactive value co-creation, SLA and service contracting

These data were created during the history of delivering the requested service, weighted by the current market and compliance (legal, financial) context.

So, the value proposition, as output of the value co-creation process, is derived from the current service specification proposition AND the cost proposition. When both of these two propositions are accepted by the customer AND the provider, the service order (SO) sent by the client to the provider will trigger the definition of the SLA, subject to:

- an interactive clarification process involving the customer and the provider, and
- a consulting process about environmental (context) and compliance (legal, financial constraints) conditions (using the ENV, REG databases).

The SLA ends by registering the Service Contract.

From the point of view of the core service activities list (represented in Fig. 1), the **Customer Order Management** (COM) stage is associated with the *Service Design and Development* phase (and hence with the SLA and *Service Set-Up and Configuring* components in the SSyst activity model). Once the SLA is reached, the *Service Delivery & Monitoring* component of SSyst activity model is launched during the **Service Operations Management** (SOM) lifecycle stage and it involves two core service activities, namely *Service Delivery* and *Service Operations Monitoring*. The service follow-up and performance evaluation activities fall into the **Service Operations Management** (SOM) core activity set.

The *Service Performance Evaluation* (SPE) activity component selects and uses consolidated data about the particular requested service, i.e.

- value-type data (e.g. quality, cost, timeliness, a.o.) and
- perception-type data   (e.g. degree of satisfaction, market share, innovation, a.o.).

This consolidated data is obtained by integrating the value and perception data and information measured and calculated at current service deliveries and post service delivery over the whole history of that particular service offered and repeatedly delivered (and enhanced) by the provider.

Service Marketing *uses specifications* created within the *Service Set-Up and Configuring* component, value and perception data collected during *Service Delivery & Monitoring* and *Service Follow-up* components and processed by the *Service Performance Evaluation* (SPE) component. It also exploits events and strategy used in the service customer – service provider interaction process of reaching the SLA.

3   Service Operations Management and the SSyst Activity Model

This section describes the activities integrated in the SSyst activity model that counts for the Service Operations Management stage associated to the SSyst lifecycle model. The **Service Operations Management** stage of a service's lifecycle involves two core service activities [24]: *Service Delivery* and *Service Operations Monitoring*, which are represented in Fig. 4.

In addition, while monitoring service operations a number of *value measures* are performed (timeliness, quality, cost - reflecting the service value, SV), which are stored in the Current Service Value database (CSV) for the service being currently delivered [25]:

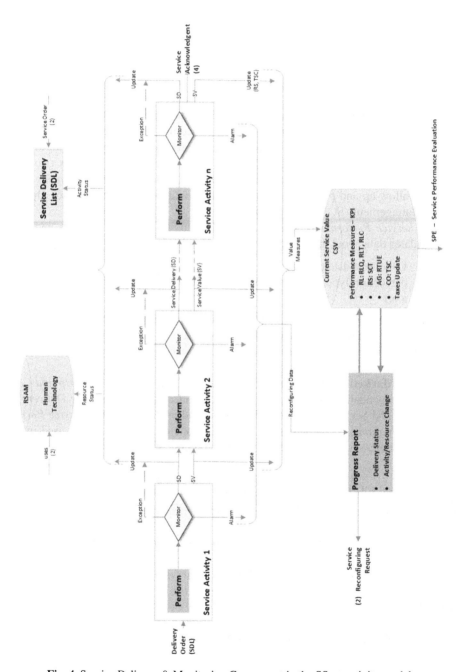

Fig. 4. Service Delivery & Monitoring Component in the SSyst activity model

- *Reliability* (RL), the value of the Service System to answer right the customer needs:

    o   RLQ: *quality* of the service - appreciation collected from the customers at several delivery stages (including the final one) of a composite service about   the standards at which the service was delivered;

    o   RLT: *timeliness* of the performed service (reflects all delivery stages of a composite service);

    o   RLC: *consistency* of the service (reflects the way in which the delivered service answers the client's requirements);

- *Responsiveness* (RS), the speed at which the client's requests were resolved by the Service System:

    o   SCT: the measured *service cycle time* (per activity, and per total composite service);

- *Agility* (AG), the reaction speed of the Service System *to respond to market changes* to maintain / gain competitive advantage - statistical data introduced by the provider's staff combined with data collected from customers, and with statics data from the service market (about competitors) and the delivery environment (new parameters, rules, directives, laws and application instructions)

    o   RTUE: *reaction time to unplanned events*, this data is partially collected during service delivery;

- *Cost* (CO), the cost reported by the provider's departments while operating the Service System to effectively resolve customer requirements:

    o   TSC: partial service costs measured during delivery stages and finally aggregated to evaluate *total service costs*.

In parallel with service delivery, the financial data about *taxes update* is also collected, for post-service delivery analysis. The Current Service Value (CSV) data thus reflects the value of the service at different delivery stages; this data is gathered, primarily processed and used, after completion of the service delivery, for *Service Performance Evaluation* during the *Service Follow-up* activity of the **Service Operation Management** post-service stage.

The management of the delivery and monitoring operations represented in Fig. 4 is designed for the most general case of composite services, which feature a number of $n \geq 1$, $n \in N$ activities. These activities have been planned and scheduled in an optimal manner relative to the customer's requirements in the *Service Configuring and Set-up* (SCSU) module of the **Service Management** stage.

During each Activity i, $1 \leq i \leq n$, a task (or job) is performed and monitored. At the end of each task two types of data are collected (directly measured or introduced manually by the operator of that current activity):

1. Data about *service delivery* (SD):

    - *Resource status*: status of the resources used, degree of occupancy of the human resource, technological resource and service capacity. This data updates the status of the technology and HR in the global database of the service provider: *Resource Service Access Management* (RSAM);
    - *Service delivery status*: current status of delivery for the composite service, after completion of the current activity i, $1 \le i \le n$.

2. Data about *service value* (SV):

    - Value measures about the *quality* and *cost* of the currently performed activity i, $1 \le i \le n$.

The monitoring activity during service delivery generates two other types of information in case of abnormal events occurring while performing the current activity:

- *Exceptions* with respect to normal service delivery while performing the current activity i (delayed completion of activity i, blocked or unavailable resources for next activities $i+1$, $1 \le i \le n-1$);
- *Alarms* raised upon an interruption or failure in performing the current activity i; these alarms create a progress report which signals the necessity of new resource allocation or activity redefinition to the *Service Configuring and Set-up* (SCSU) module, by generating a *Service Reconfiguring Request* (SRR).

Once the last activity n of the composite service terminates, a *service acknowledge* message (*service_ack*) is issued for the *Service Follow-up* activity component of the **Service Operations Management** stage, signalling the completion of the composite service delivery. This message will trigger the *Service Performance Evaluation* (SPE), as post-service delivery computational activity.

Fig. 5 describes the organization proposed for the *Service Follow-up* activity component with *Service Performance Evaluation*. Two types of services were considered, with delivery just acknowledged by *service_ack*:

1. Pure service (S), no product (good) is sold: the completion of the service delivery is signalled by *service_ack* which, based on the service value data already measured during the Service Delivery and Monitoring activities and stored in the CSV database, triggers the calculation of:

    - *Value KPIs* for the (pure) service describing the *Asset Management* (AM), i.e. the provider's effectiveness in managing fixed and working capital assets to resolve clients' requirements:

        o RSWC: *return of service working capital*, for the recently delivered service;
        o RSA: *return on assets* for the same service;
        o SCCCT: service *cash-to-cash cycle time*.

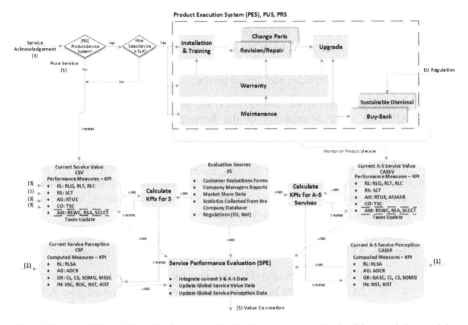

Fig. 5. Service Follow-Up & Performance Evaluation component in the SSyst activity model

- *Perception KPIs* for the (pure) just delivered service - these KPIs are added to the service perception data retrieved during the Service Contracting activity - describing:

  ➢ The provider's *Growth* (GR) after the recently delivered service, i.e. the company's ability to grow along the time and generate a net income on a consistent and sustainable basis:

  - o CL: *customer's loyalty*, estimated in terms of client identification in the provider's customer database;
  - o CS: *customer's satisfaction*, retrieved from the evaluation sheets the clients fill in after service delivery;
  - o SOMG: *service operating margin growth*, estimated post service delivery by the provider's staff.

  ➢ The provider's *Innovation effort* (IN) by considering the way in which the currently delivered service has contributed to the company's capacity to innovate and create new and knowledge-intensive services:

  - o IISC: number of incremental innovations (improvements) identifiable for the just delivered service;
  - o RISC: number of radical innovations registered for the just delivered service;
  - o NST: number of new services created and sold relative to the recently delivered composite service;
  - o KIS: number of Knowledge Intensive (KI) - services created and integrated in the recently sold and delivered composite service.

2. <u>After-sale service</u> (A-SS) or *Product-Extension Service* (PES) enhancing the utility that the ownership of the product delivers to the customer (e.g. repair, maintain and upgrade, take-back etc.) [26] - see top right part of the representation in Fig. 5. After the product is sold to the client, the A-SS is monitored during the whole product lifecycle, and, based on the:

    ➢ *Service value data*: RL (RLQ, RLT and RLC), RS (SCT) AG (RTUE and AIUASR - the adaptability of the Service System to the increase of unplanned A-SS requests), and CO (TSC), and
    ➢ *Perception data*: RL (RLSLA),
    which are already collected during the Service Contracting activity, the calculation of specific KPIs for A-SS is performed, and the following data bases are updated (see Fig. 5): the Current A-SS Value database (CASSV) respectively the Current A-SS Perception database (CASSP):

    • *Value KPIs* for the A-SS describing, in addition to the *Asset Management* indicators AM (RSWC, RSA, and SCCCT), the *Growth* indicators:

        ○ GASC: *growth of A-SS contracts*, by adding the last one contracted;
        ○ SOMG: *service operating margin growth*, estimated post A-SS delivery by the provider's staff

    • *Perception KPIs* for the A-SS describing, in addition to the service perception data RL (RLSLA) and AG (ADCR) retrieved during the Service Contracting activity, the Growth indicators GR (CL, CS) and the *Innovation effort* indicators IN (NST and   KIST).

The value and perception data measured and collected during service delivery and the KPIs for current service value and perception calculated during the Service Follow-up activities are stored in the CSV and CSP databases for the current (pure) service, respectively in the CASSV and CASSP databases for the current after (product) sales service.

These data are then integrated with the historical data kept for that types of services in the provider's Global Service Value Database (GSVD) and Global Service Perception Database (GSPD), during the *Service Performance Evaluation* (SPE) activity; these two global databases are then checked against the customer's requirements in case of a new request of a service of that type during the process of value co-creation.

The Service Performance Evaluation (SPE) component is used in order to:

    • analyse the company's (service provider) financial results in terms of: costs, revenues, operating profit, Return On Assets (ROA) and cash flow;
    • analyse the company's competitive performance in terms of: market share, customer satisfaction, customer loyalty and ranking among competitors (results);
    • analyse the company's innovativeness;
    • to support the value co-creation process.

4   Extending Products with Services in Product-Service Systems (PSS)

As discussed in section 3, the ***Service Follow-up*** activity is extended to services to be performed after a product is sold to the customer. Such services, i.e. the After-Sales Service (A-SS) or Product-Extension service (PES) represent a category of Product Service System (PSS) and are characterized by the customer ownership of the physical good [26], [27].

The concept of "*Product-Service System*" (PSS) was first defined in [28] in order to identify a "marketable set of products and services capable of jointly fulfilling a user's needs". A PSS uses a physical product as vehicle for delivering generic or specific services related to that product. The transition from pure-product to pure-service providers is a continuum and manufacturing firms move along this axis as they incorporate more product-related services (see Fig. 6).

Three categories of PSSs where considered for the Service System (SSyst) activity model, for further management analysis and design, according to who owns the PSS and who uses it:

- *Product-Extension services* (PES), also called *After-Sales services*: this category is characterized by the customer ownership of the physical good [25];
- *Product-Utility services* (PUS): this category refers to two main areas of services which are connected with rentals and leasing;
- *Product-Result services* (PRS): this category is related to situations where a provider supplies a complete solution to an on-going need for a customer.

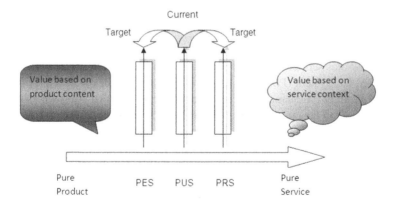

Fig. 6. The continuum-type transition from pure-product to pure-service business

Table 1 below proposes KPIs for performance evaluation (value and perception) of After-Sale services; some of the service perception indicators are used to analyse the firm's innovativeness and capacity to create new and knowledge-intensive services.

Table 1. KPIs proposed for A-SS value and perception evaluation

A-S Service Performance Evaluation	Performance category			KPIs
	Type	Description	Estim.	
Service Value (SV)	Reliability (RL)	Performance of the SSyst. to answer right to customer needs	meas.	RLQ: quality of service
			meas.	RLT: service timeliness
			meas.	RLC: provide the right answers to client enquiries (consistency)
	Responsiveness (RS)	Speed at which client enquiries are resolved by the SSyst.	meas.	SCT: service cycle time
	Agility (AG)	SSyst. agility to respond to market changes to gain/maintain competitive advantage	meas.	RTUE: reaction time to unplanned events
			meas.	AIUASR: adaptability to the increase of unplanned A-SS requests
	Cost (CO)	Cost reported by the company to operate the SSyst. to resolve customer inquiries	meas.	TSC: total service costs
	Asset Management (AM)	Company's effectiveness in managing fixed and working capital assets to resolve client inquiries	comp.	RSWC: return on service working capital
			comp.	RSA: return on assets
			comp.	SCCCT: service cash-to cash cycle time
Service Perception (SP)	(RL)		meas.	RLSA: generate the right SLA-contractual agreement in place
	(AG)		meas.	ADCR: adaptability to customized requests
	Growth (GR)	Company's ability to grow along the time & generate a net income on a consistent and sustainable basis	comp.	CL: customer loyalty
			comp.	CS: customer satisfaction
			meas.	GASC: growth of A-SS contracts
			comp.	SOMG: service operating margin growth
			comp.	MSSC: market share per service category
	Innovation (IN)	Company's capacity to innovate and create new and KI services	comp.	NST: no. of new services created and sold from total services
			comp.	KIST: no. of KIS from total A-S services

An After-Sales service (A-SS) or Product-Extension service (PES) is a category of Product Service System (PSS) characterized by the customer ownership of the physical good. Product-Extension services enhance the utility that the ownership of the product delivers to the customer (e.g. repair, maintain and upgrade, take-back etc.).

A Product-Utility service (PUS) is a category of Product Service System (PSS). This category refers to two main areas of services which are connected with *rentals and leasing*. The provider is still the owner of the product but the customer uses directly the product and the related service (e.g. car leasing, car rental, property sharing, etc.), see Fig. 7.

As compared to the activities associated with the Product-Extension services (or A-SS), the activities related to PUS impose a minimum of compulsory insurance that can be optionally extended to cover all situations possibly occurring during the product, monitor in more details whether the product is properly used, and provide complete technical assistance to the customer over the entire product's life cycle.

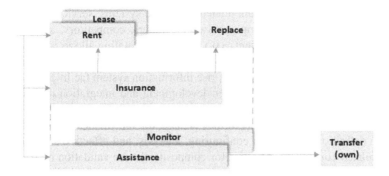

Fig. 7. PUS activities proposed for the SSyst activity model

A Product-Result service (PRS) is a category of Product Service System (PSS) related to situations where a provider supplies a complete solution to an on-going need for a customer. The client does not own and use the product, but uses instead only the product's functionality and the results created (such as energy service contracting, voicemail, etc.), see Fig. 8.

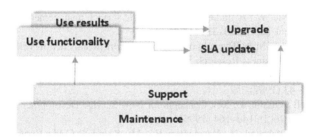

Fig. 8. PRS activities proposed for the SSyst activity model

A permanent activity, diversified by the customer's perception on the product's functionality, degree of satisfaction and fidelity is the upgrade of the technical conditions and performances in which the product's functionality can be used and the Service Level Agreement update (upon the provider's proposal, agreed with the customer). The same KPIs for the A-SS and PUS evaluation can be used for the PRS.

A key issue is to monitor and control all the processes and activities which are carried out to provide a product-service business: service measures need to be implemented and applied consistently by all the parties involved in the service network in order to enhance its overall effectiveness.

5   Conclusions

The activity-based model of the service system (SSyst) proposed in this paper is organized in a standard representation oriented towards the business process management point of view. The proposed model highlights roles and interactions between the four categories of stakeholders related to service systems: service providers (with suppliers and service outsourcing), service customers, competitors, and compliance bodies (law, financial, environment, a.o.). This representation allows a rapid customization of the SSyst model for a range of services, leading to easy configuration, simulation and validation through the enterprise information system facilities.

This approach is suitable for further development and integration in the organization's information system. Based on the analogy with BPM, a business-oriented standardized representation has been further created for the SSyst activities' model, to allow the definition, set up and configuring, analysis, and effective simulation of its inter correlated component activities for composite service validation by means on an information system, and the proposal of innovation patterns and solutions.

References

1. Vargo, S.L., Maglio, P.P., Melissa Archpru Akaka, M.A.: On value and value co-creation: A service systems and service logic perspective. European Management Journal 26, 145–152 (2008)
2. Spohrer, J.C., Demirkan, H., Krishna, V.: Service and Science. Service Science: Research and Innovations in the Service Economy. Springer (2011)
3. Kellogg, D.L., Nie, W.: A framework for strategic service management. Journal of Operations Management 13(4), 323–337 (1995)
4. Mohr, L.A., Bitner, M.J.: The role of employee effort in satisfaction with service transactions. Journal of Business Research 32(3), 239–252 (1995)
5. Shostack, G.L.: Service Positioning through Structural Change. Journal of Marketing 51(1), 34–43 (1987)
6. Edvardsson, B., Olsson, J.: Key Concepts for New Service Development. The Service Industries Journal 16(2), 140–164 (1996)
7. Nijssen, E.J., Hillebrand, B., Vermeulen, P.A.M., Kemp, R.G.M.: Exploring product and service innovation similarities and differences. International Journal of Research in Marketing 23(3), 241–251 (2006)
8. Fitzsimmons, J.A., Fitzsimmons, M.J.: Service Management: Operations, Strategy, and Information Technology, 7th edn. Irwin Professional Pub. (2010)
9. Bitner, M.J., Ostrom, A.L., Morgan, F.N.: Service blueprinting: a practical technique for service innovation. California Management Review 50(3), 66–94 (2008)
10. Gronroos, C.: Service Marketing and Management: A Customer Relationship Management Approach. John Wiley & Sons, Ltd., Chichester (2000)
11. Seyring, M., Dornberger, U., Suvelza, A., Byrnes, T.: Service blueprinting handbook. International SEPT Program. University of Leipzig (2009)
12. Tien, J.M., Berg, D.: A case for service systems engineering. Journal of Systems Science and Systems Engineering 12(1), 13–38 (2003)
13. Lopes, A.J., Pineda, R.: Service Systems Engineering Applications. Procedia Computer Science 16, 678–687 (2013)
14. Pineda, R., Lopes, A., Tseng, B., Salcedo, O.H.: Service Systems Engineering: Emerging Skills and Tools. Procedia Computer Science 8, 420–427 (2012)
15. Slack, N., Chambers, S., Johnston, R.: Operations management. Hall Financial Times, 6th edn. Prentice Hall (2010)

16. Dabholkar, P.A., Overby, J.W.: Linking process and outcome to service quality and customer satisfaction evaluations: An investigation of real estate agent service. International Journal of Service Industry Management 16(1), 10–27 (2005)
17. Hammer, M.: The Superefficient Company. Harvard Business Review, 81–91 (2001)
18. Muehlen, M.Z., Danny, T., Ho, D.T.: Service Process Innovation: A Case Study of BPMN in Practice. In: Sprague Jr., R. (ed.) Proceedings of the 41st Hawaii International Conference on System Sciences, Waikoloa, HI, January 7-10 (2008)
19. Ponsignon, F., Smart, A., Maull, R.: Service delivery systems: a business process perspective. POMS College of Service Operations 2007 Meeting. London Business School, London (2007)
20. Peng, Y.: Modelling and Designing IT-enabled Service Systems Driven by Requirements and Collaboration. PhD Thesis, L'institut national des sciences appliquées de Lyon (2012)
21. Böttcher, M., Fähnrich, K.-P.: Service Systems Modeling: Concepts, Formalized Meta-Model and Technical Concretion. The Science of Service Systems, Service Science: Research and Innovations in the Service Economy, 131–149 (2011)
22. Golnam, A., Regev, G., Wegmann, A.: On Viable Service Systems: Developing a Modeling Framework for Analysis of Viability in Service Systems. In: Snene, M., Ralyté, J., Morin, J.-H. (eds.) IESS 2011. LNBIP, vol. 82, pp. 30–41. Springer, Heidelberg (2011)
23. Baines, T.S., Lightfoot, H.W., Benedettini, O., Kay, J.M.: The servitization of manufacturing: A review of literature and reflection on future challenges. Journal of Manufacturing Technology Management 20(5), 547–567 (2009)
24. Aurich, J.C., Fuchs, C., Wagenknecht, C.: Life cycle oriented design of technical Product-Service Systems. Journal of Cleaner Production 14(17), 1480–1494 (2006)
25. Legnani, E., Cavalieri, S., Gaiardelli, P.: Modelling and Measuring After-Sales Service Delivery Processes. In: Borangiu, T., Thomas, A., Trentesaux, D. (eds.) Service Orientation in Holonic and Multi agent, SCI, vol. 472, pp. 71–84. Springer, Heidelberg (2013)
26. Legnani, E., Cavalieri, S., Ierace, S.: A framework for the configuration of after-sales service processes. Production Planning & Control: The Management of Operations, Special Issue: Manufacturing and Service Operations Networks 20(2), 113–124 (2009)
27. Gaiardelli, P., Saccani, N., Songini, L.: Performance measurement systems in after-sales service: an integrated framework. International Journal of Business Performance Management 9(2), 145–171 (2007)
28. Goedkoop, M.J., Halen, C.V., Te Riele, H., Rommens, P.: Product Service Systems: Ecological and Economic Basics. The Hague, Netherlands (1999)

Product Specification for Flexible Workflow Orchestrations in Service Oriented Holonic Manufacturing Systems

Francisco Gamboa Quintanilla, Olivier Cardin, and Pierre Castagna

LUNAM Université, IUT de Nantes – Université de Nantes, IRCCyN UMR CNRS 6597
(Institut de Recherche en Communications et Cybernétique de Nantes),
2 avenue du Prof. Jean Rouxel – 44475 Carquefou
{francisco.gamboa,olivier.cardin,
pierre.castagna}@univ-nantes.fr

Abstract. Holonic Manufacturing Systems are a response solution for the emergent need of flexible, reactive and productive manufacturing systems. This paper relies on PROSA, a classical holonic reference architecture, which makes use of a product specification, a process specification and a means to determine a resource's production abilities and capacity, but does not define a specific method for representing such. This paper proposes an approach to define a product's process specification model that integrates the principles and advantages of Service-oriented Architectures, Petri-Nets and Product Families. Then, a re-definition of the basic holons is given to have a glimpse on a possible exploitation of this new approach, together with a short-term forecasting strategy, for the flexible orchestration of workflows. Finally, it is shown how the proposed product's process specification model enhances the HMS's flexibility, reactivity and productivity giving rise to a Service-oriented Holonic Manufacturing System.

Keywords: Holonic Manufacturing, PROSA, Petri-Nets, Product specification, Manufacturing-Services, Process families, Workflow exploration.

1 Introduction

For the last few decades it has been seen an evolution in the goods market trend which manifests with an increasing demand of customized products. This evolution has been boosted by the rise of the e-commerce market which makes available customization platforms to customers via internet. Companies, in their search to compete in the marketplace, have been looking for ways to expand their production lines and differentiate their offer with the belief of improving their sales [1]. However, as [2] pointed out, as variety increases the law of diminishing returns does not keep pace. Thus, the problem of Customization i.e. reaching Production Efficiency (PE), relies on process design, whose main concern is manufacturability and cost. For this reason, to attain PE and respond to product variety, the next generation

T. Borangiu et al. (eds.), *Service Orientation in Holonic and Multi-Agent Manufacturing and Robotics*, Studies in Computational Intelligence 544,
DOI: 10.1007/978-3-319-04735-5_12, © Springer International Publishing Switzerland 2014

Manufacturing Execution Systems (MES) should provide increased levels of flexibility, re-configurability and intelligence [3].

Holonic Manufacturing Systems (HMS) has been recognized as a paradigm providing to MES the above mentioned attributes by means of a decentralized architecture. Such attributes are obtained thanks to the identification and recognition of autonomous intelligent entities, each one attributed with subset of the various responsibilities in the system. These entities, called holons, are capable of cooperating with other entities and organize themselves for the achievement of a specific goal. PROSA [5], a reference holonic architecture, identifies and classifies three main holon roles, i.e. product, resource and order holons, each one in charge of managing a part of the production control system such as: product and process specifications, resources' capacity utilization and logistics, respectively.

PROSA, in its definition, structures the design principles of a HMS. Such definition recognizes the existence of a product specification, process specification and the capacity of determining a resource's production abilities and capacity. However it does not establish specific tools to implement these. The objective of this paper is to propose a modelling strategy for a product's process specification that welcomes product customization and enhances the HMS' flexible, reactive and productive potential to attain production efficiency. A second objective is to propose a methodology to determine the Resource Holons' (RHs) production capabilities that can interface with the proposed product's process specification.

The second section of this article gives a brief description of the type of system of application. The third section of this article deals with the specification of the product, leading to a model including product and process families' specification. This section ends up with the proposal of using Petri Nets in order to represent product recipes. Finally the fourth section is intended to show how these new concepts can be integrated into a HMS with the use of SOA's principles for flexible workflow orchestration.

2   Description of System of Application

Before going further in introducing the work presented in this paper, it is important to have a look on what kind of systems this work is addressing. This will introduce the reader into the context, in order to have a better understanding of the ideas and concepts discussed in this paper.

This work is mainly directed to companies needing to implement new production systems with enough flexibility to produce a great product variety that the new trend of product customization implies. Such need of flexibility comes from the idea that such flexibility will translate into a greater competitiveness in terms of product quality, speed of product delivery, greater product offer, and the ability to introduce faster new products into production and market. Hence, this work is intended for those companies looking to push to the limit the efficiency of their production systems while keeping a high degree of flexibility in opposition to companies seeking for high volume productions where the system's physical configuration inclines more towards continuous flow production lines which favour high production flows by scarifying flexibility.

Due to product customization and the great product variety that it engenders, there is a high uncertainty in product demands. For this reason this work is mainly directed to production systems implementing a push strategy or Make-to-Order (MTO) strategy where orders can arrive at any moment during production time, requesting estimations on delivery dates. These so called emergent orders impose a dynamic behaviour to the systems, changing its state with each new arrival. Such dynamism makes the implementation of traditional scheduling systems not a viable solution as it makes its calculations based on a static state therefore having to recalculate with each new arrival. The degree of the MTO strategy can be either an Assemble-to-Order (ATO) or a Build-to-Order (BTO) strategy where product parts are already available or they can be ordered as production orders arrive. Orders can come in small batches or individual products as it is the customer who submits them.

The intended system of application owns a physical topology resembling that of a Flexible Job-Shop (FJS). This is natural, as a job-shop is typically the initial production floor configuration for manufacturers willing to offer a variety of products thus; needing of flexibility. The constraints remain the same: all jobs are formed of a certain number of manufacturing operations that can be executed by one or more of the resources available in the production floor.

The production floor is composed of three major components: a set of workstations, a transportation system and a set of work-in-process (WIP) products. It is a multi-station manufacturing system where there exists more than one workstation or machine capable of doing one same operation. Such workstations can count with stocks of materials or sub-products that will be needed to provide a certain manufacturing transformation to products in order to allow the implementation of an ATO and/or BTO strategy.

The transportation system gives physical interconnectivity between the different workstation. Due to the FJS characteristic of having more than one production sequence, the transport system is considered to be a multi-routing system where products can follow jumbled routings among the different workstations. Such routing system might not be designed to have full reachability, thus it is considered the possibility of non-reachable physical states.

Due to the great number of product variants that can exists in the work-in-process, products are considered to possess an appropriate identification for its proper treatment and in order to keep track of its production evolution. A product in the WIP can use of auto-identification technologies as proposed in [14], in order to communicate its identification to the system so that its environment can interact with it accordingly.

Taking into account system integration, the possibility of a system integrating all types of different technologies will be considered. For instance, a work station could well be an automated machine, an automated work cell or an operator in a manual or semi-automated workstation. Although re-configurability is out of the scope of this paper, this aspect will later be demonstrated for adding flexibility in the system's reconfiguration process.

Here are some characteristics or assumptions on the manufacturing process:

- Operations are non-preemptive. Once a manufacturing operation has started it cannot be interrupted unless the product in question is going to be fully discarded.
- There is no parallelism in the execution of manufacturing operations for one product. Parallelism is only present in an indirect way with the simultaneous production of composing sub-products.

Productivity in a Job-Shop production system is strongly linked to its physical layout. One of the main challenges is to design a layout that minimizes material handling costs, process inventories, idle times and that attaints full reachability. The proposed product's process specification, presented in this paper, intends to exploit to a maximum the intrinsic flexibility of the manufacturing process itself according to its precedence rule with no regard on the physical resources. The strategies on how this flexibility will be exploited for the formation of workflows in terms of sequence and providers is out of the scope of this paper, however section 4 will give a slight insight on a possible solution.

3   Product Specification

3.1   Product Families

In their attempt to achieve mass customization, companies face the problem of an increased internal complexity due to a gain in product variety which raises production cost [2]. In order to solve this complexity problem and achieve economy of scale while satisfying Customer Needs (CN), companies have been adopting the development of product families, which seems to be a well-recognized solution to keep competiveness in the marketplace [6].

The principle of product families' development is based on the exploitation of the inherent commonality between different product variants. A product family refers to a set of individual products that share a set of common structural characteristics and yet are differentiated one from another by certain specific features [5]. Such commonality among product structures inside a product family inherently enables commonality as well in the corresponding production process [6]. This gives origin to process families, which in turn takes advantage on the existing commonality in operations and sequences among the different family members.

A process family is therefore a collection of manufacturing tasks that respond to the realization of the corresponding structural modules within a modular product architecture. Process families, in the same way as product families, carry the attributes of commonality, modularity, reutilization and scalability [1] [5] [7]. This process specification is actually the production recipe of a product family member and thanks to its modular nature, it can be reconfigured into different sequences/workflows, also called process orchestrations.

All in all, a product data model, in a customization production system implementing product families, comprises a product family specification (physical domain) and a process family specification (manufacturing domain).

3.2   Product Model

In the context of product customization and product families design, there are two main challenges in the organization of a product's production data model. First, instead of being a collection of individual product variants, the model should explain the relationships between these variants. Second, an individual product variant should be defined out of the selection of the parameters related to the product family. Such parameters are the result of a customer specification process, i.e. product customization. Thus, a specific description of a product's variant production process is a function of both parameters specification and a process family description. For such, the following Product Manufacturing model, Fig.1, is proposed for representing the process family of a specific product family.

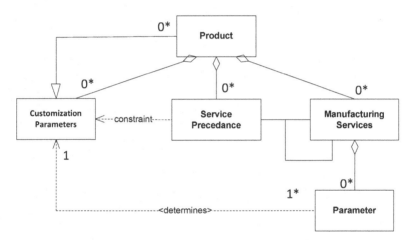

Fig. 1. UML Product Manufacturing-Model

This model is based on the product model described in ISA SP-95 standard, which contains all the necessary information for the manufacture of a product. To adapt this model to product families and product customization, the information is clustered in three main classes:

• *Customization Parameters:* Collection of variables to be defined by the customization process for a given product family. They possess an identifier and a range of allowed values corresponding to a structural module. The level of customization for a given product family is reflected by the cardinality of this class: more customizable parameters exist, more product variants a family represents. The instantiation of all customizable parameters of a product family results in a list of defined parameters for a product variant. Examples of these parameters are: laptop colour, hard disk capacity, type of screen, type of Wi-Fi antenna, type of keyboard, optional Bluetooth, etc. These will then be mapped into process parameters which are explained later with the Manufacturing-Service class.

- *Manufacturing-Service:* Represents a manufacturing-task or group of tasks forming part of the products production process; it results from the mapping of a given structural module from the physical domain into the process domain. It is a manufacturing process module describing manufacturing transformation ability with no regard of the methods and technology for its implementation. This class also contains the class *Parameters* which correspond to certain variables needed to be determined for the correct execution of the manufacturing task. These service parameters are determined according to the choices made for the customization parameters in order to reproduce such specifications.
- *Manufacturing-Service Precedence / Production Conditions:* Information explaining the relation and interdependencies between the different Manufacturing Services forming the given process family. It represents the precedence rules between services for the orchestration of production workflows. Its cardinality can be zero, considering the possibility of the existence of a non-decoupled production process characterized by a single manufacturing service (no need of precedence).

It is therefore the instantiation of these three elements that completely determine the information required for the realization of a product variant. Such specification is independent of the physical platform as manufacturing services are mere operation descriptions with no consideration of the resources or methods implementing them. This quality makes the product manufacturing-specification compatible with all types of resource models as long as they can provide the required services.

Product differentiation is then achieved by both; parameter specification and configuration of the different manufacturing service modules through the addition, subtraction and/or substitution of these. In this manner, the model explains the process family description through services and their interdependencies, customer specification through customization parameters specification, and the bill of manufacturing services for configurable product families as some manufacturing services can be held out for some family members.

3.3   Manufacturing-Services

As mentioned before, manufacturing services are the result of the direct mapping of structural features of a product family into the process domain. Such services, as stated in [9] for Service-oriented Architectures, represent a single operation or a series of operations of more or less intangible nature, that normally take place in the interactions between a customer and a provider, given as a solution for a customer problem. Services can then be standardized and a bank of these reusable services (operations) can be created for further reutilization in case of existing commonality with other product families.

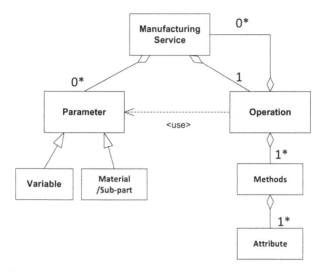

Fig. 2. UML Manufacturing-Service Model; (provider's perspective)

As the Manufacturing-Service model shows, Fig. 2, a **manufacturing service** is composed of two elements:

- *Operation*: Represent the activities related to the service. From the consumer perspective, these are descriptions of the transformations made on the product, more generally, a service identifier. From the provider's perspective, this is the function with the algorithms executing the service, as shown in Fig 2. Contrary to service description, service algorithms are proprietary to the provider and therefore dependent on the resource's technology. The service has a list of attributes on which its performance can be evaluated. Such attributes values are different for each of the methods that implement the service as it is seen in the aggregation relation with the Method class. For instance, an evaluation attribute could be the operation energy consumption. A machine using newer technology could consume less energy than other machines when providing the same manufacturing service.
- *Parameter*: They can be of two forms: variables and materials. In the first case, they are variables with a range of allowed values corresponding to a design parameter from the physical domain (e.g., element X positioned by coordinates). In the latter case, they indicate the category of the component to be added to the main product by the operation. The selection of the material or sub product is done inside a range of component variants belonging to the same category (e.g., *Category*: hard disk, *Range*: {200 GB, 300 GB, 400 GB, 1Tb}). It is determined after the customization process and used to fully determine the operation description.

It is the decoupling of parameters from the operation that allows bringing product customization to the process domain. Therefore, as product customization is based on the reutilization of structural features, the manufacturing services used to produce such features can be adapted to the different product variants and/or families through service parameterization. The group of services that can be derived from this model can be seen as a family sharing a same operation description but differentiated by the values in their parameterization.

3.4   Sequence Modelling through Petri-Nets

As mentioned in the Product Manufacturing-model, a product specification should express all the information needed for the manufacture of a specific product: parameters, manufacturing services and service precedence conditions. Its modelling tool should be capable of expressing all the possible service choreographies (workflows) that can produce the specified product. In addition, it must also facilitate the online edition, be easy to understand and program and should have low memory footprint for potential embedded applications.

Traditionally in process design, static models are implemented by specifying only one single predefined production plan. Mendes et al. in [13] enrich the process model by considering the existence of different alternative services for a given production state. They used the Petri-Nets formalism as modelling tool with the objective of involving decision motors in the system. The approach proposed in this article has the intention of going farther in enriching the process model by increasing the decision area by questioning the order of execution of tasks. Petri Nets, a well-known modelling formalism in the academic and industrial domain, turns out to be a very good candidate for this purpose. This is mainly due to its characteristic ability to capture the synchronous and asynchronous aspects between the manufacturing services involved in the production process. Thanks to this and to their evolution mechanisms, Petri-Nets have the capacity of representing a great number of sequence combinations with a single net. This is of great importance as the main goal is to design a product manufacturing information model that allows the exploitation of the HMS's inherent flexibility, which in turn will project through the exploration of all the possible alternative production workflows that produces a specific product.

The proposed product production model is represented using a Petri-Net extension which includes the inhibitor arcs and the test arcs. For more information on the Petri-Net formalism, refer to [12]. In this model, manufacturing services are represented by the net's transitions. The different production states are indicated by the net markings. These markings, implicitly determine the manufacturing services that have been executed at a given point. The service interdependencies/precedence rules are inherently defined by the collection of arcs relating places and transitions and the evolution rules of the Petri-Nets formalism. Test arcs together with the inclusion of permission places are used to indicate the selection of the optional modules that differentiate versions of the product i.e. product sub-families. A token is added to the permission places of those optional services that have been selected to be included in the process. Inhibitor arcs on the other hand are used to include more complex precedence conditions like those of mutual exclusion and negation which can be especially present in chemical processes. Finally, the set of parameters is added as attributes associated to the manufacturing services (transitions).

To better understand the Petri-Net approach, Figs. 3-5 show an illustrative example of a theoretical process family using Legos. It consists of a Lego platform representing the transporter of the product, and a set of blocks each one standing as an instance of a theoretical service type. Three types of Legos are used to represent such types and are differentiated according to their sizes as can be seen in Fig.3. Manufacturing service type 1 is represented by a 2x2 Lego block, a service type 2 by a 2x4 and finally a service type 3 by a 2x6 Lego block. Each service has a set of parameters for its instance from which a subset is not customizable and is defined by the process designers. The other subset of parameters is used to capture the customers' choices in the process domain; they include the scalable aspects of customization into the production process while test arcs and permissions places capture the configuration aspects of product customization. The idea behind this example is to illustrate the structural interdependencies in the Lego structure and to illustrate the process family precedence conditions and how this can be represented with a Petri-Net.

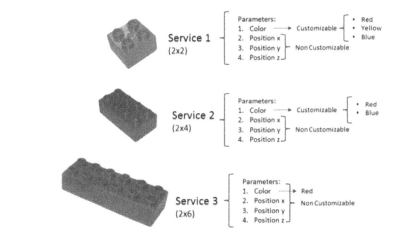

Fig. 3. Types of Services

Fig. 4. Product Base Configuration

In this example, the colour of the Lego block is the customizable parameter while its coordinates are set a priori during process design creating an instance of the service type. The base configuration of the product family is exposed in Fig.4. Such base sub-product can then be transformed into a member of one of the different product sub-families, as shown in Fig.5, belonging to the process family to be modelled. As can be seen, it can result in two different versions. Version 1 includes two service-type 1 instances which are differentiated by their coordinates from those instances in version 2. This shows a possibility to configure (available for the customer), which in this case would be of mutual exclusion. Therefore, this product family is customizable in colour; each of the blocks of the structure can take one of the colours available, as indicated in Fig.3, plus a certain customization level in the product structure defined by the optional modules issuing a version 1 or a version 2 product.

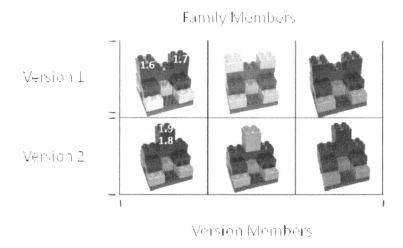

Fig. 5. Product Variants of a Family

As it is illustrated in Fig.4, the base configuration of the Lego product is realized by the application of 5 instances of service type 1 {1.1, 1.2, 1.3, 1.4, 1.5}, two instances of service type 2 {2.1, 2.2} and one instance of service type 3 {3}. Then parting from such base configuration, versions 1 and two can be realized by the application of two service type 1: {1.6, 1.7} and {1.8, 1.9}. Table 1 contains the bill of manufacturing services involved in the realization of the given product family and the precedence conditions for each one of them. Using the Lego structure to demonstrate the interdependencies between services, it can be seen that services 1.1 through 1.5 are independent of any other service - other than the existence of a platform. This same independence is expressed in Table 1 with a zero precedence condition. Service 2.1 and 2.2 are directly dependent on the previous application of services 1.1&1.2 and 1.3&1.4 respectively. However, in order to make a richer example, a more complex precedence condition is used as indicated in Table1. In this case the main interest is to avoid the situation of applying services 2.1 and 2.2 before service 1.5 has been applied. This can represent a scenario with a physical limitation like that of a robotic arm, in an assembly process, being unable to place block 1.5 once blocks 2.1 and 2.2 have been placed. Hence, the possible sequences (with respect to this three services) are {1.5→2.1→2.2}, {1.5→2.2→2.1}, {2.1→1.5→2.2} and {2.2→1.5→2.1}. It is important to note that complex conditions like this should be expressed in canonical form as a sum of minterms. Service 3, on the other hand, has a compound condition. It depends on the previous application of services 2.1 and 2.2, followed by the rest of the services in the list with its single precedence conditions.

Table 1. Example: Manufacturing Services Precedence Table

Manufacturing Service Type	Precedence Condition
Service 1.1	-
Service 1.2	-
Service 1.3	-
Service 1.4	-
Service 1.5	-
Service 2.1	$(1.1*1.2*\overline{2.2}) \mid (1.1*1.2*1.5*2.2)$
Service 2.2	$(1.3*1.4*\overline{2.1}) \mid (1.3*1.4*1.5*2.1)$
Service 3	$(2.1*2.2)$
Service 1.6	3
Service 1.7	3
Service 1.8	3
Service 1.9	1.8

Serving from the precedence table and a series of modelling rules, the following Petri-net structure can be derived (Fig.6). As mentioned before, the production state of the product in question is given by the net's marking which enables the triggering of certain transitions. Thanks to this, at a certain production state, the allowed manufacturing services to be next executed, that will respect the service precedence rules, can be known. This allows the exploration of the alternatives given by the asynchrony between certain services. The selection of production modules, resulting from the personalization process, is done by adding tokens to those services forming part of the different product versions. From the example, if version 1 is to be done, a token will be added to the permission place linked to both Services 1.6 and 1.7 by the test arcs which will avoid the rehabilitation of those transitions.

In short, a single Petri-Net can generate a state-automaton representing the arborescence of all possible production workflows while consuming a small amount of memory and a more straight forward programming and design.

4   Integration into SoHMS

The proposed approach for modelling product specification through services and Petri-Nets gives origin to a Service-oriented Holonic Manufacturing System (SoHMS). This takes the core of a Service-oriented Architecture (SOA) with the provider and customer entities having a holonic behaviour with roles defined by PROSA. Next, there will be explained some of the activities of the PROSA's basic Holons, indicating how to integrate the proposed approach and how this gives answer to the paper' objectives: enhancing the HMS flexibility, reactivity and productivity and a way to determine the RHs production capabilities.

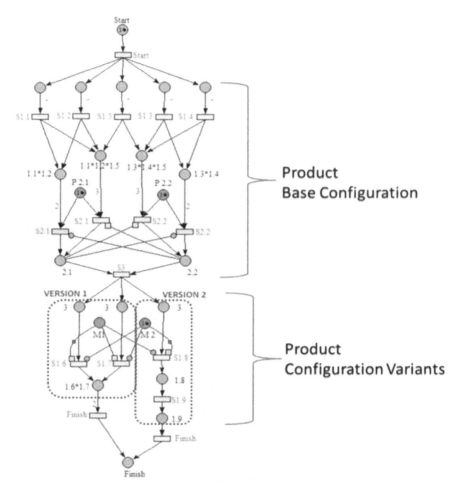

Fig. 6. Petri-Net Product Manufacturing Model

4.1   Holons' Roles

Product Holon (PH)

The PH, as in PROSA, contains the product specification using the present approach through Petri-Nets, Services and parameters specification. However, instead of standing as just an informational server, the PH leaves its passive character and adopts a more active one by involving itself in the decision process. Its main responsibility is then the exploration of the best possible production solutions according to the rules inherently expressed in the Petri-Net production recipe. It is also responsible for the evaluation and the selection of the best explored solution according to a certain criterion, e.g. the due date of the service. Exploration is done in two stages: prior to the order launching and during order production, with the intention of re-evaluating the system's state and react to changes by proposing new solutions based on the present state. Such decision is then communicated to the OH for their execution.

Order Holon (OH)

The OH contains the solution selected by the PH in the form of an execution table with all the information related to their proper execution, i.e. time constraints and service contractors. Its main responsibility is to ensure the proper execution of a task or series of tasks in the manufacturing system.

It is in charge of the product routing through the production plant (factory) from one resource to another according to the physical ports indicated in an execution table issued by the PH after workflow exploration. As production evolves, it notifies the production state to the associated PH for a continuous evaluation of new alternatives. Similarly, as production goes on, it sends intention confirmations to all the contractors to maintain contracts valid (reservations) and waits for their acknowledgments. Such reservations have a limited lifetime and become invalid in case of not being updated with a certain frequency. This is done in order to detect changes in the system's state as first stated in [11].

Resource Holon (RH)

A RH is a virtual representation of the physical resources that provide production capabilities in the factory floor. Such virtualization can be of one or of a group of physical resources for which manufacturing functions have been pre-programmed according to their internal models. In the same way as the agents in the operator level defined in [10], the RH can offer services that involve the interaction of various machines with a shared physical environment. The main idea behind this is that, by the unification of the individual physical resources' abilities, more complex manufacturing services could be offered, thus augmenting the manufacturing abilities of the SoHMS.

In contrast to PROSA's initial definition, it does not contain the controller of the resource. Its main activity in the HMS platform is the exposition of services and the negotiation for the allocation of the resources' activities according to a specific criterion (e.g. maximize resource utilization). On a lower level, there is a corresponding RH that can be called the "Operator-RH" (inspired by [10]), which contains the list of pre-programmed functions for the cluster of physical resources forming an RH. This operator-RH contains the utilization time-table of each of the physical resources involved, that the higher level RH accesses to manage its allocation. This separation is important as both activities - resource allocation and service execution - require execution environments with different time constraints.

4.2   Holonic Interaction

Fig.7 shows a UML sequence diagram to express, in a general way, the production lifecycle of a single product. It all starts with the exploration of the possible production sequences according to the product specification Petri-Net.

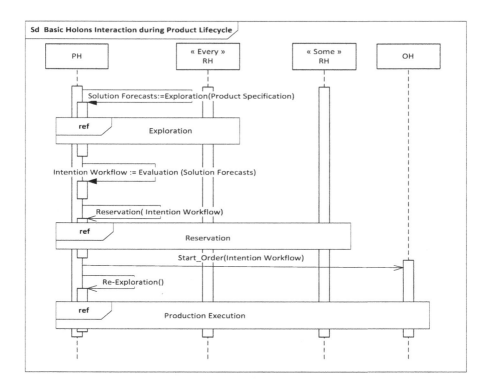

Fig. 7. Order Launching in a SoHMS

The issue of such exploration is a set of different trajectories that can achieve the production of a terminated product. Such trajectories are differentiated one from another by the order of execution of their services according to their precedence rules, the candidate service providers and the estimates of their execution. Once the exploration is concluded, the PH evaluates the alternative solutions according to a specific criterion (depending on interests) and selects one solution, making it the intention workflow. After defining the production workflow, the PH attributes contracts to establish reservations based on the RHs' service proposals and waits for contract establishments confirming the validity of the production plan. Once the intention has been established, the PH passes the confirmed intention to the associated OH for managing its execution through the production factory. At the same time, the PH enters in a mode of reactivity to re-explore and re-establish a new solution in the case of a disturbance on the system that invalidates the original production workflow.

Exploration of Alternative Workflows
Exploration starts by attributing an initial marking to the product specification Petri-Net. The marking, as stated before, represents production possibility states, i.e. states that do not depend on history but on the permissible future actions. At a given marking, the set of enabled transitions represent the permissible services that can be executed for that production state. The PH then sends tasks announcements to the cloud

of RH requesting for proposals coming from the RHs capable of providing such services. These are then added to the solution graph and continue exploration in time of each of these proposals in terms of sequence and of provider selection. This process continues until all possible trajectories are explored, generating an automaton with solution forecasts, as illustrated in Fig. 8. Such exploration process is based in the emergent short term forecasting approach proposed in [11].

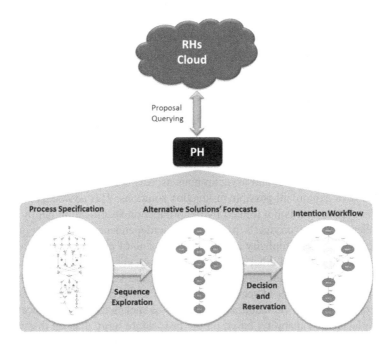

Fig. 8. Workflow Exploration and Selection[1]

In this way, the exploration process concludes with an arborescence of possible solutions according to the systems present state. Later on, by the use of graph search algorithms, the evaluation of the best solution can be done for their reservation and subsequent execution as presented above.

Continuous Exploration
One of the main requirements for SoHMS is a good reactivity to production disturbances. For this reason the PH enters in a reactive mode that enables it to re-start the exploration for new possible workflows, again in terms of service sequence, provider and time of execution, which could have a better performance than the original solution proposed. This re-exploration for new solutions can be done in a periodic manner and/or in an event-driven manner starting from the present production state:

---

[1] The state diagram arborescence shown in this image does not correspond to the Petri-Net in the example. This was simplified for visual reasons.

- *Periodic Re-Exploration*: The exploration of new alternatives is made every fixed period of time in order to detect changes in the system, that are not directly related to the PH in question. Such changes have an indirect impact in the product production plan and aren't directly notified to the PH's to trigger exploration. Examples are: changes of intention of other PH, breakdowns of RHs out of the PHs holarchy (non-contractor RHs), etc.
- *Event-Driven Re-Exploration*: Exploration is triggered by the notification of a disturbance involving one of the holons directly associated to the PH production plan (holons belonging to the same holarchy). In this case if an associated RH suffers a disturbance, the PH can immediately start exploration of new feasible alternatives.

5   Conclusion and Perspectives

Using Petri-Nets and the concept of Services in an HMS, creating a SoHMS brings several advantages to the control system. The Petri-Nets, thanks to its great expressiveness of the synchronous and asynchronous aspects between manufacturing tasks, allows the PH to explore all possible production solutions. This advantage adds flexibility to the system as it is not limited by the production specification but by the constraints inherent to the product's production interdependencies. Equally important, they express with great simplicity a potential explosion of production trajectories that would be difficult to model otherwise, as such arborescence depends on the combinatorial nature of manufacturing services and on contractor selection. On the other hand, the inclusion of the concept of services, giving rise to the SoHMS, introduces a clear and unified way to describe manufacturing tasks as means to determine accurately a RH's manufacturing capability, based on the task nature more than on the resources' model. This facilitates the introduction of different resource technologies as it is independent of the technology used. Moreover, it welcomes the customer involvement in the manufacturing specification and imports the advantages of service reutilization (as in product families) for cost reduction and faster design to production time. Finally, the short-term forecasting approach for the exploration of production alternatives represents a step for augmenting the systems productivity and bringing it close to optimality, by augmenting the vision of the system for the whole production lifecycle of the product.

On future work, more detail will be added on the reactive mechanism for the re-exploration of new solutions due to disturbances. For such, also social behaviour rules will be defined in order to avoid chaotic interactions in the system [11].

Work is also to be done in defining the RH model and behaviour algorithms for scheduling local resources such as to maximize their utilization or other criteria. Due to the potential explosion of alternative solutions, the exploration of all these can be time consuming and difficult to compute. In such case, machine learning algorithms could help in identifying those trajectories that give the best results according to the evaluation criterion used and limit the exploration of new solutions to a reasonable number around these solution areas.

References

1. Jiao, J., Simpson, T.W.: Product family design and platform-based product development: a state-of-the-art review. Journal of intelligent Manufacturing 18, 5–29 (2007)
2. Child, P., Diedrich, R., Sanders, F.H., Wisniowski, S.: Sloan Management Review. SMR forum: The Management of Complexity 33, 73–80 (1991)
3. Molina, A., Rodriguez, C.A., Ahuett, H., Cortes, J.A., Ramirez, M., Jiménez, G., Martinez, S.: Next-generation manufacturing systems: key research issues in developing and integrating reconfigurable and intelligent machines. International Journal of Computer Integrated Manufacturing 18(7), 525–536 (2005)
4. Van Brussel, H., Wyns, J., Valckenaers, P., Bongaerts, L., Peeters, P.: Reference architecture for holonic manufacturing systems: PROSA. Computers in Industry 37, 255–274 (1998)
5. Meyer, M., Lehnerd, A.P.: The power of product platform- building value and cost leadship. Free Press, New York (1997)
6. Martinez, M.T., Favrel, J., Fhodous, P.: Product family manufacturing plan generation and classification. Concurrent Engineering: Research and Applications 8(1), 12–22 (2000)
7. Simpson, T.W., Maier, J.R.A., Mistree, F.: Product platform design: Method and application. Research in Engineering Design 13(1), 2–22 (2001)
8. Jiao, J., Tseng, M., Duffy, V., Lin, F.: Product Family Modeling for Mass Customization. Computers and Industrial Engineering 35(3-4), 495–498 (1998)
9. Grönroos, C.: Service Management and Marketing. A customer relationship management approach, 2nd edn. Wiley, Chichester (2001)
10. Nagorny, K., et al.: A service- and multi-agent-oriented manufacturing automation architecture. Computers in Industry (2012),
http://dx.doi.org/10.1016/j.compind.2012.08.003
11. Valckenaers, P., Karuna, H., Saint Germain, B., Verstraete, P., Van Brussel, H.: Emergent short-term forecasting through ant colony engineering in coordination and control systems. Advanced Engineering Informatics 20(3), 261–278 (2006)
12. Cassandras, C., Lafortune, S.: Introduction to Discrete Event Systems. Springer Science (2008)
13. Mendes, J.M., Leitão, P., Restivo, F., Colombo, A.W.: Process Optimization of Service-Oriented Automation Devices Based on Petri Nets, pp. 274–279. IEEE (2010), doi:978-1-4244-7300-7
14. Wong, C.Y., McFarlane, D., Zaharudin, A.A., Agarwal, V.: The Intelligent Product Driven Supply Chain. In: IEEE Systems Man and Cybernetics, Hammammet, Tunisia (2002)

A Multi-Agent Architecture for Compensating Unforeseen Failures on Field Control Level

Christoph Legat and Birgit Vogel-Heuser

Institute of Automation and Information Systems,
Technische Universität München, Germany
{legat,vogel-heuser}@ais.mw.tum.de

Abstract. Technical failures not considered during plant engineering result in a reduced productiveness due to unscheduled system downtimes. Dynamic reconfiguration of field control software offers an alternative in contrast to the traditional way of immediate maintenance stops. In this contribution, a novel agent-based architecture for handling unforeseen failures is proposed. The reconfigurability of the field control software is achieved by implementing real-time agents in an IEC 61131-3 compliant language. The flexibility of the control software the agents rely on is achieved by applying the service-oriented paradigm to the field level control functionality. The realization of the concept on a laboratory plant with standard automation components is described and the advantageousness of dynamic reconfiguration is critically discussed.

Keywords: Discrete Field Level Control Software, Dynamic Reconfiguration, Fault-tolerant System, Agent-based Systems.

1 Introduction

The availability of production systems is one of the major challenges in operating production systems since failures in technical systems cannot be avoided completely [1]. Typically, exhaustive analysis of potential malfunction sources during engineering provide information to field control engineers about failures for which a compensation mechanism has to be realized. Failures which have not been identified and explicitly considered during engineering can entail various consequences. Firstly, failures affecting the precision of control execution cause minor product quality and possibly cost-extensive wastage. Since the failure has not been considered, manual monitoring is required and consequently, the problem arises. Secondly, some failures cause a complete loss of some control operations and finally result in a breakdown of a whole production line. This lead to a loss of until the system has been recovered. In both cases, the system has to be stopped for recovery. Since the failure is not documented, the process of manual failure diagnostics and maintenance is comparative time-consuming and results in drastically decreasing productiveness.

For this reason, a novel knowledge-based, multi-agent architecture is proposed which can keep up production by detecting and compensating failures not considered

T. Borangiu et al. (eds.), *Service Orientation in Holonic and Multi-Agent Manufacturing and Robotics*, Studies in Computational Intelligence 544,
DOI: 10.1007/978-3-319-04735-5_13, © Springer International Publishing Switzerland 2014

in advance. Missing documentation of failures also affects automated detection mechanisms because only little information is available. For this reason, we focus on failures which result in unexpected but observable temporal behaviour, i.e. an expected result is not achieved after a specific time. Automated compensation presumes the adaptability of field control functionality. But today's practice of implementing discrete field level automation software in accordance to the IEC 61131-3 standard result in a rather monolithic software structure. Hence, the adaptability of the control software is limited. Despite this drawback, IEC 61131 is the most common standard in industrial practice, reason for which the proposed approach has to be applicable for programmable logic controllers (PLCs) only supporting this standard. To handle this challenge, the multi-agent system relies on control services which encapsulate control functionality in a well-defined manner but are implemented in IEC 61131 compliant languages. Furthermore, parts of the multi-agent system are also implemented in accordance to this standard.

The remainder of this contribution is structured as follows: In Section 2, state of the art is discussed and shortcomings of existing approaches are identified. The main contribution of this paper comprises the agent architecture for dynamically reconfiguring manufacturing control systems in case of unforeseen failures. This will be presented in Section 3. Subsequently, its realization on standard automation software and one concrete failure scenario is presented in Section 4 to demonstrate the benefits of this approach. In Section 5, the advantage of dynamic reconfiguration of field level control software compared to the traditional way of immediate maintenance stops is investigated with respect to its effects on manufacturing systems' throughput. The paper is concluded in Section 6 with a brief summary.

2   Related Work

Agent technology is increasingly adopted for flexible automation software in order to realize different functionality required for dependable production automation (see [2, 3] for comprehensive overviews). Whereas in discrete open-loop automation, control capabilities are rather limited due to the system design [4], robotic systems provide higher degrees of flexibility for reconfiguration [5]. Leitão et. al [6] presents an approach to combine the adaptability offered by self-organizing holons while considering the optimization of performance. Self-organizing agents controlling tasks of system components in order realize required production processes are explored within the efforts towards Evolvable Assembly Systems [7]. Whereas all these concepts provide increased control flexibility, they either do not deal with component failures, especially ones not considered in advance or are not applicable for IEC 61131 compliant PLCs.

In recent years, much research has been conducted on reconfiguring control software based on the novel IEC 61499 standard for distributed automation software [8]. Its application for adapting control behaviour of inner logistic systems during operation has been proposed in [9]. The compensation of machine breakdowns under varying throughput conditions is described in [10]. These approaches are applicable for

PLC-based control and can adjust the control behaviour in case of failures. But they focus exclusively on the IEC 61499 standard, which is rarely supported by controllers used in industrial practice.

The application of agent technologies to design fault-tolerant control systems requires detailed knowledge about the technical system in order to build up required knowledge models [11]. Instead of building physical models of the system and the technical process, the service-oriented paradigm (SOA) provides a way to encapsulate functionalities offered by field devices [12]. In order to compose services to production processes, e.g. Mendes et. al [13] proposed to use a Petri Net formalism which can be used as knowledge-base for holonic agents controlling a manufacturing system.

Furthermore, service-orientation support the agile integration of field devices in a Plug & Produce manner [14] and enables describing behavioural aspects of the devices in a well-structured way [15]. Consequently, SOA is a promising way in order to provide required flexibility of control functionality but no approach exists yet which exploits SOA for fault-tolerant control systems in an IEC 61131 compliant manner.

In a nutshell, approaches on dynamic reconfiguration which are applicable with standard field control hardware and established standards are rare. Furthermore, no approach can deal with failures not considered explicitly in advance.

3   Multi-Agent System for Compensating Unforeseen Failures

In this section, a multi-agent based architecture for compensating unforeseen failures during production operation is presented. The functionality of compensating unforeseen failures is realized by agents, each implementing one or more roles as depicted in Fig. 1.

Based on a role's real-time requirements, agent(s) implementing them are allocated to adequate environments. An arrow connecting between two roles designates interaction between the implementing agents. The sequence of interactions for operating the system in case of a malfunction is denoted by the numbering of interactions. Since knowledge-based systems have been identified as major requirement for future production control systems [1], Fig. 1 also shows the models which have been associated to the roles.

In order to provide the required flexibility of the field control software and avoid monolithic software structures, we rely on fine-grained, real-time, and self-contained control operations whose execution results in a well-defined state after a finite execution time.

The applicability of this concept for developing and describing field control functionality has been presented in [16, 17]. Such control operations are referred to as *control services*; the sequence of operations, i.e. control services, is referred to as *strategy* henceforth.

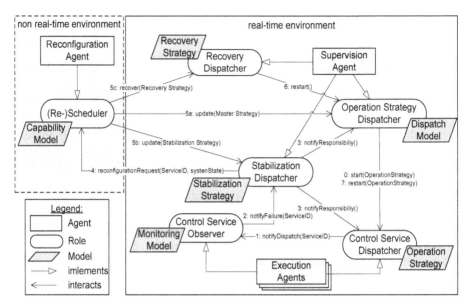

Fig. 1. Agents, roles, interactions and associated models

3.1   Control Service Dispatcher (CSD)

A control service dispatcher (CSD for short) is responsible to execute the schedule of control services defined by its operation strategy in real-time. An Operation Strategy defines the sequence of required control service invocations in order to manufacture a single good. Consequently, a single CSD is allocated to each good handled by a manufacturing system in parallel. A successful execution of a control service is indicated by achieving the post condition associated to a respective control service within the operation strategy. This indicates that the next control service has to be dispatched.

3.2   Operation Strategy Dispatcher (OSD)

Since more than one good has to be manufactured in parallel on a manufacturing system, the operation strategy dispatcher (abbreviated OSD), is responsible to schedule the CSDs. The dispatch model contains on the one hand the operation strategy which has to be executed by the CSDs and on the other hand a pair (STATE, t) which encapsulates the information when to start a CSD. Whereas the first OSD is initialized after system start up, the OSD awaits at least the time t and subsequently monitors the system state to initialize a further CSD if the state STATE is reached. This ensures that a specific control service is finished (given by the STATE) on a specific good (given by time t) before starting the process of manufacturing the next good.

3.3   Control Service Observer (CSO)

Whereas the OSD and CSDs are responsible for executing operation strategies and consequently realize the production process, control service observers (CSOs) monitor the execution of the control services dispatched by CSDs in order to detect a malfunction of the service. Due to the unforeseenness of failures, no concrete monitoring model is available. Following the maximum of everything not expected is unexpected, a malfunction is benchmarked based on an estimated duration of service execution. Consequently, if a post condition of a service recently executed is not manifested within a given temporal horizon, the service is assumed to be defective. Information about the service recently executed by a CSD is given to the CSO by the notifyDispatch message. The CSO is also required to be executed in real-time in order to avoid system damage.

3.4   Stabilization Dispatcher (SD)

In case of a detected malfunction of a service, the stabilization dispatcher (SD) is informed by the CSO which has detected the failure via the notifyFailure message. Based on the information about the malfunctioned service, the SD executes a predefined sequence of services (stabilization strategy) which is associated to a respective service. Such stabilization strategies are required in addition to the operation sequences since even if executions are stopped by the CSDs, probably some additional control service executions might be required in order to prevent system and/or product damage (cp. Section 4). In order to avoid responsibility conflicts between the SD and the CSD, since both dispatch control services, the CSD is informed about the control take-over of the SD by the notifyResponsibility message. When the manufacturing system is stabilized, i.e. a state is reached in which the manufacturing system can remain without any criticality, information about the malfunctioned service and the current, stable state are passed to the (Re-)Scheduler when requesting for reconfiguration.

3.5   (Re-)Scheduler

The central task of the overall system is associated with the (Re-)Scheduler. It is responsible to compute the models required by the OSD (and consequently the CSDs), the SD, and the RD. Since the computation complexity of problem solving algorithms is rather high, the stabilization offered by the SD provides required temporal flexibility to the (Re-)Scheduler.

The capability model of the manufacturing system contains the knowledge about control services, like e.g. their execution restrictions and effects to the system and the product. Based on the set of control services (except the malfunctioned service), the (Re-)Scheduler is able to determine the optimal operation strategy (if such exists). Detailed information about a knowledge model and its processing to determine the optimal Operation Strategy can be found in [17]. To support multiple goods operated by the manufacturing system in parallel, the additional restart condition to generate the dispatch model (cp. Section 3.2) is calculated and passed to the OSD. For each control service, which is used in the operation strategy, a stabilization strategy is computed in addition and passed to the SD.

Furthermore, a recovery strategy is required because the stable state in which the system remains until reconfiguration is not necessarily the starting state of the novel operation strategy (cp. Section 4). Consequently, a sequence of control service invocations manipulating a manufacturing system's state towards the starting state of the operation strategy, referred as recovery strategy, is computed and passed to the recovery dispatcher.

3.6   Recovery Dispatcher (RD)

The recovery dispatcher (RD) is responsible for transferring a manufacturing system into a state which allows restarting the OSD. Similar to the strategies used by the other roles, the recovery strategy, which was received from the (Re-)Scheduler, defines a sequence of control services to be executed. After successful execution of the recovery strategy, the control is passed back to the OSD and the manufacturing system can continue operation despite an unforeseen failure has occurred previously.

4   Realization

In this section, we briefly introduce a laboratory plant and some implementation details about the realization of the proposed agent architecture for dynamically reconfiguring its field control software. Subsequently, a specific failure scenario is described in detail to exemplify the benefit of the proposed approach.

The Hybrid Laboratory Plant is a research lab for practical testing of new technologies and automation solutions. Since it is built with real state of the art automation systems and software, it provides a good base for evaluating novel methods for development, specification and operation in a real plant under realistic conditions.

The plant demonstrates a production process for commissioning, transporting, and filling of bottles (see Fig. 2, centre). A five-axis-robot is used for warehousing the bottles and is connected to a grid of conveyors and rotary switches which are used to transport bottles between the warehouse robot and two filling stations. Both filling stations are equipped with pneumatic separators for bottle positioning and enable to fill solid pellets of different colours into the bottles.

The Hybrid Laboratory Plant is controlled by a single Siemens SIMATIC S7-400 programmable logic controller for running the field level control software. These PLC series are programmed using IEC 61131-3 compliant languages. It hosts the Supervision Agent who implements the OSD, SD, and RD roles and the Execution Agents implementing CSO, CSD roles (cp. Fig. 1). A SIMATIC Rack PC which is connected to the programmable logic controller hosts the Reconfiguration Agent. We rely on a proprietary agent platform currently under development.

The benefit of the presented approach is demonstrated by a single scenario: the malfunction of switch 2's control service "turnTo0()". This service is responsible for positioning rotary switch 2 for material take over from conveyor 3 (see Fig. 2). In failure-free operation, bottles are separated after filling and transported via switch 2. An excerpt of the corresponding operation strategy is given in Fig. 2.

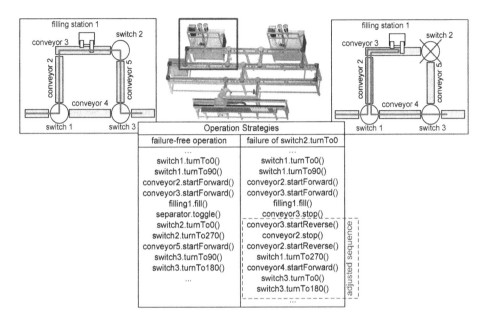

Fig. 2. Hybrid Laboratory Plant (centre), operation strategy for failure free mode (left) and malfunctioned switch 2 (right)

At some time of operation, the malfunction occurs and is detected by the Execution Agent since the turning operation does not result in the expected angle of the rotary switch – the expected post condition - in time. In this case, the stabilization is initialized by the Supervision Agent. The stabilization strategies comprise in this case solely the stoppage of the conveyors. After successful stabilization, the Supervision Agent requests for support by the Reconfiguration Agent who determines an alternative operation strategy without the malfunctioned service (cp. Fig. 2, right). The recovery strategy, to be executed by the Supervision Agent in his role as RD, depends on the state in which the system has been stabilized – especially on the states of the products. If, for example, multiple bottles are on the way towards the filling station (conveyors 2 and 3), the recovery strategy has to contain control service invocations which transport all bottles except one beyond this section in order to enable restarting the system by executing the novel operation strategy.

As shown in this section, the proposed approach for dynamically reconfiguring a manufacturing system's field control software enables to keep up production in case of (unforeseen) failures. Relying on the concept of control services provides more flexibility to the (control) system and enables to utilize functionality of a manufacturing system's components which are available but were not originally engineered for this purpose (e.g., a separator used as a stopper).

5   Comparative Study on Throughput Effects

Most manufacturing systems are designed in order to maximize their throughput, especially in Mass Production. Consequently, the effect of failures on the throughput, i.e. loss of production, is a major goal of dynamically reconfiguring a manufacturing system. A priori engineered redundancies of manufacturing systems are designed to provide the same or at least minimal loss of throughput in case of a specific failure. But reconfiguring a manufacturing system in case of unforeseen failures focuses on keeping up production with available system capabilities which are not originally designed for compensating failures, especially not for specific ones. For this reason, a throughput decrease by reconfiguration is supposable. Considering the scenario of the Hybrid Process Plant (malfunctioned switch 2) which was presented in the previous section, the throughput decrease is ascribed to the fact that after reconfiguring the system only a single bottle can be on the way from switch 1 to the filling station and backwards (cp. Fig. 2). Consequently, there are reconfigurations which have negative effect on the overall throughput of the manufacturing system but the plant is already operating. In contrast, an immediate, unscheduled maintenance stop manifests directly in high production loss. The decision to be taken by a plant's operator whether a reconfiguration should be used to keep up production or the manufacturing system should be directly shut down for maintenance is difficult to appraise. For this reason, an analytical study about the advantage of reconfiguring a manufacturing system in case of unforeseen failures is presented in this section.

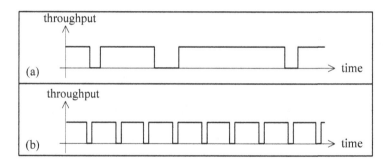

Fig. 3. Comparison of operating modes of manufacturing systems with respect to realized throughput: (a) twenty-four-seven operation and (b) shifting work

In the following, two possible ways of operating a manufacturing system are considered: twenty-for-seven operation and shifting work. In both cases, the manufacturing systems, if operating failure-free, realize a (more or less) constant throughput of d. As depicted in Fig.3a, a manufacturing system's operating plan in twenty-four-seven operation contains some few scheduled maintenance periods in which all required maintenance tasks are executed. A maintenance period's point in time depends on its scheduled necessity e.g. due to predictive replacement of equipment. Therefore, the time between these periods is not equidistant in contrast to shifting work where frequent operation and resting phases are scheduled (cp. Fig. 3b). Scheduled maintenance periods in twenty-four-seven operation can be seen as resting phases like in shifting operation (henceforth both are referred to as resting period with duration T_{rest}).

In the remainder of this section, we firstly discuss the effects of immediate maintenance stops and dynamic reconfiguration on the overall throughput of a manufacturing system. Subsequently, the section is concluded with a comparison of both approaches.

5.1   Throughput Effects of Immediate Maintenance Stops

One alternative, the operator can choose in case of an unforeseen failure is to immediately shut down the manufacturing system for maintenance. If a failure occurs at time point *ToF* (time of failure), the period between the *ToF* and the beginning of the subsequent resting period is referred to as Δ (cp. Fig. 4). The duration between production stop and restart of the system in case of a specific failure is called mean time to repair (MTTR).

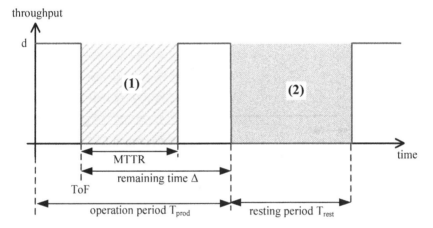

Fig. 4. Exemplary throughput realization in case of an immediate maintenance stop due to a failure (1) and a planned maintenance stop (2)

We assume that the repair of failures take at most $T_{rest} + \Delta$, i.e. a manufacturing system's failure can be maintained within the failure-affected operation period and a subsequent resting period. The function d_M (given in equation 1) defines the throughput $d_M(\Delta)$ which can be realized in case of an immediate maintenance stop with respect to the remaining production time Δ. The throughput loss in this case solely depends on the MTTR. When a failure occurs during the operation period and the remaining production time Δ is less than the MTTR, production cannot be restarted in a failure-affected operation period (second case of eq. 1).

$$d_M(\Delta) = \begin{cases} (\Delta - MTTR) \cdot d & if\ MTTR \leq \Delta \\ 0 & otherwise \end{cases} \qquad (1)$$

5.2   Throughput Effects of Dynamic Reconfiguration

When reconfiguring a manufacturing system dynamically, time elapses while executing the reconfiguration procedure, i.e. executing the stabilization strategy, calculating a new operation strategy, reconfiguring the agents' models and executing the recovery strategy (cp. Sect. 4). This period without any throughput is referred to as mean time to configure (*MTTC*). The throughput which is realized by a new operation strategy after reconfiguration is referred to as d_{new} (cp. Fig. 5) which is less or equal to the throughput d of the original one. The throughput d_{new} will be typically not higher than the original throughput d because a manufacturing system is designed for throughput optimization and consequently, the operation strategy with maximized throughput will be chosen as default one.

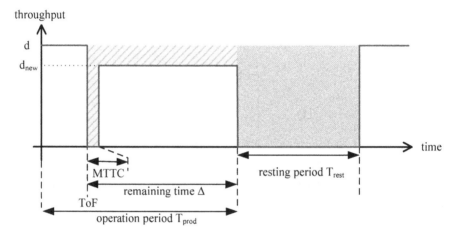

Fig. 5. Exemplary throughput realization when reconfiguring the control software in case of a failure

With function d_R given in Equation 2, the resulting throughput $d_R(\Delta)$ can be calculated if a manufacturing system is reconfigured dynamically in case of an unforeseen failure. If the MTTR of the failure is smaller than the resting period T_{rest}, maintenance can be done completely within the scheduled resting period. Consequently, after reconfiguring the system, i.e. after MTTC, the new operation strategy with throughput d_{new} can be executed until the scheduled resting period begins and the system can be maintained (first case in Equation 2). In shifting work, resting periods are often significantly shorter than in twenty-four-seven operation. If the MTTR is longer than the resting period and not as long as $T_{rest} + \Delta$ (in accordance to the initial assumption), the failure cannot be fixed solely in the resting period. Consequently, the remaining time duration for maintenance, i.e. MTTR-T_{rest}, is required to be performed during an operation period. The throughput of the operation strategy after reconfiguration is at most as good as the original operation strategy's throughput. For this reason, the operator will prefer the reconfigured operation period for maintenance additionally to T_{rest}, i.e. the resting respective maintenance period has to be scheduled earlier (cp. second case in Equation 2). Otherwise, reconfiguration doesn't make sense, since at least a subsequent operation period is affected and (partially) required for maintenance (third case of Equation 2).

$$d_R(\Delta) = \begin{cases} (\Delta - MTTC) \cdot d_{new} & if\ T_{rest} \geq MTTR \wedge MTTC < \Delta \\ (\Delta - MTTC - (MTTR - T_{rest})) \cdot d_{new} & if\ T_{rest} < MTTR \leq \Delta + T_{rest} \wedge MTTC < \Delta \quad (2) \\ 0 & otherwise \end{cases}$$

5.3   Conclusion: Immediate Maintenance Stop vs. Dynamic Reconfiguration

In the previous sections we have analysed the effects of immediate maintenance stops and dynamic reconfiguration on a manufacturing system's throughput and identified that the decision depends in both cases on the time of failure, i.e. remaining operating time Δ, and on failure-specific parameters MTTR for immediate maintenance as well as MTTC, d_{new} in case of dynamically reconfiguring the manufacturing system. Furthermore, parameters which depend on the way a manufacturing system is operated, i.e. especially T_{rest}, influence an optimal decision.

Fig. 4 visualizes an exemplary evolution of the throughput over time. The ToF is at time point t_1; the MTTR is less than Δ. In case of a failure the overall throughput dc_{max} cannot be achieved but there is a break-even point at time t_2, where an immediate maintenance stop outperforms a dynamic reconfiguration.

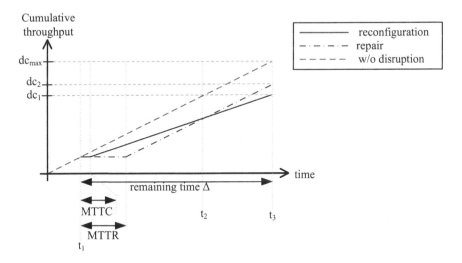

Fig. 6. Exemplary cumulative throughput function in case of a failure (MTTR < Δ; MTTR < T_{rest})

In a nutshell, dynamically reconfiguring a manufacturing system is a promising option to increase manufacturing systems' availability, especially in case of unforeseen failures. Its advantage compared to the traditional way of immediate maintenance stops is not universally valid. In some cases, e.g., if a failure can be fixed quickly, maintaining the system immediately outperforms a dynamic reconfiguration. In shifting work, frequent but significantly shorter resting periods exist. Consequently, there is a higher risk that the resting period is too short for fixing a failure. In contrast, twenty-four-seven operation is characterized by significantly longer operation periods than in case of shifting work. Therefore, if a failure with limited MTTR occurs very early in an operation period, an immediate maintenance might outperform a dynamic reconfiguration approach. Smaller the throughput after a reconfiguration, longer can be the MTTR so that an immediate maintenance stop has the edge over a dynamic reconfiguration. But a variety of influencing factors, as exemplarily discussed in this section, with their complex dependencies have to be considered towards an optimal operator decision. In practice, the decision is much more complex as described within this case study. For example, without assuming that solely a single sequence of operation and subsequent resting phase is affected by a failure, the formulas' complexity increase drastically. Furthermore, in order to enable an optimal decision, business conditions have also to be considered, e.g. availability of required maintenance personnel and spare parts or costs for the tardiness of some production jobs (rush jobs, contract penalties, etc.). For this reason, an integrated approach on a decision support system considering all relevant aspects and hiding the complexity while providing necessary information to the operator is mandatory. Towards this vision, research on automation technology, especially architectures and engineering approaches on field level control software, and aspects of operations research (e.g. approaches on dynamic rescheduling) have to be investigated together and their interdependencies and relationships have to be studied in detail.

6   Summary and Outlook

We proposed an agent-architecture for robust production which can handle failures not considered during engineering. The handling of unforeseenness is achieved by dynamic reconfiguration based on a model describing all capabilities of a control system. The required flexibility of the field control software is achieved on the one hand by applying the concept of control services, and on the other hand by the dispatching capabilities of implemented real-time agents.

It has been shown in an effect analysis, that dynamically reconfiguring a manufacturing system is not universally the optimal way with respect to a manufacturing system's throughput, although the availability is increased. In order to handle the decision complexity an operator is faced with, the vision of an integrated support system considering field level control and operations management aspects is derived.

While the proposed approach can be used to ensure the productiveness of manufacturing systems, some kinds of redundancy are already required to be engineered in the system. The proposed concept of control services provide more flexibility and might reduce the required hardware redundancy by increased control intelligence. In order to support the control intelligence of a manufacturing system, an engineering approach based on the concept of control services will be developed as a next step.

References

1. Legat, C., Lamparter, S., Vogel-Heuser, B.: Knowledge-based technologies for future factory engineering and control. In: Borangiu, T., Thomas, A., Trentesaux, D. (eds.) Service Orientation in Holonic and Multi agent, SCI, vol. 472, pp. 355–374. Springer, Heidelberg (2013)
2. Leitao, P., Marik, V., Vrba, P.: Past, present, and future of industrial agent applications. IEEE Transactions on Industrial Informatics (2012) (accepted and published online)
3. Brennan, R., Vrba, P., Tichy, P., Zoitl, A., Sünder, C., Strasser, T., Marik, V.: Developments in dynamic and intelligent reconfiguration of industrial automation. Computers in Industry 59, 533–547 (2008)
4. Vrba, P., Marik, V.: Capabilities of dynamic reconfiguration of multiagent-based industrial control systems. IEEE Trans. on Syst., Man, Cybernetics, A. 40(2), 213–223 (2010)
5. Borangiu, T., Gilbert, P., Ivanescu, N.A., Rosu, A.: An implementation framework for holonic manufacturing control with multiple robot-vision stations. Engineering Applications of Artificial Intelligenc 22, 505–521 (2009)
6. Barbosa, J., Leitão, P., Adam, E., Trentesaux, D.: Self-organized holonic manufacturing systems combining adaptation and performance optimization. In: Camarinha-Matos, L.M., Shahamatnia, E., Nunes, G. (eds.) DoCEIS 2012. IFIP AICT, vol. 372, pp. 163–170. Springer, Heidelberg (2012)
7. Frei, R., Serugendo, G.D.M.: Self-organizing assembly systems. IEEE Trans. Syst., Man, Cybern. C, Appl. Rev. 41(6), 885–897 (2011)
8. Vyatkin, V.: IEC 61499 as enabler of distributed and intelligent automation: State-of-the-art review. IEEE Trans. Ind. Informat. 7(4), 768–781 (2011)
9. Lepuschitz, W., Zoitl, A., Vallée, M., Merdan, M.: Toward self-reconfiguration of manufacturing systems using automation agents. IEEE Trans. Syst., Man, Cybern. C, Appl. Rev. 41(1), 52–69 (2011)
10. Khalgui, M., Mosbahi, O., Li, Z., Hanisch, H.M.: Reconfiguration of distributed embedded- control systems. IEEE/ASME Trans. Mechatronics 16(4), 684–694 (2011)
11. Romanenko, A., Santos, L.O., Afonso, P.A.: Application of agent technology concepts to the design of a fault-tolerant control system. Control Engineering and Practic 15(4), 459–469 (2007)
12. Jammes, F., Smith, H.: Service-oriented paradigms in industrial automation. IEEE Transactions on Industrial Informatic 1(1), 62–70 (2005)
13. Mendes, J.M., Leitao, P., Colombo, A.W., Restivo, F.: High-level Petri nets for the process description and control in service-oriented manufacturing systems. Int. J. Prod. Res. 50(6), 1650–1665 (2012)
14. Zühlke, D., Ollinger, L.: Agile automation systems based on cyber-physical systems and service-oriented architectures. In: Lee, G. (ed.) ICAR 2011. LNEE, vol. 122, pp. 567–574. Springer, Heidelberg (2011)
15. Mendes, J., Restivo, F., Colombo, A., Leitao, P.: Behaviour and integration of serviceoriented automation and production devices at the shop-floor. International Journal of Computer Aided Engineering and Technology 3(3), 281–291 (2011)

16. Schütz, D., Legat, C., Vogel-Heuser, B.: On modelling the state-space of manufacturing systems with UML. In: Borangiu, T., Dolgui, A. (eds.) IFAC Papers Online Proceedings of the 14th IFAC Symposium on Information Control Problems in Manufacturing, Bucharest, Romania, pp. 469–474 (2012)
17. Legat, C., Schütz, D., Vogel-Heuser, B.: Automatic generation of field control strategies for supporting (re-)engineering of manufacturing systems. Journal of Intelligent Manufacturing (accepted and published online, 2013)

Integrating Agents and Services for Control and Monitoring: Managing Emergencies in Smart Buildings

Monica Pătraşcu and Monica Drăgoicea

University Politehnica of Bucharest, Department of Automatic Control and Systems
Engineering, 313 Spl. Independentei, Bucharest, 060042-Romania
{monica.patrascu,monica.dragoicea}@acse.pub.ro

Abstract. The present work introduces a research perspective on developing *Smart Building* control and monitoring solutions using a service-centric conceptual framework in which agents and services are integrated in order to solve both the problem of comfort and the issue of safety. The proposed conceptual framework relies on the service oriented architecture approach and its related supporting technologies, tools, mechanisms that facilitate discovery, integration, processing and analysis of datasets collected from various ubiquitous appliances. At the same time, agents can take, based on environmental data, decision for control, monitoring, fault diagnosis and maintenance of more and more complex systems. In order to further develop the above mentioned service-centric conceptual framework, this paper proposes an extensive integration of emergency protection systems that take into account a varied range of hazards and disasters, from small fires to earthquakes, with *a priori* defined Intelligent Operations Centre for *Smart Cities*. In this respect, the CitySCAPE development framework is exploited, as being the architectural style of thinking in terms of *Smart Building* integration on different control levels, monitoring and safety intervention, meeting basic requirements of seismic protection at city level.

Keywords: agents, service orientation, smart buildings, Operations Centres for Smart Cities.

1 Introduction

Over the past two decades a major interest was dedicated to innovating building performance evaluation methods. In a larger perspective, "whole building" approaches to the operation of buildings were intended to be developed, where building components (construction materials) and systems (ambient components, like heating, lighting, ventilation etc.) are supposed to be integrated, not only to support the "green building" development, but also to educate users towards a sustainable use of planet resources.

There are many terms used today to describe different levels of device integration in a building, all of them enabling new "intelligent" building automation. *Home Automation* [1], *Smart Home* [2], *Smart Energy* [3] are only some few names to define

T. Borangiu et al. (eds.), *Service Orientation in Holonic and Multi-Agent Manufacturing and Robotics*, Studies in Computational Intelligence 544,
DOI: 10.1007/978-3-319-04735-5_14, © Springer International Publishing Switzerland 2014

current design of intelligent building management perspectives able to allow *real time monitoring* and *data collection* across different infrastructure components, centralization of real time events and data building in order to enable infrastructure-wide analytical and optimisation capability.

Smart Building is a term that defines a broader set of approaches, technologies, methods, tools and devices that crystallize European citizens and business increasing awareness towards *environmental, safety* and *comfort of living* issues [4]. It can be approached on different levels, through legislation, local initiatives of citizens and business organizations to better insulate and to install renewable energy sources, but also through better monitoring and control of building energy performance and *safety*. However, it was definitely recognized that none of these initiatives would be fully successful without the implication of the ICT.

Smart Building refers in fact to a new paradigm that has been developed in the last years, trying to define, develop and deliver intelligent building management solutions for energy optimization and facilities management. It offers real opportunities to innovate services based on the computational power of the Internet. Smart buildings are designed to run more efficiently and to communicate with and about their various systems assuring the *interoperability* of *functionalities exposed as services*.

In this respect, the research topic proposed in this paper defines CitySCAPE [5] as an *architectural style* of thinking in terms of *Smart Building* integration on different levels and control, monitoring and safety intervention meeting basic requirements of seismic protection at city level. An implementation of an inSCAPE type supervision system integrated in CitySCAPE has been tested, both in simulation and in hardware-in-the-loop configurations in [6]. As part of CitySCAPE, a building-level Critical Systems Emergency Protocol has been blended into the SOA based decision support system [7] of an integrated Intelligent Building Management solution.

The paper is organized as follows. Section 2 presents a novel perspective on the development of smart solutions for intelligent building management based on the CitySCAPE architecture. Section 3 presents the solution that integrates the two key aspects of CitySCAPE at building level, eSCAPE and inSCAPE, making use of *agents* and *services*, respectively. Section 4 includes two case studies, while section 5 offers final conclusions and further development perspectives.

2   Smart Buildings, Services and Agents – A Solution Development Framework

This section introduces a framework in which *agents* and *services* are *composed* to define a *smart product* – the *smart building* here. It is based on the following common observation. The most valuable features of a smart product aren't contained entirely within that product itself, but are delivered as a result of interactions with other products or services within an ecosystem that needs to collaborate and share information. In a smart building, system functionalities are distributed over the various *intelligent, interconnected, instrumented devices* in the building environment [8].

Today, a new generation of intelligent building applications are shifting from the centralized, local desktop computer application software towards the provision of distributed geo-spatial services and components that foster software modularity and reusability. On the perspective of the *smart* attribute, the next generation of IT development deals more with the integration of the existing software and infrastructures, than with creating new applications.

The smart building might contain a whole range of sophisticated software intensive systems, designed to make living comfortable while improving safety and optimizing energy consumption (Fig. 1).

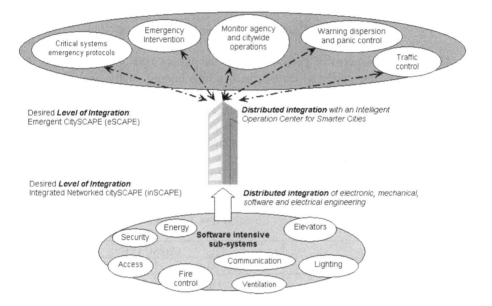

· **Fig. 1.** Smart building – a system of systems in a larger ecosystem

At the same time, the smart building system might interact with other systems external to the building itself. For example, the building security systems might be engineered to interact with emergency response centres (such as the IOC – the Intelligent Operation Centre for Smart Cities) in order to deliver incident details to first responders based on data collected from sensors within the building.

CitySCAPE *(a Synergic Control Architecture for Protection against Earthquakes)* was proposed as an architecture dedicated to the control and monitoring of urban systems [5]. It has an hierarchical structure that implements a decentralized character at lower levels and centralized components at higher levels, that deal with the integration of its subsystems into the whole. CitySCAPE ensures structural integrity, implements and supervises social protection norms, and ensures emergency response in case of disasters.

This paper intends to propose o solution that integrates two key aspects of CitySCAPE at building level, eSCAPE and inSCAPE (Fig. 2), making use of *agents* and *services*, respectively. Each of these systems and their role in the protection of human life are described in the next paragraphs.

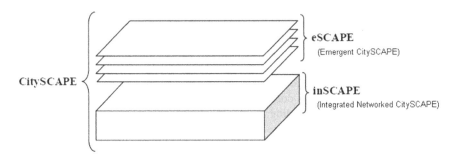

Fig. 2. CitySCAPE - a macroscopic level perspective

The two main aspects that are considered in the present solution are: *The Right To Life* (ensuring safety during an earthquake is a high interest, high complexity problem, that has attracted attention ever since the development of seismic structural design) and *The Right To Comfort* (in what concerns social protection and ensuring high living standards, one must take into account the requirements for energy consumption, evolution of social groups, view on living standards and, of course, green energy generation and Earth friendly living solutions).

Table 1 presents a comparison between different levels of integration in the Smart Building paradigm and the CitySCAPE architecture.

Table 1. Levels of integration – CitySCAPE vs. Smart Building perspective

	Smart Building	**CitySCAPE**	
		inSCAPE	eSCAPE
Building level integration	Distributed integration of electronic, mechanical, software and electrical engineering	Hierarchical integration of services, devices and their associated control systems for structural integrity	Emergent integration of intelligent control agents for human safety norms and protocols (during emergencies) and comfort (during normal operation)
IOC level integration	Distributed integration with underlying agency such as emergency management, public safety, social services, transportation or water	Hierarchical integration of services and associated control systems for city-wide seismic and disaster protection	Emergent integration of intelligent control agents for city-wide protection and comfort of social systems (for emergencies and normal operation)

2.1   The Service Side

Integrated Networked CitySCAPE (inSCAPE) is a CitySCAPE subsystem which integrates and interconnects a hierarchical system for structure integrity (including, for instance, seismic vibration control).

At city level, inSCAPE is wholly defined by two concepts: vertical interconnection and horizontal integration. In Fig. 3.a (**vertical interconnection**) the top-down decomposition is defined (for a city comprised of q clusters, there is a hierarchical structure *HS* that can perform the monitoring of the disaster protection systems). In Fig. 3.b (**horizontal integration**) the bottom-up composition is accepted (for a city comprised of q clusters, over which a hierarchical structure *HS* has been defined, on each layer of the *HS* there is a distributed structure through which structural control can be obtained).

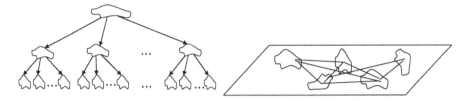

Fig. 3. a. Vertical interconnection (left). **b.** Horizontal integration (right).

Scaled down to the building level, this subsystem can be built as a supervision-type sub-architecture. Although the modules necessary to shape up inSCAPE at this level can be implemented in various ways and using various techniques (from heuristic supervisors to multivariable controllers integrated in intelligent control systems), this paper presents a service oriented solution. Past work for this approach includes an implementation of an inSCAPE type supervision system that has been tested, both in simulation and in hardware-in-the-loop configurations, in [6]. Moreover, a building-level Critical Systems Emergency Protocol has been blended into the SOA based decision support system [7] of an integrated Intelligent Building Management solution.

Let a structure $S = \{f(struct, em), Ds, Css, Drs, Comm\}$ [5]. The structure S is comprised of a function that describes the structural behaviour during emergencies and normal operation $f(struct, em)$, the set *Ds* of devices (both safety and comfort, as well as any transducers and actuators necessary for operation), the structure's control supervisor *Css*, a disaster response service *Drs* and a communication module *Comm*.

The function $f(.)$ can be as complex as necessary, depending on structure, seismic zone, building materials, age and wear, installed devices and so on. From a systems engineering point of view, $f(.)$ describes the controlled plant. From a service science point of view, the analysis of $f(.)$ yields information necessary for the development of building specific services, be they safety or comfort oriented. For instance, the type of structure (number of floor, destination, use of space etc.) will generate specific evacuation paths.

The *Css* component is a supervisory-type service that manages the control and monitoring of all devices throughout the building, as well as all maintenance and fault diagnosis related tasks, while the *Drs* module is a service in charge of disaster response and protection of human life during such events. Both the *Css* and the *Drs* are high level services, operating on an elevated abstracting and containing building wide protocols and schedules. These two services access medium and low level agents and sub-services in order to ensure structure integrity during emergencies, optimal operation of its systems in normal conditions, and, of course, human safety. A detailed view of the connection between the aforementioned services and agents can be observed in section 3 of this paper.

2.2   The Agent Side

The emergent CitySCAPE (eSCAPE) subsystem models, controls and monitors the protection of living beings. This component manages the social system and human behaviour during disasters, with a flexible communication network. With a main role of protecting human life and critical systems, eSCAPE is the decision making entity that will decide, for example, what are the best evacuation paths, which are the high risk areas, which are the zones that need clearing for emergency intervention teams and so on. eSCAPE is defined as comprised of cells and tissues, the devices that form these components being implemented as agents in what follows, exploiting the natural reorganizing and emergent properties of multi-agent systems. The high diversity of the eSCAPE components can thus be properly implemented by making use of the properties and versatility of intelligent agents. Thus, a generic cell $< \cdot > = \{D_C, D_P, D_A, U_{comm}\}$ is a set of control modules D_C, perception modules D_P and actuating modules D_A, as well as a communication unit U_{comm}, while a generic tissue $\ll \cdot \gg = \{< \cdot >_i \mid i=1,n\}$ is a set of n cells of the same type.

Cell properties: *flexibility*: each cell can be dissolved and aggregated depending on the context in which it needs to operate; *modularity*: each cell is independent of other cells; *self-reorganizing*: each cell can reorganize around a nucleus represented by the communication agent U_{comm}, according to priority lists and/or proximity.

In order to build the eSCAPE components as multi-agent systems, two operations need to be defined, as follows: *Collaboration* - the operation of grouping cells, represented by their communication units as nuclei, into tissues: $\ll \cdot \gg = Co(< \cdot >_1, < \cdot >_2, ..., < \cdot >_n)$; and *Aggregation* - the operation of grouping modules into cells: $< \cdot > = Ag(module_1, module_2, ..., module_n \mid U_{comm})$ around a communication unit as nucleus. Each module of a cell can be an agent in itself, while cells as a whole function based on the generic intelligent agent architecture proposed in [9].

A series of tissue properties are to be noted: each cell of a tissue is interconnected with the others, giving the tissue a global emergent behaviour; cells of different tissues can occupy the same physical space; two or more cells of different type which occupy the same physical space are connected through their communication units.

At building level, eSCAPE is comprised of 4 tissue types (fig. 4). Each of these manages one aspect of the protection system: (a) emergency response; (b) evacuation protocols; (c) warning dispersion and panic control; (d) critical systems protocols.

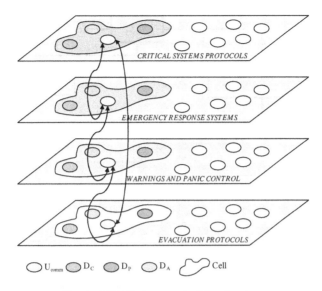

CRITICAL SYSTEMS PROTOCOLS

EMERGENCY RESPONSE SYSTEMS

WARNINGS AND PANIC CONTROL

EVACUATION PROTOCOLS

U_{comm} D_C D_P D_A Cell

Fig. 4. eSCAPE tissues at building level

For example, an emergency response cell <ERC> = $\{D_C, D_P, D_A, U_{comm}\}$ is formed of: D_C - agents for combined behaviour generation of intervention personnel (such as fire fighters, paramedics, police etc.) on a specific floor; D_P - agents for information processing from the building supervisor *Css* and emergency response service *Drs*; D_A - agents for intervention requests' transmitted to building clusters' supervisors and/or city-wide emergency response centres, like the IOC.

Thus, eSCAPE appears as a multi-agent system variation with emergent components specialized on specific problems, synergistically interconnected with the service oriented inSCAPE.

3   Services and Agents in Intelligent Building Control and Monitoring

A proposed further implementation of the Control and Monitoring solution discussed in this paper is depicted in Fig. 5. This holistic view of the architecture permits, on the one hand the integration of services and agents, and on the other, the coordination of safety and comfort specific subsystems.

Thus, the DRS^{HL} and CSS^{HL} services represent their counterparts from the in-SCAPE side of CitySCAPE, as described in section 2.1 of this paper. These modules are high level (HL) components that access the Floor Agents (FA^{ML}) and the Control & Monitoring Agents (CMA^{LL}). In turn, they are coordinated by a general supervisor agent SupA^{HL}, whose role is to connect the building to the city-wide implementation of the architecture, through a communication unit *Comm*. This function allows structures to be integrated in the city level instances of eSCAPE and inSCAPE, and, ultimately, a complex large scale CitySCAPE.

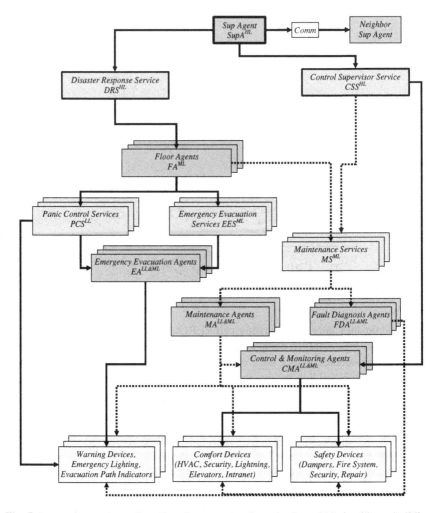

Fig. 5. Integrating agents and services for control and monitoring within intelligent buildings

The medium level (ML) agents and services deal, in turn, with disaster response on each floor (Floor Agents FA^{ML} and Emergency Evacuation Services EES^{ML}), and with optimal device operation throughout the building (Maintenance Services MS^{ML}) – these two aspects emphasize the horizontal interconnection and vertical integration of inSCAPE. For visualization purpose, in Fig. 5, the maintenance branch has dotted line connectors, while the flow of information for the control branches is represented using continuous connectors.

The low level (LL) agents and services, as well as some medium level (ML) agents are in charge of control and monitoring of physical devices ($CMA^{LL\&ML}$), specific maintenance tasks ($MA^{LL\&ML}$), fault diagnosis ($FDA^{LL\&ML}$), panic control ($PCS^{LL\&ML}$) and emergency evacuation ($EA^{LL\&ML}$).

The proposed solution includes three types of agents: low, medium and high level, each with its own level of intelligence. This perspective allows for the composition of heterogeneous groups of agents into several MAS, which, in turn, could group into a higher level MAS. Fig. 5 presents the three different types of agents according to their goal inside the building, taking into account their relation to the services considered for proper functioning, both from a structural point of view and a social/human perspective.

The LL agents (low level) are usually persistent software tools that perform basic tasks, such as the role of control algorithms. These agents are directly connected to the physical devices, either through their own network (on a seismic damper, for instance), or through basic web services (for example, to monitor and/or control the ventilation system). These agents are mostly reactive or are based on reflexive rules, sometimes incorporating very simple inference modules (for example, to allow various functioning points to be reached, such as night and day energy consumption levels or lighting services and so on).

The ML agents (medium level) can be either embodied entities or software tools that include higher reasoning than the LL agents. The ML layer performs like a heterogeneous agent society, in which several interactions types can be observed, goals are either defined or communicated, there can be cooperation, and different levels of autonomy can be identified. Considering these points, a more extensive classification is required for the medium layer [10]. Thus, ML agents are discussed based on:

— level of intelligence:

  o reactive agents: reflexive agents, either simple or with internal rules
  o deliberative agents: reasoning, intelligent inference systems, adaptive behaviours are only a few points that can describe such agents
  o composed reactive-deliberative agents

— composition:

  o singular: these agents exist on their own in the system and usually perform specific tasks; they receive directives from higher level agents;
  o emergent: these agents are usually the cell-type agents that are comprised of entities performing various roles (as described in section 2); the agents that group to form an emergent agent can be either reactive or deliberative, embodied or software etc.

— entity type:

  o embodied: these agents have a physical body
  o pure software: these agents exist only in the virtual world and interact only with each other through various communication systems

— human interaction

  o with strong social skills: these agents are required to interact with humans; their environment is highly non-deterministic and they require a high degree of autonomy to perform their goal
  o with weak social skills: these agents are only required to interact with other agents

— goal:

   o control & monitoring: these agents perform all the actions required for the control of the devices throughout the building
   o maintenance & fault diagnosis: these agents perform all actions required for maintenance & fault diagnosis; their goal is intrinsically different from the control & monitoring ones', due to the particularities of decisions and actions that these agents must take
   o protection: these agents are tasked with protection of human life, from hazard detection to hazard mitigation, including management of panic control devices and of the human component during emergency

— interaction protocols:

   o communicative agents: these agents are usually cell-type agents that have a communication component; the information transmitted between these agents has been described in section 2.
   o cooperative agents: these agents are usually the ones that perform the grouping into emergent agents; they have to cooperate to achieve a goal and are usually required to be of different types in order to compose a higher level emergent agent (as described in section 2)

— action type:

   o autonomous agents: these agents are usually the higher level ones, that are capable of making abstract decisions
   o dependent agents: these are usually the lower level agents that receive goals, directives or tasks from higher level entities

In this paper, the HL agent (high level) is only one software entity, with a high level of intelligence. This agent has a supervisory role, it interacts with human decision making entities, it communicates with adjacent building, integrating the structure in the more complex acceptation of CitySCAPE.

4   Case Studies

4.1   Fire Event Evacuation Scenario

A version of EES^{ML} has been implemented in [6], while several groups comprised of FA^{ML}, EES^{ML} and MS^{ML} are works in progress. As part of a more complex modelling and simulation of CitySCAPE, this case study illustrates a fire event scenario and subsequent evacuation on a floor of a smart building.

The first case study's interface included in this work can be observed in Fig. 6. An evacuation protocol cell is dynamically modelled, along with a emergency response cell for fire hazard, on a building story. The program includes a set of embodied evacuation agents that gather the floor occupants and guide them to safety, fire detection sensing elements, fire control systems for offices, a particular CO_2 fire control system for an archive space with an agent that dispenses oxygen masks for the human occupants. This simulation includes three different fire scenarios, with an interface that displays the transducers and actuators for the fire suppression system, the human entities, and the embodied agents.

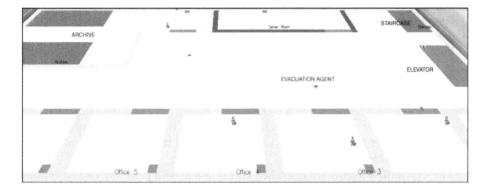

Fig. 6. Fire event evacuation: floor view

The world model of this application is a floor map comprised of walls, elevators (with suspended activity due to fire hazard), stairway access and a hallway. Depending on space destination, the behaviour of the agents involved is particularized, described in what follows.

The fire suppression agent for offices is an emergent cell agent tasked with monitoring fire events and suppressing flames in office spaces. It is a cell agent that includes several instances of the same types of lower level agents:

— sensing agents: the fire detection transducers are reactive, singular, embodied, with control & monitoring goals, communicative and dependent agents
— control agents (not visible in interface): the control algorithms that analyse the information received from the sensing agents and transmit commands to the acting agents
— acting agents: the fire suppression actuators (sprinklers) are reactive, singular, embodied, with control & monitoring goals, communicative and dependent agents
— communication inside this cell has been provided by the AOP medium used for implementation of this application [11]

This agent controls & monitors the fire suppression system for the entire building.

The fire suppression agent for archive and/or server room is an emergent cell agent tasked with monitoring fire events and suppressing flames in special spaces, like archives and server rooms, which require particular fire suppression systems. In this example, the suppression agent is CO_2 (lethal to humans). It is a cell agent that includes:

- sensing agents: the fire detection transducers are reactive, singular, embodied, with control & monitoring goals, communicative and dependent agents
- control agents (not visible in interface): the control algorithms that analyse the information received from the sensing agents and transmit commands to the acting agents
- acting agents: the fire suppression actuators are reactive, singular, embodied, with control & monitoring goals, communicative and dependent agents
- communication inside this cell has been provided by the AOP medium used for implementation of this application [11]

This agent also transmits information to the evacuation system and deploys oxygen masks for humans caught inside the archive at the time of fire hazard event. Thus, an intrinsic connection between this fire emergency protocol cell and the evacuation cell (as viewed from the eSCAPE perspective) is built, the two cell-type agents *working together to protect human life*. This is an illustration of the natural emergent properties of the eSCAPE system.

The evacuation agents are embodied agents tasked with gathering human entities and guiding them to the staircase doors. These agents' degree of intelligence is in the deliberative-reactive combined category. Apart from being embodied (for example a mobile robot type of embodiment), the evacuation agents of this simulation are singular, with strong social skills, have protection as a goal and are agents with a medium-to-high degree of autonomy.

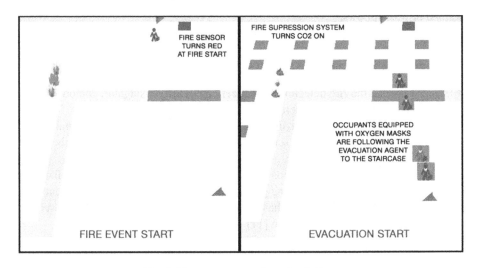

Fig. 7. Fire event evacuation: simulation

Fig. 7 presents a screenshot taken during simulation of a fire event inside the office, as well as inside the archive room.

Last, but not least, the fire hazard protection protocols and evacuation command are generated at the superior Emergency Evacuation Service EES^{ML}, that contains, for example, the entire set of evacuation protocols for that particular floor, as described by the authors in [6], and can commission, for instance, the evacuation agents with different goals in what concerns paths taken, staircases used and so on.

4.2   Chemical Spill in Laboratory Scenario

The second case study's interface included in this work can be observed in Fig. 8. An evacuation protocol cell is dynamically modelled, along with a emergency response cell for chemical spills [12], in a laboratory that spans on a building story. The world map includes nine lab spaces, an emergency shower room, one elevator and two stairwell access points.

The program includes a set of embodied evacuation agents that gather the floor occupants and guide them to safety, cell agents tasked with chemical spill control and with fire events in laboratory spaces. This simulation includes different chemical spill and fire scenarios, with an interface that displays the transducers and actuators, the human entities, and the embodied agents.

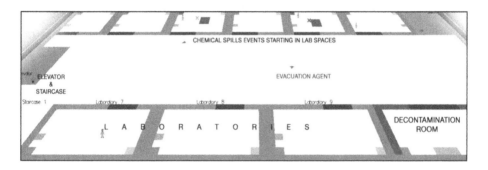

Fig. 8. Chemical spill in laboratory: floor view

The chemical spill control agent is an emergent cell agent tasked with monitoring chemical spills and containment in laboratory spaces. It is a cell agent that includes several instances of the same types of lower level agents:

— the sensing agents are air toxicity detection transducers (reactive, singular, embodied, with control & monitoring goals, communicative and dependent)
— the control agents (not visible in interface) include the control algorithms that analyse the information received from the sensing agents and transmit commands to the acting agents
— this particular cell includes three different types of acting agents that are used to implement the neutralization of toxic substances and protection of human life, as follows:

o embodied agents, such as robots for chemical spill clean-up (deliberative-reactive, singular, with weak social skills, with protection as a goal and a medium-to-high degree of autonomy)

o the shower decontamination agents (reactive, singular, embodied, with control & monitoring goals, communicative and dependent)

o sealing doors for isolating non-exposed personnel in adjacent laboratory spaces, in absence of fire or when whole floor evacuation is not necessary (reactive, singular, embodied, with control & monitoring goals, communicative and dependent)

— the communication inside this cell has been provided by the AOP medium used for implementation of this application [11]

The fire suppression agent is an emergent cell agent tasked with monitoring fire events and suppressing flames in office spaces. It is a cell agent that includes several instances of the same types of lower level agents and is similar in nature with the fire suppression agent for archive and/or server room described in section 4.1 (CO2 fire suppression acting agents, control agents, fire sensing agents). The main difference is the protection protocol that implements a different behaviour:

• in the affected room: don't perform evacuation, but guide the lab technicians toward the emergency shower room
• for the rest of the floor: perform evacuation

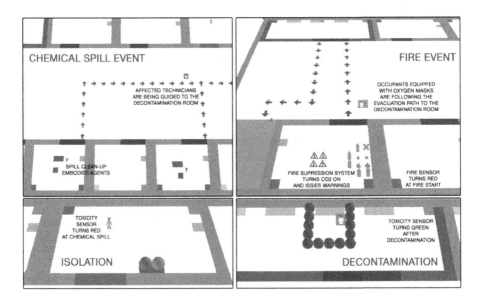

Fig. 9. Chemical spill simulation

These agents also transmit information to the evacuation agent and deploy oxygen masks for humans caught inside highly toxic spaces. Again, an intrinsic connection between the chemical spill control cell agent, the fire suppression cell agent and the evacuation agent is obtained. These agents are *working together to protect human life*, illustrating of the natural emergent properties of the eSCAPE system.

The evacuation agents are embodied agents similar to those presented in section 4.1. They are tasked with gathering human entities and guiding them to the staircase doors (in case of fire) or elevators (in case of spill). These agents' degree of intelligence is also in the deliberative-reactive combined category. They are embodied, singular, with strong social skills, have protection as a goal and a medium-to-high degree of autonomy.

Fig. 9 presents a screenshot taken during simulation of a fire event inside the office, as well as inside the archive room.

The chemical spill control protocols, the fire event protocols and evacuation command are generated at the superior Emergency Evacuation Service EES^{ML}, that contains for example the entire set of decontamination and evacuation protocols for that particular floor, as described by the authors in [6], and can commission different actions in what concerns paths taken, decontaminants used and so on.

5   Conclusions

It is in the perspective of the above mentioned paradigm – the *Smart Building* – that the solution presented in this paper is described. It is part of, CitySCAPE, a larger control and monitoring architecture. Primarily built to deal with earthquake protection and emergency response, this system can been expanded to encompass other life aspects and disasters, such as comfort and fire protection. Moreover, this framework can be applied to other fields, such as manufacturing, where service orientation and multi-agent system integration is a forthcoming direction with broad prospects.

CitySCAPE incorporates two subsystems, inSCAPE and eSCAPE, each with its own particularities and scopes of action. The two coordinate protecting the physical structures and the population, at both city-wide level, or, as it is the case of this work, at building level.

The case studies presented in this paper implement three evacuation agents, a chemical spill control agent and three fire suppression agents. Their architecture is the cell type proposed in this paper, that includes sensing elements, an inference system (reasoning), acting elements and an internal world model. Thus, the evacuation agents include the floor plan, position of offices/laboratories, staircase access etc., and they lead the human groups toward the evacuation points on the safest routes possible. The chemical spill control agents deal with neutralizing toxic substances and request the human occupants to move toward the decontamination chambers. The fire suppression systems sense the fire events, deploy oxygen masks when necessary and then activate the suppression systems. All protocols are included in the Emergency Evacuation Service, the can generate both low level and medium level requests, commands or behaviours, as much for evacuation, as for fire suppression and chemical spill protocols.

Future work will include implementation of the other services described in this paper and the particular communication between agents and services required by the

Smart Building. At the same time, communication with the outside world will be considered, taking into account emergency responders, outside environment before evacuation planning, neighboring buildings and their seismic structural stability and other such factors that can influence the operations of the Smart Building, be it in emergency management or in normal operation.

The proposed architecture integrates services and agents, making use of their advantages within the control and monitoring of intelligent buildings. In addition, this solution, as part of the CitySCAPE framework, allows further integration of structures as systems in the city level ecosystem, in a modular and visionary perspective over systems-of-systems and their role ensuring human safety and comfort.

References

1. Bucur, L., Tsai, W.T., Petrescu, S., Chera, C., Moldoveanu, F.: A Service-Oriented Controller for Intelligent Building Management. In: Proceedings of the 18th International Conference on Control Systems and Computer Science, CSCS 18, Bucharest, Romania, vol. 2, pp. 665–670 (2011) ISSN 2066-4451
2. Smart Home. Smart Home Learning Center, http://www.smarthome.com/learningcenter.html (accessed February 2013)
3. Smart Energy. Smart Energy, Web-based Energy Modelling Software, http://www.smartenergysoftware.com (accessed February 2013)
4. European Commission, Advisory Group and the REEB Consortium On the Building and Construction sector. ICT for a Low Carbon Economy Smart Buildings (2009), http://ec.europa.eu/information_society/activities/sustainable_growth
5. Patrascu, M.: Advanced Techniques for Seismic Vibration Control. PhD Thesis, University Politehnica of Bucharest, Faculty of Automatic Control and Computers (2011)
6. Drăgoicea, M., Bucur, L., Pătrașcu, M.: A Service Oriented Simulation Architecture for Intelligent Building Management. In: Falcão e Cunha, J., Snene, M., Nóvoa, H. (eds.) IESS 2013. LNBIP, vol. 143, pp. 14–28. Springer, Heidelberg (2013)
7. Drăgoicea, M., Pătrașcu, M., Bucur, L.: Service Orientation For Intelligent Building Management: an IOT and IOS Perspective. In: UNITE 2nd Doctoral Symposium, R & D in Future Internet and Enterprise Interoperability, Sofia, Bulgaria, pp. 79–86 (2012)
8. IBM White paper. A mandate for change is a mandate for smart, http://www.ibm.com (accessed February 2013)
9. Meystel, A.M., Albus, J.S.: Intelligent systems: architecture, design, and control. Wiley & Sons, New York (2002)
10. Monica, P., Dragoicea, M.: Integrating Services and Agents for Control and Monitoring: A Smart Building Perspective. Preprints of the Int. Workshop on Service Orientation in Holonic and Multi Agent Manufacturing and Robotics, SOHOMA 2013, Valenciennes, France, June 20-22, pp. 978–973 (2013) ISBN 978-973-720-490-5
11. Wilensky, U.: NetLogo. Centre for Connected Learning and Computer-Based Modeling. Northwestern University, Evanston (1999), http://ccl.northwestern.edu/netlogo/
12. American Chemical Society's CEI/CCS Task Force on Laboratory Waste Management: Guide for Chemical Spill Response Planning in Laboratories (2013), http://www.acs.org/content/acs/en/about/governance/committees/chemicalsafety/publications/guide-for-chemical-spill-response.html

State of the Art and Future Perspectives for Smart Support Services for Public Transport

João Falcão e Cunha and Teresa Galvão

Faculdade de Engenharia da Universidade do Porto, Porto, Portugal
{jfcunha,tgalvao}@fe.up.pt

Abstract. This paper summarizes existing systems and research on information transport services, and proposes a hypothetic scenario for future travellers using public transport. Increased distributed intelligence in pervasive mobile smart devices and in sensor networks in public transport vehicles is enabling a new approach for enhancing the experience of public transport customers. Such environment could be modelled through a distributed multi-agent service system. This paper presents advanced information services already available on such environments, in particular the MOVE-ME smartphone application, and indicates a possible service environment where people's feedback may benefit all transport service stakeholders. Mobile computing and crowdsourcing are key enablers for enhancing user experience in the transport services, and also for enhancing overall public transport services. Better experience leads to increased usage of shared mobility modes, and therefore to more sustainable cities in the future. Concerns about data security, and anonymity of travellers will need to be adequately addressed in the future scenarios presented.

Keywords: mobile, integrated, multimodal, information systems.

1 Introduction

Pervasive mobile smart devices and sensor networks in public transport vehicles are enabling a new approach for enhancing the experience of public transport customers. The experience may actually start before a journey takes place, by planning a trip, and travellers could still be providing relevant information after reaching their destinations. Transport services and associated information services are closely coupled, and this relation needs to be better understood to improve overall service.

We claim that pervasive mobile services can be used for enhancing user experience in the transport services, and also for enhancing overall public transport services, including providing accurate feedback to management. Research going on in public transport information services at FEUP, the School of Engineering of the University of Porto for the past ten years, and at Imperial College London for the past three years has been aiming at providing users with real time information, but also finding ways to use the user feedback for benefiting all other transport service stakeholders in particular usual or sporadic travellers. Mobile computing is a key enabler for making such vision possible [1].

T. Borangiu et al. (eds.), *Service Orientation in Holonic and Multi-Agent Manufacturing and Robotics*, Studies in Computational Intelligence 544,
DOI: 10.1007/978-3-319-04735-5_15, © Springer International Publishing Switzerland 2014

The transport infrastructure, networks and vehicles, is being continuously monitored using a variety of sensors with variable degrees of intelligence. The information that they produce combined with the information available from the user's devices and feedback creates vast amounts of data and an opportunity. With adequate management it is possible that patterns will be derived from such data in order to help transport users to have an improved experience, and also other transport stakeholders such as control staff in the transit authorities. For instance, the smart device of a regular user of some transport service may be monitoring the network information and be able to advise if changes are required, or if some special promotions are being offered on the usual route or in an alternative route. The smart device of a tourist arriving in a new destination and already aware of the user needs, and her behaviour patterns, may provide advice on alternative travel modes, points of interest, and prices.

This paper will introduce the MOVE-ME service, application and infrastructure, and then refer to work in progress on extending it in several directions. Experiments with emotional and operational feedback from travellers, in a crowdsourcing approach, are presented, and also a prototype mobile payment and validation service. These prototypes have being tested in real situations in Porto and London with voluntary users. Finally some ideas are presented of how a distributed multi-agent system could be used to model and improve the overall service. An imaginary future scenario of traveling in a city with the support of an intelligent application is presented. This application relies on bringing together existing information and services, using new infrastructure technologies and business models.

2   The MOVE-ME Service and Architecture

The MOVE-ME project (www.move-me.mobi) has developed an infrastructure and a mobile application enabling users to access public transport information in real time. From 2012, this application enables travellers to plan their journeys based on real time or planned data from metro, bus, coach, and train schedules (see Fig 1 and 2). The infrastructure brings together geo-referenced data from different transport companies, Google map data, and also relevant touristic locations enabling multimodal journey planning. Real time multimodal travel planning can be done in a 60 minute time horizon. When real time data is not available or when the time window is larger than 60 minutes and shorter than 3 days, travel planning is done based on available published schedules over a 3 days horizon [1]. With such time horizons the current available infrastructure has been able to guarantee acceptable user response time in the order of less than 20 seconds in 95% of cases. Interaction is also supported by a location based service that enables answers to be calculated taking into consideration the user location.

The MOVE-ME service with advanced real time functionalities was launched in May 2012 in the Metropolitan Area of Porto, and is now also available in the Metropolitan Area of Lisbon, and in other regions of Portugal. Over 25 distinct metro, bus, coach, and train companies share information on their service, and over 30.000 Android and iPhone users are now benefiting from such service. Its user interface adapts to the default language of the device, and offers Portuguese and English details. There is also a Web interface, but with limited functionality.

 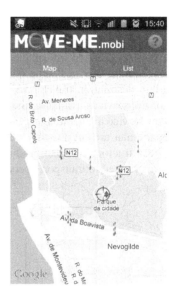

Fig. 1. Main MOVE-ME user services: home application screen options, and location based identification of nearby transport stops

Fig. 2. The map shows a multimodal public transport route from Vila do Conde to Porto (blue, in the original) and its stops/stations (red and white poles, in the original). Users can follow a trip in real time.

Most of the system architecture of MOVE-ME, see Fig. 3, evolved from previous transport information services, both for planned and real time situations. Its main data base collects detailed information from transport companies, including network details, schedules, and real time data, and algorithms maintain an updated large matrix of origin destination travel possibilities with associated travel times. MOVE-ME is one of the services available, the most demanding one. Fig. 3 shows also InfoBoard, another service available using the same data for displaying travel information related to a particular area of a town, such as a transport interface, a hospital, or a university. The infrastructure can also produce other information, for example printed materials to be distributed to transport users, or to be available at bus or metro stops.

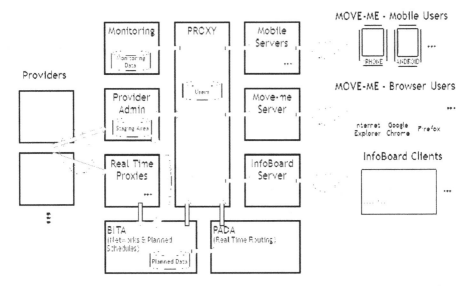

Fig. 3. MOVE-ME system architecture: the core of the system is shared with several other information services such as the InfoBoard large display information service

The MOVE-ME mobile service, application, and infrastructure was developed by OPT www.opt.pt, with research support from FEUP www.fe.up.pt, INEGI www.inegi.up.pt, and UGEI at INESC TEC www.inescporto.pt, in the context of the CIVITAS Elan European project www.civitas-initiative.org. MOVE-ME won the 2012 CIVITAS European Technical Innovation Award for the Porto municipality, Portugal.

3   On-Going Research on Smart Mobile Traveller Information Services

Helping the user to plan his journey is an important service to improve experience of public transport services, but several other components need to be addressed in order to fulfil the most demanding expectations.

More advanced mobile computing research with the objective of enhancing public transport user experience is under way at FEUP for the past ten years, and in collaboration with Imperial College London for the past three years. The research group at Imperial College London is focusing in capturing affective and emotional aspects of the traveling experience in public transport.

This section presents on-going work on enabling users to benefit from shared travel information in a crowdsourcing environment, and in suitable ways to manage payments for the travel service. It introduces experiments with emotional and operational feedback from travellers, and also a prototype mobile payment and validation service under evaluation in STCP www.stcp.pt, the public service bus company in the Metropolitan Area of Porto. Prototypes are being tested in real situations in Porto and also London with voluntary users.

3.1 Measuring and Sharing Users Emotional State

A prototype cloud-based mobile service to assess the relationship between affective state and travelling context has been developed and then tested with commuters of the Porto transport network [2] (see Fig. 4). User's affective state was captured using a simple emotional model of travelling mood, with cognitive pleasure and physical arousal dimensions based on Russell's circumplex of emotion [3]. Travelling context included noise, saturation, smoothness, ambience, speed and reliability. The findings show a strong correlation between mood and context, dependent on the user.

Fig. 4. Measuring public transport users' affective state – prototype screens tested with Porto commuters [2]

3.2   Sharing Users Information On Operational Situation

Mobile devices are also being used to measure users' affective state in real public transport contexts, and how to share and disseminate real time information through social networks for the benefit of other travellers and also for the benefit of the control rooms of transport companies [4]. Preliminary experiments have been conducted in Porto and in London transport services.

A prototype crowdsourcing application to share information has been developed and tested with commuters of the London transport network. The model borrows key principles from Internet based services. It strives to intensify win-win relationships between public transport passengers and operators. The structured exchange of information is sustained by a validation mechanism for data reliability, and an incentive mechanism to encourage passenger participation. Passengers benefit from rich real-time data to ease their journeys and improve travel experience, in exchange for their own participation providing and validating information. Operators gain access to rich customer generated data, which in an aggregated format may provide a real-time assessment of customer experience and of local performance across the entire network operation.

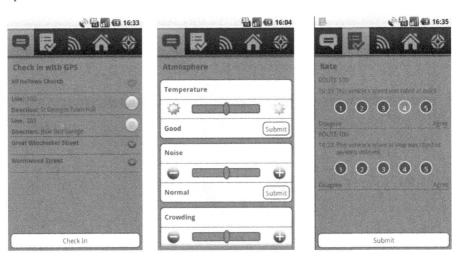

Fig. 5. Measuring public transport operational state – prototype screens tested with London commuters [4]

3.3   Managing Payment and Travel Authorization

Current research is also addressing ways of extending MOVE-ME to enable travellers to pay for their journeys using their mobile devices. The Metropolitan Area of Porto has an open transport network with no barriers. In the future it is possible to envision an entirely pay-as-you-go transport service based on users' mobile devices. As Web enabled smartphones become pervasive, ticket selling machines and validating machines could become redundant and unnecessary [5].

4   Perspectives of Evolution – The MOVE-US Service Scenario

The evolution of technology and the needs to have improved transport could result on the following extreme scenario:

Fig. 6. Home screen of mobile payment service for STCP, zone ticket options (showing Z2-Z5), and ticket validation example [5]

- Every transport user will have a smartphone fully connected to the Internet, and location enabled through GPS and alternative technologies.
- Ticketing systems and existing associated fixed physical infrastructure at depots, stops and vehicles will disappear, being completely replaced by user mobile devices and pay-as-you-go services.
- Every vehicle becomes fully intelligent, aware of its driver (if there is one), passengers, and ticket supervisors (when they enter the vehicle).

Security concerns will need to be fully addressed in such scenario, and it is possible that anonymous passengers will be allowed in the system provided they have the necessary authorizations for travelling.

In such a scenario all available transport information will be collected and handled with artificial intelligence. Usage patterns will be identified, including the ones related with unplanned situations.

It is possible that software will be available in the user smartphone that integrates or collects his/her usage patterns and travel plans in order to provide suggestions or help making decisions. Such software would also benefit from information available from the other relevant users and naturally from all the relevant network data. Regular users would mostly benefit from such support in unusual transport situations, such as delays or accidents that would disrupt the normal plans. Users in new situations would benefit from counselling regarding travelling options.

The following scenario does not exist yet, but we can imagine the technology required to make it real in a few years.

*Monica and Peter are planning their next trip to Porto. They have now booked
their flights and hotel. Their **Move-Us** app knows that they prefer to use public transport services in most countries. They will only use a taxi when there is heavy rain,
and when they would need to walk for over 5 minutes, or when a taxi would take them
to their destination in half the time of public transport, and would not be paying more
than twice the price. **Move-Us** has access to all the relevant travel and context information in Porto, like whether forecast, and suggests the way to get to the hotel on the
date and time of arrival (and also on return).*

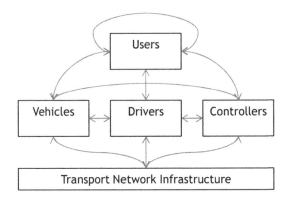

Fig. 7. Transport users, drivers and controllers will have access to vast amounts of data and will
be able to share real time information. Intelligent treatment of data is critical for providing
relevant feedback.

*When Monica and Peter arrive in Porto, flight delayed, **Move-Us** provides updated
real time travel proposals, including trip cost and estimated arrival time. They accept
the public transport suggestion.*

*Move-Us then guides them to the Metro stop at the airport using their location and
the available online maps. **Move-Us** knows exactly where the Metro stop is located,
when the next vehicles will arrive, and it can also calculate the cost of the trips. It
also has suggested that they pay-as-they-go during their stay in Porto. The payment
service will automatically chose the cheapest available fare. It also keeps them informed of the taxi alternative, its cost and time to arrive at the hotel.*

*Monica and Peter don't need to buy tickets from a machine or a shop. When the
next metro arrives at the airport, **Move-Us** connects to the public transport payment
service in Porto, informs it of the travel intentions, and will then charge their credit
card. Every payment transaction could be automatic, and fully transparent for
Monica and Peter, but **Move-Us** always requires validation before entering and when
exiting the vehicles. It also provides other relevant information about cost, position,
directions, and approaching commuting stops, if needed. This is possible because
vehicles and user mobile devices are geo referenced with precision, and intelligent*

software matches the routes. In the very unlikely event that such system fails, the user can manually override it or complain later on.

*During their trip **Move-Us** may be collecting relevant data and sharing it with other travellers. Monica and Peter may also provide feedback about their trips. When they interact with their friends using Facebook or Twitter (e.g.) **Move-Us** will collect mood information (if permission is given) and other emotional status and relevant travel statements relating to the current trip. It may share some of this information with other users, or with travel controllers. If they were travelling in a bus, relevant information about driving experience could also be relevant.*

*****Move-Us** could also automatically connect Monica and Peter into a temporary social network of users to enable sharing of relevant information. For instance, travellers ahead of them could be sharing some interesting information about a shoe shop offering large discounts near the metro stop near the hotel. This information would be made available as **Move-Us** also knows that Monica and Peter like to buy high quality shoes every time they are in Porto.*

*In general, whenever Monica and Peter intend to go to some place **Move-Us** will provide assistance for planning the trips, for guiding them during the trips, and will ask for any comments they may like to add during or after reaching their destination. This is done in non-intrusive ways, and their comments may be used to improve service. They may also get some rewards, given to the most active travellers in the Porto public transport system.*

In fact, most of the technology required already exists, and the possibility of such scenario becoming real depends on socio-organizational evolution, and on making the investments that are needed. The scenario cannot be real while the great majority of people travelling miss a smartphone, but mixed scenarios are also possible during the evolution.

5   Conclusions

It is expected that ubiquitous availability of high quality Web enabled mobile devices and services will improve public transport user experience, both functionally and emotionally. Public transport operators may also benefit from collaboration of users feedback to enhance the transport service levels.

The MOVE-ME service will start incorporating some of the functionalities that have been shown, but it is not yet clear which ones will be adopted. Current research is trying to define priorities.

It is also envisaged that the availability of distributed computing power could enable the capture of usage patterns of users and such information could be used to improve interaction and also improve the understanding of transport patterns. A distributed multi-agent system could be used to model and improve the overall service, improving the transport experience of travellers.

Better experience leads to increased usage of shared mobility modes, and therefore to more sustainable cities in the future.

Acknowledgments. The authors would like to acknowledge the work of colleagues, PhD students, and collaborators at OPT. Funding from the following projects is also acknowledged: OneStopTransport QREN TICE no. 13843, MOBIPAG QREN TICE 13847, EU CIVITAS Elan, IBM CAS Portugal, INEGI, IDMEC Polo FEUP, UGEI - INESC TEC, PhD grants from FCT, Portugal. This paper is the evolution of work presented at the 3rd ERRIC Workshop on Service Orientation in Holonic and Multi-Agent Manufacturing and Robotics, in Valenciennes, 2013.06.20-21 [6].

References

1. Falcão e Cunha, J., Galvão Dias, T., Pitt, J.V.: Mobile Real Time Applications for Enhancing Public Transport User Experience - The MOVE-ME project. ERCIM News (93), 41–42 (2013) ISSN 0926-4981
2. Costa, P.M., Pitt, J.V., Falcão e Cunha, J., Galvão Dias, T.: Cloud2Bubble: Enhancing quality of experience in mobile cloud computing settings - A framework for system design and development in smart environments. In: MCS 2012 Proceedings of the 3rd ACM Workshop on Mobile Cloud Computing and Services, pp. 45–52 (2012)
3. Russell, J.A.: Circumplex Model of Affect. Journal of Personality and Social Psychology 9(6), 1161–1178 (1980)
4. Nunes, A.A., Galvão Dias, T., Falcão e Cunha, J., Pitt, J.V.: Using social networks for exchanging valuable real time public transport information among travellers. In: Proceedings of the IEEE Conference on Commerce and Enterprise Computing, pp. 365–370 (2011)
5. Ferreira, M.C., Nóvoa, M.H., Dias, T.G.: A Proposal for a Mobile Ticketing Solution for Metropolitan Area of Oporto Public Transport. In: Falcão e Cunha, J., Snene, M., Nóvoa, H. (eds.) IESS 2013. LNBIP, vol. 143, pp. 263–278. Springer, Heidelberg (2013)
6. Falcão e Cunha, J., Galvão Dias, T.: Smart services for public transportation: state of the art and perspectives. In: Proceedings of SOHOMA 2013, 3rd ERRIC Workshop on Service Orientation in Holonic and Multi-Agent Manufacturing and Robotics (2013)
7. Costa, P.M.: Enhancing Quality of Experience in Affective Pervasive Environments. PhD Thesis, Imperial College London (2013)
8. Hocová, P., Falcão e Cunha, J.: A Service Science and Engineering Approach to Public Information Services in Exceptional Situations - Examples from Transport. In: Morin, J.-H., Ralyté, J., Snene, M. (eds.) IESS 2010. LNBIP, vol. 53, pp. 65–81. Springer, Heidelberg (2010)

Part IV
Intelligent Products
and Product-Driven Automation

QLM Messaging Standards: Introduction and Comparison with Existing Messaging Protocols

Sylvain Kubler, Manik Madhikermi, Andrea Buda, and Kary Främling

Computer Science Building, Aalto University,
Department of Computer Science and Engineering,
P.O. Box 15400, FI-00076 Aalto, Finland
sylvain.kubler@aalto.fi

Abstract. Recent advancement in web technology enabled the development of new Business-to-Business (B2B) infrastructures, e.g. based on the concept of Service Oriented Architecture (SOA). These infrastructures enable seamless information exchange among different stakeholders and complex business procedures. However, there is still a lack of sufficiently generic and standardized application-level interfaces for exchanging the kind of information required by such infrastructures. These interfaces must be as complete and flexible as possible to support changing organization needs and structures. Their development is an essential step to design future SOA services and to enhance product lifecycle management. The Quantum Lifecycle Management (QLM) messaging standards are proposed as a standard application-level interface that would fulfil such requirements. This standard is introduced in this paper and compared to existing ones. Several real-life implementations are presented to show why such messaging standards are needed and how flexible QLM messaging standards are.

Keywords: Service Oriented Architecture, Internet of Things, Quantum Lifecycle Management, messaging protocols, Intelligent Product.

1 Introduction

Today's enterprise architectures are based on several computing paradigms that have evolved over the years with the advent of the ubiquitous network era and the so-called Internet of Things (IoT) [2]. This gave way to complex heterogeneous enterprise system combinations to meet the enterprise expectations. However, as the system complexity increases, its infrastructure and maintenance costs also increase. Organizations are therefore looking for ways to minimize Information and Communications Technology (ICT) costs, while offering better products and services to users [16]. Finding the right balance (cost minimization *vs.* product/service enhancement) is a challenging task for organizations since their IT infrastructure has to continually provide the infrastructure and systems that support their business needs, even when they drastically change. This is particularly difficult because infrastructures and systems are often designed in isolated factions, thus limiting their openness and collaboration [26].

T. Borangiu et al. (eds.), *Service Orientation in Holonic and Multi-Agent Manufacturing and Robotics*, Studies in Computational Intelligence 544,
DOI: 10.1007/978-3-319-04735-5_16, © Springer International Publishing Switzerland 2014

A first step has already been taken to make systems more adaptable through the use of Electronic Data Interchange (EDI) technologies that allow B2B integration [15]. However, such systems are still limited in terms of *flexibility* and require higher maintenance costs [11]. Recent advancements in web technology (e.g. web 2.0) provided the opportunity to develop new B2B infrastructures that make it possible to perform both seamless information exchange among different stakeholders and complex business procedures by integrating different internal and external services [24]. In this regard, the concept of SOA is a good example since it enables to turn business applications into individual business functions and processes [21].

SOA is an architectural paradigm and discipline that may be used to build infrastructures enabling those with needs (consumers) and those with capabilities (providers) to interact via services across disparate domains of technology and ownership [23]. A SOA architecture is a form of distributed system architecture whose main properties are service abstraction, orientation and autonomy [8]. Although the potential of SOA has been widely recognized, there are still fundamental questions and issues that need to be addressed, e.g. no proper agreement on a common standard for data exchange between organizations, whether in terms of data structure or data communication, has yet been reached; new flexible business and communication interfaces for supporting the many changes of organization infrastructures and needs throughout their lifetime has yet to be designed [11]; developing new strategies for context-aware services, i.e. services able to self-adapt autonomously depending on current management conditions is becoming a glaring demand [26]. These challenges have to be addressed so as to increase the scope of SOA architectures, i.e. to provide architectures sufficiently flexible to be used in any domain of the Product Life Cycle (PLC), regardless the product context and environment. The notion of *flexibility* is the watchword to design future and dynamic SOA services and to enhance the product data management (i.e. product data interoperability, visibility, sustainability, manageability and security) [3].

This paper introduces a new messaging protocol named Quantum Lifecycle Management (QLM) messaging that aims to enable all kinds of intelligent entities[1] in and between organizations to exchange IoT information in ad hoc, loosely coupled ways. This proposal, currently under standardization, complements the existing portfolio of messaging protocols (e.g. JMS, oBIX...), which are all potential solutions for supporting SOA features. Each protocol has *pros* and *cons* that may limit their openness to new applications and new organization needs. In this regard, section 2 gives a more comprehensive view of the necessity to implement an appropriate messaging protocol to ensure data exchange interoperability, which is of the utmost importance to efficiently support SOA infrastructures. Section 2 also presents four messaging protocols (in addition to QLM) well recognized in the literature, whose primary goal is to meet the data exchange interoperability requirements, as QLM. Section 3 defines a framework used in our paper to compare QLM with these messaging protocols. Section 4 provides greater details on the two standards that compose QLM. Section 5 details

---

[1] An "entity" might be a product, a device, a computer, a user or any other type of system that consumes or provides information.

two real-life industrial applications where product information are collected, exchanged and processed in and between various organizations based on the QLM messaging standards. These applications help to show how important and flexible these standards are.

2   Data Exchange Interoperability

2.1   Two-Level Challenge

Data exchange interoperability can be achieved at the application level, which is a two-level challenge as illustrated in Fig. 1:

- i) *Communication level integration:* necessity to provide infrastructures and mechanisms that support communications and transactions between distinct organizations, networks, applications and IT technologies,
- ii) *Data level integration:* necessity to handle the changes in data media and formats that occur throughout the product and information life cycle,

Over the last decades, techniques, concepts and standards have emerged at both levels [4,20]. For instance, the ISO 16739 or UDEF are standards defined at the level of data integration. The focus of this paper is not given at this level but at the communication level.

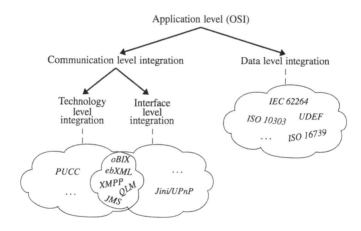

Fig. 1. Data exchange interoperability at the "Application level"

The communication level integration is in turn a two-level challenge as depicted in Fig. 1:

- i) *Technology level integration:* necessity to provide a framework for integrating heterogeneous hardware and software platforms and to support the wide range of IT technologies (proprietary solutions, extended networks, *etc.*),
- ii) *Interface level integration:* necessity to support different types of transactions that must be based on shared business references [6]. To gain time and

efficiency and to avoid re-defining cooperation rules and software supporting it each time, these references must be based on business standards or norms. The business standards must be independent and loosely coupled with the technology level integration so as to support openness.

It is not an easy task to develop solutions supporting both levels of integration. As a matter of example, the Peer-to-peer Universal Computing Consortium (PUCC) [18] mainly focuses on the technology integration aspect by designing a general peer-to-peer networking platform that enhances the communication capabilities of various devices via various networks. The solutions developed by this consortium do not integrate the set of interfaces directly at the communication level but requires the deployment of the PUCC middleware, thus limiting the integration with other middleware. Another example is the Jini/UPnP interoperability framework developed in [1] that allows clients to use pre-defined interfaces, but requires the use of Java technologies. This makes technology integration with non-Java applications quite daunting. A few solutions have nevertheless emerged for supporting both levels of integration such as oBIX (Open Building Information Exchange), ebXML (Electronic Business XML), XMPP (eXtensible Messaging and Presence Protocol), QLM (Quantum Lifecycle Messaging) or still JMS (Java Message Service). This paper only focuses on such protocols. The next section briefly introduces these five messaging protocols, which are then compared in section 3.

2.2   Candidate Messaging Protocols

JMS might be the most known messaging protocol among the five to be compared. The creation of JMS was joint effort of several corporations supported by SUN Microsystem [17]. JMS is an Application Programming Interface (API) for enterprise messaging that provides the set of interfaces enabling point-to-point communications in distributed enterprise applications. JMS defines a single, unified message API that enables loosely coupled, asynchronous and reliable communications, and support the development of heterogeneous applications that span operating systems, machine architectures, and computer languages. JMS relies on the publish/subscribe (Pub/Sub) model and supports the subscription mechanism. In this model, publishers and subscribers are generally anonymous and may dynamically publish or subscribe to specific data. A centralized system takes care of distributing subscribed data coming from the publishers to all subscribers.

oBIX is an industry-wide initiative in the building area that enables mechanical and electrical control systems to communicate with enterprise applications. oBIX defines XML and Web Services-based mechanisms and the information is represented as a concise object model, where objects with any structure and contents can be freely exchanged [22]. A client-server architecture is used for accessing oBIX information over the network. Software can act as a client, as a server or eventually as both simultaneously. Three types of requests/responses referred to as "oBIX verbs" are defined:

- *Read*: it returns the current state of an object,
- *Write*: it updates the state of an object,
- *Invoke*: it triggers an operation on an object.

oBIX also provides mechanisms called *watch objects* that enable clients to subscribe to object events. More concretely, *watch objects* are created based on client requests, then the server provides the client with a URI to enable it to poll and get the latest status updates. It makes it possible to append new history records, query historical data or do basic roll up of numeric data. Subscriptions are cancelled either when the client unregisters the *watch object* or when the predefined lease time for polling action is exceeded. The oBIX alarm feature is a cornerstone of this protocol, which enables to raise an alarm when an incident occurs and to notify users or related applications. Broadly, oBIX is a suitable communication interface for data acquisition and control of a wide range of devices.

XMPP is a communication protocol for message-oriented middleware based on XML [27]. XMPP was introduced by the Jabber open-source community in 1999 and has been improved by the XMPP Working Group to provide Internet Engineering Task Force (IETF) Instant Messaging and presence technology. The XMPP RFC defines this messaging protocol as a robust, secure, scalable, internationalization-friendly architecture for near real time messaging and structured data exchange. In XMPP, the communication of XML messages is based on the XML stream rather than the XML document between two network entities. The XMPP technology uses decentralized client-server architectures in which clients open a stream to the server, which in turn opens a stream back.

ebXML is an initiative of the OASIS group and the United Nations Centre for Trade Facilitation and Electronic Business, whose foundations rely on a former B2B solution called ooEDI (object oriented Electronic Data Interchange) [14]. This consortium promotes an open and XML-based infrastructure supporting the exchange of electronic business information in an interoperable, secure, and consistent manner. One objective for ebXML is to lower the barrier of entry to *e*-business, particularly in the context of small and medium enterprises. The ebXML specifications cover business processes and heterogeneous data shared between a wide range of actors and corporations.

QLM messaging standards emerged out of the PROMISE EU FP6 project[2] in which the real-life industrial applications required the collection and management of product instance-level information for many domains involving heavy and personal vehicles, household equipment, phone switches, *etc.* Information such as sensor readings, alarms, assembly, disassembly, shipping event, and other information related to the entire PLC needed to be exchanged between several organizations. The project consortium set out to find suitable standards for exchanging such information. PROMISE created two main specifications that fulfilled the necessary requirements: the PROMISE Messaging Interface (PMI) and the PROMISE System Object Model (SOM). At the end of the PROMISE project, the work on these standards proposals was moved to the QLM workgroup of The Open Group[3]. The QLM messaging protocol is the continuity of PMI and consists of two standards proposals [12], namely the QLM Messaging Interface (QLM-MI) that defines what kinds of interactions between

[2] http://promise-innovation.com
[3] http://www.opengroup.org/qlm/

entities are possible, and the QLM Messaging Format (QLM-MF) that defines the structure of the information included in a QLM message. Both standards are presented in section 4.

3   Messaging Comparison Framework

In our study, several key properties are considered to compare the five messaging protocols previously introduced. These properties mostly come from the framework defined in [28], which are divided into three main categories: *i)* message delivery model, *ii)* message processing model, *iii)* message failure model. The next three paragraphs respectively introduce the set of criteria defined for each of these categories.

3.1   Message Delivery Model

This category defines the properties related to the delivery of the message to the intended recipient. The majority of the criteria composing our framework comes from this category:

1. *Representation:* it indicates the message representation adopted by the messaging protocol. It can be represented either as a *i) Data element* - the message consists of a messaging header and body that respectively contains information about the interface and the data to be conveyed (e.g. an XML message); or as an *ii) Object* - the message is specified as an object, which is nothing more than a programming method where its arguments are the set of information that compose the message. The *Object* representation assumes that this object/method is held by both the sender and the recipient of the message, while the *Data element* representation does not,

2. *Messaging API:* the protocol might be application-specific or -independent,

3. *Initiation:* it defines how the message is transmitted between two nodes, which can be either server initiated (i.e. *push* mechanism) or client initiated (i.e. *pull* mechanism). In the *pull* mechanism, a client queries the server whenever it needs the data. The messaging protocol might support one or both mechanisms,

4. *Intermediation:* it specifies whether the messaging protocol offers the possibility to rely on an intermediate party to complete its transaction,

5. *Persistence:* it specifies whether the messaging protocol has a provision for data persistency. Two modes are defined: *i) Persistent* - the data is stored in the system until and unless it is retrieved, *2) Transient* - the data is stored in the system as long as the message is valid (i.e. until TTL expires),

6. *Subscription:* it specifies whether the messaging protocol proposes subscription mechanisms. Two mechanisms are considered: *i) interval-based* and *ii) event-based*. In some of these protocols, a callback address can be specified when creating the subscription, which is used as address pointer for sending new values of the subscribed data to the subscriber node,

7. *Self-Contained:* it specifies whether the message contains all the necessary information to enable the recipient to appropriately handle the message. In more concrete terms, the message contains all the relevant information such as the actions to be performed (read, write, subscription…), the message validity period (TTL), the mode of communication (asynchronous or synchronous), the callback address, *etc.,*

8. *Protocol agnostic:* it specifies whether the messaging protocol supports multiple underlying protocols, making it possible to transport the message using most "lower-level" protocols such as HTTP, SOAP, SMTP, FTP or similar protocols. It might also be possible to transport this message using files on USB sticks or other memory devices.

9. S*ynchronicity:* it defines whether the messaging protocol supports s*ynchronous* and *asynchronous* communications,

10. *Delivery-Guarantee:* it defines whether the messaging protocol has provision for the delivery guarantee (e.g. by returning responses).

11. *Piggybacking:* it defines whether the messaging protocol allows piggybacking a new request with a response (i.e. without dissociating the new request of the response). This is a crucial property both for real-time communications and to enable communications with nodes located behind a firewall,

12. *Multiple payloads:* it states whether the messaging protocol supports multiple payload formats.

3.2   Message Processing Model

This category deals with the message reception, i.e. how messages are processed by the recipient node and how responses are returned to the requester. Two criteria are defined:

13. *Processing result:* it defines how the requested information is returned to the requestor node. Three modes of response are defined: *i) single return value* - for every request, a single response is generated, *ii) single integrated return value* - the response contains the integrated value for the requested data (mode required when supporting subscriptions without callback), *iii) set of individual return value* - multiple response at different intervals are generated (mode required when supporting subscriptions with callback),

14. *Communication:* it defines how two nodes communicate once the request message is received. Two levels of communication are defined: *i)* Separate message, *ii)* callback address.

3.3   Message Failure Model

The third category describes what provisions the messaging protocol provides in case of failures and disruptions. Only one criterion is defined:

15. *Failure notification:* it defines whether the messaging protocol has a provision for failure notifications. Three main approaches are usually implemented: *Timeout of Acknowledgement, Reply with error message* or *Exception,*

3.4   Comparison Study

The five protocols previously introduced are compared in Table 1 based on the comparison framework. Results highlight that the QLM messaging protocol covers the vast majority of the properties and sub-properties (i.e. it often offers the possibility to choose between different options where possible). It can be observed that JMS also covers many of these properties, but fundamental ones like the *piggybacking* (property 11) or still the *Callback* functionality (property 14) are missing. However, such functionalities are of great importance to increase the scope of SOA applications.

The *piggybacking* functionality, which is only covered by QLM, is an important aspect to be considered in industrial applications. Indeed, organizations today take proactive measures to protect the security, confidentiality, and integrity of their data [5]. This inevitably leads to a challenging conflict between data security and usability; security making operations harder, when usability makes them easier. It is therefore necessary, in certain situations, to enable two-way communications through firewalls (e.g. for real-time control/maintenance) with the presence of conditions. The communication model consists to piggyback new QLM requests with the generated responses[4].

An aspect of prime importance considering complex product lifecycles is the support of *Multiple payloads* for being embedded in a message. The non-support of this functionality is particularly critical when many actors, corporations and systems are involved in the PLC. Indeed, the higher the product complexity, the higher the number of formats used to store information. In this regard, all protocols support this functionality.

Another important aspect is related to how protocol returns the *processed result.* Table 1 clearly states that QLM supports all possibilities (see property 13), while others do not. Once more, the faculty of QLM to propose a panel of possibilities proves its flexibility to face SOA environments and applications with diversified characteristics.

As emphasized in property 4 of Table 1, the *intermediation* layer is mandatory in the four protocols, while QLM provides the possibility to choose between using intermediation or not. This solves the issue of intermediation dependency and reinforces the claim that QLM messaging is highly appropriate for SOA implementations.

To conclude, let us add that the conceptual framework used in the QLM messaging protocol to subscribe a data (*cf.* property 6) is the Observer Design Pattern presented by [13] and applied according to [9] which signifies that a node can add itself as an *observer of events* that occur at another node. In this sense, QLM differs from e.g. JMS, which is based on the Publish/Subscribe (Pub/Sub) model. For many applications, the Observer and the Pub/Sub models can be used in quite similar ways.

---

[4] The condition, in this case, is that the node that performs the piggybacking must first be contacted by the node located behind the firewall.

However, the Pub/Sub model usually assumes the use of a "high-availability server", which the Observer pattern does not [7]. This is why the Observer model is more suitable for IoT applications where products might communicate with each other directly.

This comparison shows that the QLM messaging protocol covers a wide range of interfaces, which inevitably plays a leading role in making SOA techniques / algorithms sufficiently generic to be instantiated in new applications and systems, throughout the organization and product lifetimes [6]. Section 4 describes in greater details the QLM messaging protocol.

Table 1. Messaging protocol comparison based on 15 criteria[5]

n°	Category	Property	Sub-property	oBIX	ebXML	XMPP	JMS	QLM
1		Representation	Object				✓	
			Data Element	✓	✓	✓		✓
2		Messaging API	Application-specific	✓	✓			
			Application-indep.			✓	✓	✓
3		Initiation	Push		✓	✓	✓	✓
			Pull	✓	✓		✓	✓
4		Intermediation		✓	✓	✓	✓	✓[4]
5		Persistence	Transient					✓
	Message delivery model		Persistent	✓			✓	✓
6		Subscription	Interval-based	✓			✓	✓
			Event-based	✓	✓	✓	✓	✓
7		Self-contained		✓	✓		✓	✓
8		Protocol agnostic				✓	✓	✓
9		Synchronicity	Synchronous	✓	✓			✓
			Asynchronous	✓	✓	✓	✓	✓
10		Delivery-guarantee		✓			✓	✓
11		Piggy backing						✓
12		Multiple payloads		✓	✓	✓	✓	✓
13		Processing Result	Single return value	✓	✓	✓	✓	
	Message processing model		Single integrated...	✓				✓
			Set of individual...	✓				✓
14		Communication	Message ID	✓	✓	✓	✓	✓
			Callback					✓
15	Message delivery model	Failure Notification	Timeout of Ack.				✓	✓
			Error message	✓	✓	✓	✓	✓
			Exception					✓

---

[5] Unlike the four other messaging protocols, intermediation is not mandatory in QLM. However, QLM offers the possibility to use such functionality.

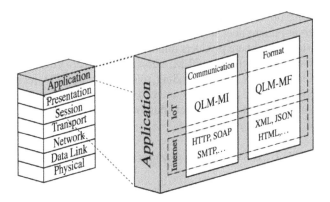

Fig. 2. QLM-MI & QLM-MF in the OSI model

4   QLM Messaging Standards

The origins of QLM messaging standards come from the work on developing a universal information system architecture for the IoT based on an open-source middleware software called DIALOG and its underlying product agent [10] and intelligent product concepts [19]. However, during the PROMISE research project, it became clear that a purely open-source based approach was not sufficient in an environment with many organizations whose existing software needed to communicate PLC data between each other. This eventually led to the creation of the QLM standards initiative.

4.1   QLM Messaging Interface

In the QLM world, communication between QLM nodes is done by using interfaces defined in QLM-MI. QLM-MI and QLM-MF are independent components as illustrated in Fig. 2. QLM messaging standards reside in the Application layer of the OSI model, where QLM-MI is specified in as a *Communication level* and QLM-MF is specified in as a *Format level*. Where the Internet uses HTTP for transmitting HTML-coded information mainly intended for human users, QLM is used for conveying lifecycle-related information mainly intended for automated processing by information systems. In the same way as HTTP can be used to transport payloads also in other formats than HTML, QLM can be used for transporting payloads in nearly any format (*cf.* Fig. 2). XML might currently be the most common text-based payload format due to its flexibility and structuring that provide more opportunities for complex data structures [26]. However other payload formats such as JSON, CSV can be used. As studied in section 3, the QLM messaging protocol offers a wide range of properties, whether regarding QLM-MI or QLM-MF. A cornerstone property of

QLM-MI is the concept of deferred retrieval of information, which corresponds to the subscription mechanisms. QLM-MI actually defines two variants of subscriptions[6]:

- *with callback address:* the data is sent using a QLM response at the requested interval. The interval "-1" means that it is event-based (i.e. the subscribed target has to push the value each time it publishes a new value),
- *without callback address*: the data can be retrieved (i.e. polled) by issuing a new QLM read query by specifying the subscription ID, which has been returned by the server when creating the subscription.

4.2   QLM Messaging Format

QLM-MF is defined as a simple ontology, specified using XML Schema, which is generic enough for representing "any" object and information that is needed for information exchange in the PLC, or in the IoT to be more generic. It is intentionally defined in a similar way as data structures in object-oriented programming as illustrated in Fig. 3. It is structured as a hierarchy with an "Objects" element as its top element (cf. row 6 of the XML message and the related-hierarchy in Fig. 3). The "Objects" element can contain any number of "Object" sub-elements (*cf.* row 7 and the hierarchy). The most important attribute of an "Object" is "type", which specifies what type of object is[7]. In this example, the messaging interface (i.e. QLM-MI) specifies that it is a write operation (see row 2), which concerns the object "Refrigerator" (see row 7) whose ID is given et row 8: *SmartFridge22334411*. In this example, only one object is contained by this message but, as mentioned previously, others might be included, such as the TV, the washing machine or any other object from the fridge's environment). "Object" elements can have any number of properties, referred to as InfoItem(s) (*cf.* row 9 and 12), which are in this example information about the fridge location and the thermostat value. An "Object" element can also have any number of sub-elements (e.g. a subpart of the fridge that should not be directly characterized by an "InfoItem" could, for instance, be characterized by an "Object").

In our example, the write request aims at changing values related to the fridge location and its thermostat, but more complex scenarios and applications could get the most out of this Format proposal due to the high generality of the Objects tree. In the QLM specification, pre-defined specialized types of "Object" are proposed as shell components, which will respect specific data structures or data models. For instance, the QLM Data Model initially conceived as the basis for the information model for the Product Data Knowledge Management/Decision Support System (PDKM/DSS) in PROMISE SOM, is now defined as a single object in the QLM specification, which can be used as an independent object in the hierarchy [25]. This makes the QLM messaging protocol truly independent of data models adopted by applications.

---

[6] A subscription request is a QLM read that includes, among others, an interval parameter in the message interface.

[7] An optional attribute called "udef" may be used for specifying the object class using the Universal Data Element Framework (UDEF; www.udef.com) taxonomy.

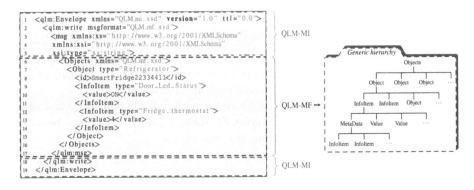

Fig. 3. QLM message format

5   Application Scope of the QLM Messaging Protocol

As stated previously, the QLM messaging protocol emerged out of real-life industrial applications where product information related to the entire PLC needed to be collected, exchanged and processed between several organizations. In this regard, QLM provides a wide range of interfaces and a flexible messaging format to let this happen. Fig. 4 illustrates how QLM enables to interconnect distinct organizations throughout the PLC, where various applications and actors take place. In this regard, two distinct scenarios are presented in sections 5.1 and 5.2. The first scenario is defined in BoL (Beginning of Life; *cf.* Fig. 4), while the second one takes place between MoL (Middle of Life) and BoL.

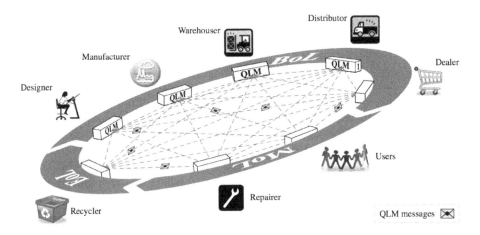

Fig. 4. QLM architectures allowing data exchange between organizations

5.1   Production Line of Car chassis

This scenario is a real case study carried out in the framework of the LinkedDesign EU FP7 project[8]. In this scenario, different actors work on a production line of car chassis as illustrated in Fig. 5: chassis are first moved from the oven to a press machine, and then to other operations. This process segment involves two robots to transfer the chassis from machine to machine (see robots 1 and 2 in Fig. 5). The actors involved in the manufacturing plan expressed, on the one hand, the need to check each chassis throughout the hot stamping process and, on the other hand, the need to define communication strategies adapted to each actor. To comply with these two requirements, first, scanners are added between each operation for the verification procedure (see scanners 1 and 2), and then the QLM messaging protocol is adopted to provide the types of interfaces required by the different actors. In more concrete terms, physical nodes from the production line have been augmented with the QLM messaging capabilities as depicted in Fig. 5, namely:

- the two scanners,
- a server in charge of collecting all the events generated by the scanners,
- the PDA of the line maintainer,
- the computer of the quality manager.

The next two sections define the appropriate communication strategies regarding each actor, which correspond to the dashed lines in Fig. 5 annotated: ① – ④.

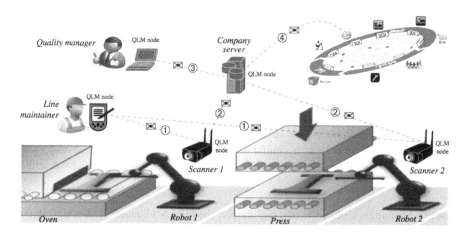

Fig. 5. Hot stamping process monitoring and control (BoL)

Line Maintainer. The line maintainer expressed the need of receiving all the verification events generated by scanners 1 and 2 so as to identify in real-time whether a problem occurs on a chassis. Accordingly, the line maintainer directly subscribes (via his PDA) to the events generated by scanners 1 and 2 (*cf.* communications ① in Fig.

---

[8] http://www.linkeddesign.eu

5). This is made possible by performing a QLM read query by *i)* setting the interval parameter to "-1" (this value means that the subscription is *event-based[9]*), *ii)* by including his own address as callback (i.e. the PDA's address) and iii) by setting the TTL parameter to "-1" (this value means that the validity period of the subscription is *forever[10]*). The parameters are defined in the message interface as shown in lines 1 and 2 of the subscription request provided in Fig. 6. Line 7 of this request indicates the name of the InfoItem that the manager subscribes, i.e. *StatusD*.

```
1   <qlmEnvelope xmlns="QLM_mi.xsd" version="1.0" ttl="-1">
2     <read msgformat="QLM_mf.xsd" interval="-1" callback="http:
          //207.46.130.1/ServletPDA">
3       <msg xmlns:xs="http://www.w3.org/2001/XMLSchema" xmlns:xsi="http:
          //www.w3.org/2001/XMLSchema-instance" xsi:type="xs:string">
4         <Objects xmlns="QLM_mf.xsd">
5           <Object type="Hot_stamping_machine">
6             <id>HotStamp1223</id>
7             <InfoItem class="StatusD"></InfoItem>
8           </Object>
9         </Objects>
10      </msg>
11    </read>
12  </qlmEnvelope>
```

Fig. 6. PDA subscribes (forever) to the InfoItem named *StatusD* on Scanner 1

The sequence diagram in Fig. 7(a) gives insight into the transactions resulting from this subscription. First, a response that contains the ID of the subscription is returned to the PDA. This ID is needed if the line maintainer wants to cancel this subscription. Then, each time a chassis passes under scanner 1, the subscribed InfoItem value (i.e. *StatusD*) is pushed to the PDA through a new QLM response. Based on these events, it is possible to develop scripts, for instance, to raise an alarm if a failure occurs on a chassis. Such an example is illustrated in Fig. 7(b) where chassis 3 has a default. However, the development of such scripts/tools is outside the scope of the QLM messaging protocol, but rely on the data obtained via that one. As the line maintainer subscribes to scanner events by specifying a callback address, the company server subscribes to these events in a similar manner (see communications ② in Fig. 5). Then, other people internal or external to the organization can access the subscribed data on the server.

Quality Manager. Unlike the line maintainer, the quality manager is not interested in receiving a continual flow of events from the scanners. His primary function is not to guarantee real-time control, but rather to deal with weekly or monthly evaluations (e.g. to estimate the failure rate over a period of time). Accordingly, the second type of subscription supported by QLM is more appropriate in this situation, which consists in retrieving (i.e. polling) one or several historical values on the server (*cf.* communication ③ in Fig. 5) by issuing a new QLM read query containing the sub-

---

[9] The node has to push the subscribed value every time this value is modified.

[10] The subscription is valid as long as the node initiator of the subscription does not cancel it.

scription ID. More concretely, the quality manager (via his computer) has to send a read query by setting the interval parameter to "-1" but, this time, without including a callback address, as illustrated with the argument "null" in Fig. 8 when the computer sends its subscription request to the server. Then, the targeted node (the server) is aware that the events generated by scanners 1[11] should be kept on the server as long as this subscription is valid. The quality manager can therefore issue a new read query by requesting one or several historical values, as depicted in Fig. 8.

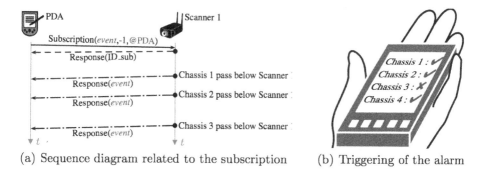

(a) Sequence diagram related to the subscription   (b) Triggering of the alarm

Fig. 7. Subscription with callback address performed by the line maintainer

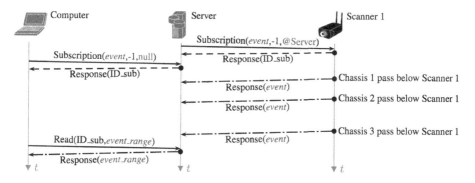

Fig. 8. Subscription with callback address performed by the quality manager

External Actors or Organizations. If other organizations in the PLC support the QLM messaging protocol and if the security rules allow them to access specific devices and information on the production line, they can further take advantage of the QLM interfaces to appropriately subscribe, modify or read specific device InfoItems. Such possibilities correspond to communication ④ in Fig. 5.

---

[11] As mentioned previously, the server subscribed the InfoItem *StatusD* to scanner 1 and 2 beforehand, which is a subscription of type *based-event* as depicted in Fig. 8.

5.2   Smart House Application

The scenario described in this section involves actors from two distinct PLC phases as depicted in Fig. 9, namely:

- in MoL: a user has equipped devices from his house (fridge, TV, *etc.*) with the QLM messaging protocol,
- in BoL: the fridge designer agreed with the user to collect specific fridge information over a certain period of the year (summer season),

The next two sections again describe the expectations of each actor and the appropriate QLM interfaces to set up.

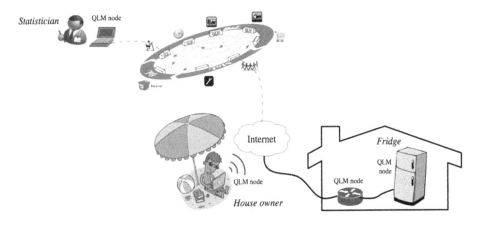

Fig. 9. Smart house (MoL) and related service provider (BoL)

House Owner. The house owner goes on vacation for a period of two weeks and would like to continuously monitor the fridge temperature during this period. Since the owner is not aware of all the features of each device, the RESTful QLM "discovery" mechanism can be used to retrieve the exact InfoItem(s) to be subscribed to. An example of how this can be achieved by using the Unix *wget* utility is shown in Fig. 10 with *wget 1*, which returns the set of devices from the house that are reachable using QLM. Then, the user can refine his research by retrieving the set of InfoItems related to a specific device (e.g. regarding *SmartFridge22334412*). Such a refinement is performed via *wget 2* in Fig. 10, which returns the list of InfoItems that can be accessed (for read, write, subscription, *etc.*) on *SmartFridge22334412*. Then, the user is able to send a subscription request to this fridge by including:

- the InfoItem to subscribe: it is *RefrigeratorIndoorTemperature*,
- the callback address: required by the fridge to send the subscribed value,
- the interval parameter: the user does not want to perform a subscription based-event as done in the previous scenario, but would like to get the indoor temperature every hour. Accordingly, the interval parameter is set to 3600 (expressed in seconds) in the message interface,

- the period of the subscription validity: this corresponds in our scenario to two weeks. Such a period is specified through the TTL parameter contained in the message interface.

Fig. 11 shows the evolution of the fridge temperature over the two weeks after subscribing it.

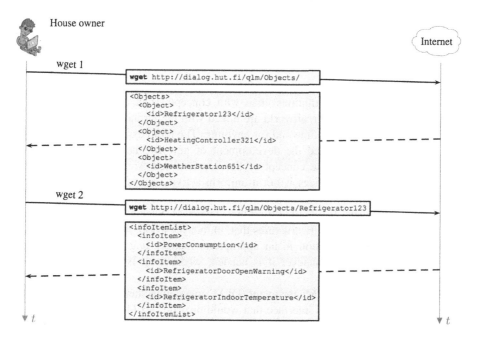

Fig. 10. RESTful mechanism used by the owner to retrieve device information

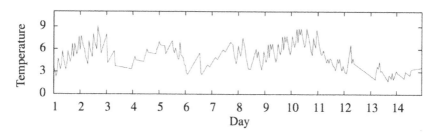

Fig. 11. Evolution of the fridge temperature resulting from the subscription

Fridge Designer. As with the quality manager in the previous scenario, the statistician is not interested in receiving a continual flow of events from the smart fridge. Rather, the statistician is interested in retrieving historical values over a period of time considering a panel of users, which will help to develop learning models and algorithms capable of representing the fridge behaviour into diversified environments and, accordingly, to enhance the design of the future generations of fridges. The statistician thus performs a subscription with a specific interval, without call back address and with TTL equal to 8035200 seconds. Indeed, as explained previously, this value corresponds to the validity period of the subscription which, in our case, corresponds to 3 months (June, July, August): $\approx 3 \times 31 \times 24 \times 3600 = 8035200$ seconds. Once the subscription is created, the statistician is able to retrieve a range of values over this period by sending a classical read query that includes the subscription ID (similar to that in Fig. 8).

6   Conclusion

New challenges and opportunities arise with concepts such as Internet of Things (IoT), where objects of the real world are linked with the virtual world, thus enabling connectivity anywhere, anytime and for anything. The rapid expansion of the IoT and the web technology enabled the development of new Business-to-Business (B2B) infrastructures based on the concept of Service Oriented Architecture (SOA). The technical provision of services is not a uniform activity but is skewed by prevalent technological solutions, which have often a lack of interfaces required by users or are designed in an isolated faction that limit their openness. Such a limitation has a direct impact on organizational infrastructures that, in today's world, need more than ever to exchange product information in an appropriate manner (e.g. to provide the right service or information, whenever it is needed, wherever it is needed, by whoever needs it).

The Quantum Lifecycle Management (QLM) messaging standards are proposed as a standard application-level interface that would provide flexible and a wide range of properties/interfaces, which aim to increase the SOA scope as well as the data exchange interoperability in the IoT. The QLM messaging properties are introduced in this paper and compared to properties of existing messaging protocols (e.g. JMS, oBIX, ebXML...). This comparison study shows that QLM covers more properties than those solutions, including fundamental ones like piggybacking, callback functionality or still the support of multiple payload formats. Several real-life implementations are also presented in this paper, which show how flexible the QLM messaging protocol is in the framework of Product Lifecycle Management.

Too often, SOA infrastructures give little attention to the hardware architecture by using "closed" technologies and applications (e.g. using private technologies or adding new software codes). Such designs both limit their interoperation with other applications and hinder their transposition to other domains/sectors. Messaging protocols such as QLM or similar ones (JMS, oBIX) are an essential step to design future SOA services and to enhance product lifecycle management. For instance, further research could be carried out to provide context-aware SOA, which will help to develop ideas for new environment-friendly products, or still to provide customized and advanced products and improve the customer experience.

References

1. Allard, J., Chinta, V., Gundala, S., Richard III, G.G.: Jini meets UPnP: an architecture for Jini/UPnP interoperability. In: Symposium on Applications and the Internet, pp. 268–275 (2003)
2. Atzori, L., Iera, A., Morabito, G.: The Internet of Things: A survey. Computer Networks 54(15), 2787–2805 (2010)
3. Bazjanac, V.: Building energy performance simulation as part of interoperable software environments. Building and Environment 39(8), 879–883 (2004)
4. Berre, A.J., Hahn, A., Akehurst, D., Bezivin, J., Tsalgatidou, A., Vermaut, F., Kutvonen, L., Linington, P.F.: State-of-the art for interoperability architecture approaches. InterOP Network of Excellence-Contract no.: IST-508 11 (2004)
5. Bishop, M.: What is computer security? IEEE Security & Privacy 1(1), 67–69 (2003)
6. Chen, D., Doumeingts, G., Vernadat, F.: Architectures for enterprise integration and interoperability: Past, present and future. Computers in Industry 59(7), 647–659 (2008)
7. Eugster, P.T., Felber, P.A., Guerraoui, R., Kermarrec, A.M.: The many faces of publish/subscribe. ACM Computing Surveys 35(2), 114–131 (2003)
8. Feuerlicht, G., Govardhan, S.: Soa: Trends and directions. In: Proceedings of the 17th International Conference on Systems Integration, pp. 149–154 (2009)
9. Främling, K., Ala-Risku, T., Kärkkäinen, M., Holmström, J.: Design Patterns for Managing Product Life Cycle Information. Communications of the ACM 50(6), 75–79 (2007)
10. Främling, K., Holmström, J., Ala-Risku, T., Kärkkäinen, M.: Product agents for handling information about physical objects. Report of Laboratory of Information Processing Science Series B, TKO-B 153(3) (2003)
11. Främling, K., Holmström, J., Loukkola, J., Nyman, J., Kaustell, A.: Sustainable PLM through intelligent products. Engineering Applications of Artificial Intelligence 26(2), 789–799 (2013)
12. Främling, K., Maharjan, M.: Standardized communication between intelligent products for the IoT. In: 11th IFAC Workshop on Intelligent Manufacturing Systems, São Paulo (Brazil), vol. 11, pp. 157–162 (2013)
13. Gamma, E., Helm, R., Johnson, R., Vlissides, J.: Design patterns: elements of reusable object-oriented software. Addison Wesley Publishing Company, Reading (1995)
14. Gibb, B.K., Damodaran, S.: ebXML: Concepts and application. John Wiley & Sons, Inc. (2002)
15. Gunasekaran, A.: Agile manufacturing: enablers and an implementation framework. International Journal of Production Research 36(5), 1223–1247
16. Haller, S., Karnouskos, S., Schroth, C.: The Internet of Things in an Enterprise Context. In: Domingue, J., Fensel, D., Traverso, P. (eds.) FIS 2008. LNCS, vol. 5468, pp. 14–28. Springer, Heidelberg (2009)
17. Hapner, M., Burridge, R., Sharma, R., Fialli, J., Stout, K.: Java message service. Sun Microsystems Inc., Santa Clara (2002)
18. Ishikawa, N., Kato, T., Sumino, H., Murakami, S., Hjelm, J.: Pucc architecture, protocols and applications. In: 4th IEEE Consumer Communications and Networking Conference, pp. 788–792 (2007)
19. Kärkkäinen, M., Holmström, J., Främling, K., Artto, K.: Intelligent products–a step towards a more effective project delivery chain. Computers in Industry 50(2) (2003)
20. Koronios, A., Nastasie, D., Chanana, V., Haider, A.: Integration Through Standards - An Overview of International Standards For Engineering Asset Management. In: 4th International Conference on Condition Monitoring, pp. 11–14 (2007)

21. MacKenzie, C.M., Laskey, K., McCabe, F., Brown, P.F., Metz, R., Hamilton, B.A.: Reference model for service oriented architecture 1.0. In: OASIS Standard (2006)
22. Neugschwandtner, M., Neugschwandtner, G., Kastner, W.: Web services in building automation: Mapping KNX to oBIX. 5th IEEE International Conference on Industrial Informatics 1, 87–92 (2007)
23. Nickul, D., Reitman, L., Ward, J., Wilber, J.: Service oriented architecture (SOA) and specialized messaging patterns. In: Adobe Systems Incorporated White Paper (2007)
24. O'reilly, T.: What is web 2.0: Design patterns and business models for the next generation of software. Communications & Strategies 65(1), 17–37 (2007)
25. Parrotta, S., Cassina, J., Terzi, S., Taisch, M., Potter, D., Främling, K.: Proposal of an interoperability standard supporting PLM and knowledge sharing. In: Prabhu, V., Taisch, M., Kiritsis, D. (eds.) APMS 2013, Part II. IFIP AICT, vol. 415, pp. 286–293. Springer, Heidelberg (2013)
26. Perera, C., Zaslavsky, A., Christen, P., Georgakopoulos, D.: Context aware computing for the Internet of Things: A survey. IEEE Communications Surveys & Tutorials (99), 1–41 (2013)
27. Saint-Andre, P.: Extensible messaging and presence protocol (XMPP): Core. Tech. rep., Core; IETF: Fremont, CA, USA (2004)
28. Tai, S., Rouvellou, I.: Strategies for integrating messaging and distributed object transactions. In: Coulson, G., Sventek, J. (eds.) Middleware 2000. LNCS, vol. 1795, pp. 308–330. Springer, Heidelberg (2000)

Proposition of an Analysis Framework to Describe the "Activeness" of a Product during Its Life Cycle

Part I: Motivations and Modelling

Yves Sallez

Univ. Lille Nord de France, F-59000 Lille, France
UVHC, TEMPO Lab, "Production, Services, Information" team
F-59313 Valenciennes, France
yves.sallez@univ-valenciennes.fr

Abstract. Recent advances in infotronics and communication have enabled the development of "intelligent" products. However, there is no clear and unanimous definition of an "intelligent" product. This paper reviews the different typologies used in this field and points out their main limitations. An analysis framework based on the concept of "activeness" is then proposed. This framework describes the situation of interaction between a collective of "active" products and a support system for a given function and a given phase of the life cycle.

Keywords: Typology, analysis framework, activeness concept, intelligent product.

1 Introduction

Over the last decade, the rise of embedded technologies (e.g. RFID, smart cards, wireless communication), as well as research in the field of ambient intelligence, have enabled the development of "intelligent" products that can fully interact with their environment. However, as outlined in reference [4], there is no clear and unanimous definition of an "intelligent" product, which is mainly due to the recent nature of the research in this area. In addition, several communities in different research fields, such as the Internet of Things, Manufacturing, Logistics and Product Life cycle Management, have worked on the concept of "intelligent" product or "smart" object which has led to many different typologies being proposed.

In this context, the aim of this paper is firstly to review the different typologies proposed in the literature and secondly to propose an adequate analysis framework.

This paper is organized in three sections. Section 2 provides a review of the different typologies associated with "intelligent" products and presents their limitations. Section 3 describes the "activeness" concept associated with a product and proposes an analysis framework available throughout the entire product life cycle. Finally, concluding remarks are provided in section 4.

T. Borangiu et al. (eds.), *Service Orientation in Holonic and Multi-Agent Manufacturing and Robotics*, Studies in Computational Intelligence 544,
DOI: 10.1007/978-3-319-04735-5_17, © Springer International Publishing Switzerland 2014

2   The State of the Art of "Intelligent" Product Typologies

As depicted in Fig. 1, two broad categories can be distinguished:

— "Individual": this category focuses on the product as an "individual" entity and is in turn divided into two major typology classes:

  (i)   mono-criterion typologies distinguishing broad classes of "intelligent" products according to their level of intelligence,
  (ii) multi-criteria typologies taking into account the different characteristics of an "intelligent" product (sensory capacities, location intelligence...).

— "Collective": this category tries to characterize the types of interactions which exist in a collective of "intelligent" products.

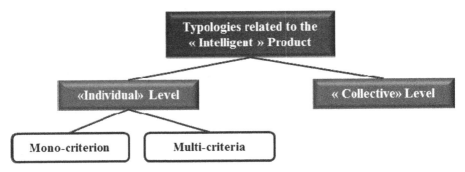

Fig. 1. Illustration of the different categories of typologies

2.1   Mono-criterion Typologies

The typologies presented below attempt to classify objects or products along a single axis / criterion corresponding to the level of intelligence.

Le Moigne typology [11]: In a very general context, Le Moigne distinguishes seven levels of intelligence, from a totally passive object (such as a raw material) to a self-completing object. Other intermediate levels can be distinguished, such as "informational" products (which have an informational capacity and can communicate) and "decisional" products (able to optimize their behaviour to achieve their goals). This first typology clearly shows that different levels of intelligence may be associated with an object. However, the decomposition into seven levels is a little tedious to apply and the concept of a self-finalized object seems a bit futuristic.

Wong typology [20]: On the basis of the definition introduced by McFarlane [12], the authors define two levels of Intelligent Product:

  • "Information-oriented Product": This type of product "provides" information for its environment. Only the first three features introduced by McFarlane are concerned (Possesses a unique identification, Is capable of communicating effectively with its environment, Can retain or store data about itself).

- "Decision-oriented product": This type of product possesses the other two characteristics mentioned in McFarlane's definition (Deploys a language to display its features, its production requirements, Is capable of participating in or making decisions relevant to its own destiny).

As outlined in [13], this typology is aimed primarily at describing objects using RFID technology in the fields of manufacturing and logistics. It does not specifically cover "intelligent" products with embedded processing capabilities.

Bajic typology [1]: Four categories of objects can be distinguished:

- "Data bearing": The object stores a minimal amount of data which is then accessed by external resources using barcode or RFID technology for example.
- "Information system pointer": RFID technology is used to identify the object and link it to an external information system (e.g. web page).
- "Service provider and requestor": This object has a dual role: passive when it responds to service requests from other players and active when it generates a query (service request) for other players.
- "Sensitive communicating object": Some sensors are associated with the object to provide information about the environment.

Kiritsis typology [8]: The author distinguishes four levels of intelligence:

- Level #1: Products with no physical embedded system (hardware or software).
- Level #2: Physical products equipped with simple sensors (e.g. thermostat) allowing them to interact with their environment.
- Level #3: Physical products with embedded sensors and data storage and processing capabilities. Embedded devices and software allow the product to adapt quickly to a changing environment.
- Level #4: Physical products equipped with PEIDS (Product Embedded Information Devices). PEIDs are devices (e.g. RFID tags, sensor networks, embedded processors) which allow products to assess internal changes, to communicate with other products and to implement decision-making processes.

This classification uses the Wong classification, but in a little more detail.

CERP-IOT typology [2]: In the book describing the work of European clusters in the field of the Internet Of Things, five levels of intelligence of "Things" are proposed. The fifth level refers to objects that can create, manage and destroy other objects and could self-replicate. Even if this futuristic vision does not seem appropriate for current applications, such a level may be useful in the promising context of nanotechnology.

2.2   Multi-criteria Typologies for an "Intelligent" Product

Several studies have introduced more detailed classifications taking into account several features of an "intelligent" product.

<u>Meyer typology</u> [13]: As illustrated in Fig. 2, the authors propose a typology using three axes describing respectively the level, location and aggregation level of intelligence.

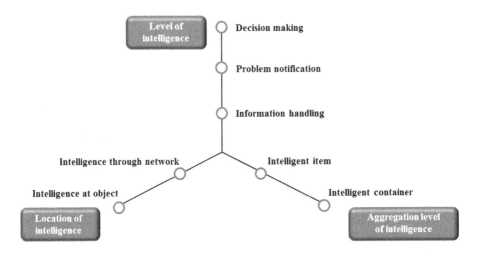

Fig. 2. Meyer Typology [13]

- The axis entitled "level of intelligence" is divided into three categories: Information handling, Problem Notification and Decision making. These three levels describe products that can "carry" information, generate alarms relating to their condition and undertake a decision-making process, respectively.
- The axis denoted "Location of intelligence" is divided into two classes depending on whether the intelligence is embedded in the product or external (accessible via a network).
- The axis named "Aggregation level of intelligence" is made up of two classes:

  (i) Intelligent Item: the product manages the information and / or decisions relating to itself. If several components are included, they cannot be distinguished as individual objects.
  (ii) Intelligent Container: the product is "aware" of its components and can serve as an intermediary (proxy device) for them.

The main interest of this typology is to clearly point out the level and the location of intelligence associated with the product. However, the various relationships that exist within an "Intelligent container" product are not sufficiently distinguished by the axis "Aggregation level of intelligence" with its two classes. The relational aspect within a group of "intelligent" products is not covered by this typology either.

Kawsar typology [7]: In the field of the Internet Of Things, the author defines a set of cooperating objects, such as an SOS (Smart Object System), and distinguishes three categories of "smart objects".

- "Stand-alone Smart Objects": they are self-contained objects independent of any infrastructure that can perceive their environment and make decisions autonomously. Awareness technologies and contextual services are typically embedded in the object.
- "Co-operative Smart Objects": they can communicate with their peers and share their "awareness" capabilities to achieve collective action. A storage application of chemical drums is given as an example of such a class [19].
- "Infra-structured Smart Objects": They are constituents of a larger system and cannot work individually. These objects are included in one or multiple applications to build a proactive system via a secondary infrastructure.

The SOS concept, and more particularly the notion of "Co-operative Smart Objects", is interesting to describe a collective of "intelligent" products. However, the relationships within a collective of "Co-operative Smart Objects" are not detailed.

Kortuem typology [10]: As shown in Fig. 3, the authors introduce a typology along three axes:

- "Awareness" defined as the ability of an object to sense, interpret and react to events and human activities that occur in the physical world.
- "Interaction" defined as the ability of an object to interact with the user in terms of input / output and control.
- "Representation" refers to the model associated with an object (in terms of programming abstraction).

Three classes of "Smart Object" are then defined:

- "Activity-Aware Smart Object": the object can store information relating to status and conditions of use. It does not have the capacity to interact with the user.
- "Policy-Aware Smart Object": it may interpret events and activities with respect to predefined organizational rules. It can generate alarms in the event of misuse by the operator.
- "Process-Aware Smart Object": it has the capacity to analyze the event and understand the organizational process to which it belongs. With a process model, it is able to provide the user with context-sensitive help.

This interesting characterization of "smart objects" is however very much centred on the interaction with the user and the relational aspects between products are not addressed.

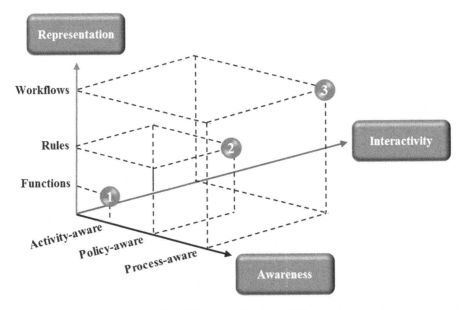

Fig. 3. Kortuem Typology [10]

<u>Sánchez López typology</u> [18]: The authors propose a classification of "smart objects" using letters. To represent a large range of "smart" objects, five letters are used to code their key features:

- "I" to identify and store all relevant data,
- "S" for the ability to sense its physical condition and its environment,
- "A" for the ability to operate internal or external devices,
- "D" for the ability to make decisions and participate in controlling other devices or systems,
- "N" for the ability to access information via a network (wired or wireless).

The "ISADN" taxonomy is a description of the functional characteristics of smart objects. As noted by the authors, all combinations of letters are not realistic. For example, an "N-smart" object has no real interest if it is not at least "IN-smart". Although in their paper the authors describe several situations of interaction between products, the "social" aspect is only represented through the "N" feature. In addition, the "D" characteristic does not distinguish different levels of decision-making. The "social" characteristic generally missing in the previous typologies is more largely dealt with in the typologies presented in the next section.

2.3   Typologies for a Collective of "Intelligent" Products

Few studies have been undertaken in the recent research field of the Internet Of Things to establish typologies for a collective of "intelligent" products.

Salkam typology [14]: The authors propose a taxonomy addressing collaborative context-aware systems. "A collaborative context-aware system (CCAS)" is defined as a "system that comprises a group of entities capable of sensing, inferring and actuating that communicate in order to achieve a common goal".

The proposed taxonomy distinguishes three axes: goal, approaches and means.

- Goal: Every CCAS aims to achieve a common goal. For example, a mission is undertaken through a form of collaboration, cooperation and/or coordination between entities.
- Approaches: Two approaches, with or without consensus, are distinguished:

  (i) In a "Consensus-free" approach, entities can make local decisions without negotiating a common decision. They can provide information (sensory data, fused information, contextual information) to facilitate local decision-making.
  (ii) In a "Consensus-based" approach, a common decision is required from the various entities involved, and so they are compelled to negotiate.

- Means: Seven ways of achieving this collaboration are proposed: ability to sense the environment, fusion of different information sources, ability to act on the environment, ability to build on, update and reason with the knowledge, ability to communicate and delegate tasks among neighbouring entities.

However, this taxonomy does not describe the links of authority that exist among entities. In our opinion, the concept of consensus remains a little too vague.

Iera typology [5]: The author has taken the typology introduced previously by the sociologist Fiske [3] and applied it to the context of the Internet Of Things. The four classes highlighted by Fiske are revisited in order to characterize the different relationships between entities:

- Communal Sharing: This class characterizes the behaviour of objects which are not relevant individually but have collective relevance when they act in a swarm.
- Equality Matching: This class can represent all forms of information exchange between objects considered as equals (as providers or consumers of information or services).
- Authority ranking: This type of relationship is established between objects of different complexity exhibiting different hierarchical levels.
- Market pricing: This class represents situations of interaction where objects work together to obtain mutual benefits.

2.4   Summary of the Different Typologies

The aforementioned typologies have several limitations:

— They are generally relevant for a specific phase of the life cycle and do not describe the impact on other phases. This limitation is important in a context of Closed-loop Product Lifecycle Management [6, 8], which requires the management of information flows upstream and / or downstream between the phases.
— They consider an object or a product as "all active" or "all passive". In our opinion, it is necessary to distinguish the level of activeness of a product according to the function studied: a product can be passive for a given function and active for another.
— Too few take into account the interactions between products and therefore do not represent the dynamics of a collective of products. This is disadvantageous for many applications.

These different limitations have motivated us to propose a more precise analysis framework which is described in the next section.

3   The Concept of "Active" Product and the Associated Analysis Framework

After a brief summary of the concept of "active" product, the proposed analysis framework is presented.

3.1   The Concept of "Active" Product in Brief

In previous studies, our team has introduced the concept of "activeness" associated with a product [15, 16]. Via an increase in its informational, communicational and decisional capacities, an "active" product (AP) is considered as an entity capable of interacting with the different support systems (e.g. manufacturing system, supply chain) during the successive phases of its life cycle. This concept contrasts with traditionally passive product behaviour where the product is not able to take the initiative in relation to a support system. The minimal activeness of an AP is to trigger events: the product is able to identify its state, compare its state with the desired one, and send information (e.g. warnings) when certain conditions are met. Obviously, more complex activities can be considered (e.g. memorization, communication, negotiation and learning).

In fact, as illustrated in Fig. 4, it is a collective of "active" products which interacts with the support system (1). To achieve this activeness, an AP can interact with other APs (e.g. information exchange, delegation of tasks) (2).

The application of the "activeness" concept leads to two main benefits:

— The improved performance of the pair "AP - Support System" for a specific function during a given phase of the life cycle;
— The improved quality of the information flows between the different phases of the life cycle, via the different forward and backwards loops, according to a closed-loop Product Lifecycle Management philosophy (3).

However, the "activeness" associated with a product must not be analysed as a whole but rather function by function. Indeed, a product can be "passive" for a function f_i and "active" for another function f_j. Taking into account this requirement and the limitations of the previous typologies, an analysis framework is proposed in the next section.

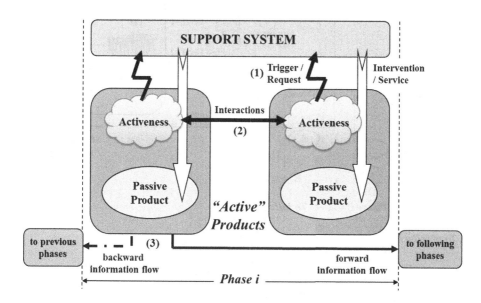

Fig. 4. Illustration of the "activeness" associated with a product for a given phase of its life cycle

3.2   Proposed Analysis Framework

The proposed analysis framework aims to represent the situation of interaction between a collective of APs and the support system for a given function. As depicted by the example in Fig. 5 (function f_i in Distribution phase), the analysis of a function is decomposed into two views (functional and organic) which are divided into five sections. The latter are highlighted by circled numbers in Fig. 5.

1. "Life cycle": This section characterizes:

   - The current phase (De: Design, M: Manufacturing, Di: Distribution, U: Use or R: Recycling),
   - Any potential link with the previous phases,
   - Any potential link with the next phases.

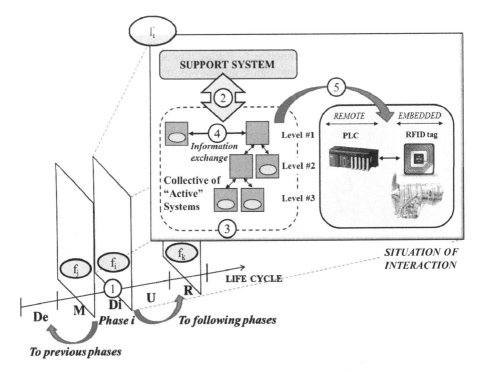

Fig. 5. Example of the five sections exhibited by the analysis framework

2. "Class": As shown in Fig. 6, four classes can be distinguished for a given function according to the share of the collective of "active" systems in the decisional process:

- Class #1 corresponds to a group of "passive" systems. The support system makes all decisions pertaining to the function. It collects the information itself and manages the entire decision-making process.
- Class #2 is representative for a group of "active" systems taking an initiator role in the decisional process. The collective plays a triggering role but the other steps of the decisional process are later assumed by the support system.
- Class #3 characterizes a group of "active" systems which manages the entire decisional process. The support system is only used to provide the services necessary to carry out the function.
- Class #4 characterizes a collective in which the "active" systems are self-sufficient. They are able to provide the services obtained from the support system in the previous classes.

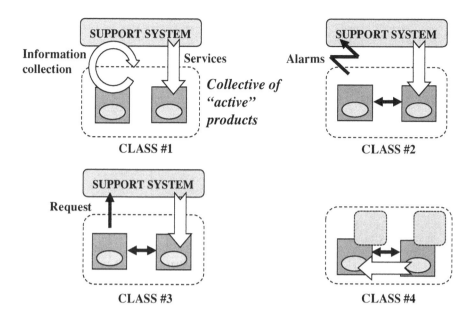

Fig. 6. The four different classes distinguished in the analysis framework

3. "<u>Description of the collective</u>": Each system is composed of a set of sub-systems which may themselves be decomposed into sub-systems. The activeness associated with a system results from the activeness of other "active" subsystems located on the successive levels. According to the holonic principle introduced by Koestler [9], this system architecture can be considered as a holarchy. For each pair (f_i, S_i) of the holonic architecture, a holon H_i is defined. For example, in Fig. 7 below, the pair (f_{13}, S_{13}) pertains to holon H_{13}.

A holon Hi is characterized by two main properties:

— Autonomy: At level #i, a holon H_i assumes the decisional process associated with the function f_i autonomously.
— Cooperation: At level #i, the H_i holon makes decisions interacting with other holons on the same level i, and with holons at levels i-1 and i+1 of the holonic architecture.

At each level of the decomposition, a type can be associated with each system:

— MOL (Molecular): decomposed into sub-systems,
— ATO (Atomic): cannot be decomposed.

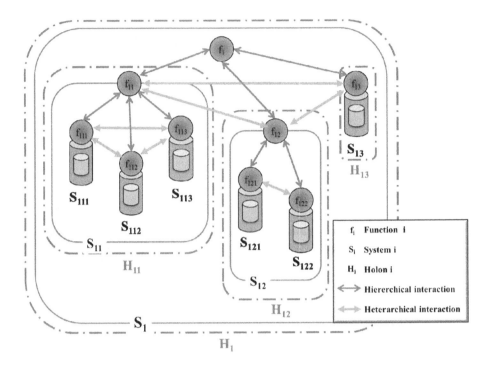

Fig. 7. Example of holarchy

4. "Description of the interactions": This section describes the interactions among the various "active" systems. Four relationships can be distinguished:

- NUL (Non-existent): there is no interaction between the "active" systems. The term "collective" only evokes a juxtaposition of individual systems.
- WAL (Without Authority Link): there are interactions between informational systems, but no authority link exists between them. For example, systems interact to exchange information on their respective contexts.
- VAL (Variable Authority Links): systems interact, with variable authority links between them. In this class, heterarchical links exist with mechanisms such as delegation or contract net protocol.
- FAL (Fixed Authority links): In this category, fixed authority links exist between the systems, exhibiting a hierarchy.

5. "Implementation": Inspired from the Meyer typology [13], two types can be distinguished in relation to the location of intelligence:

- Embedded (EMB): The decisional part is physically connected to the physical part of the system by a kinematic chain (temporary or permanent).
- Remote (REM): The decisional part is not physically connected to the physical part of the system.

4   Conclusion

After reviewing the various typologies of "intelligent" products, the different limitations were outlined. Among them, the difficulty in describing the dynamics of a collective of "intelligent" products has been highlighted. In general, the majority of the typologies proposed do not deal with the interactions that exist in a collective of products.

Following a brief presentation of the notion of "activeness", an analysis framework has been proposed to describe the "activeness" of a product throughout its life cycle. This framework focuses on the analysis of the interaction situation between a collective of "active" products and the support system for a given function. The framework contains two views (functional and organic), which can be divided into several sections that describe the different facets of the interaction situation.

However, this analysis framework needs to be tested in several applicative contexts to demonstrate its genericity. Its ability to describe the relationships between the collective of products and the support system, and to describe the behaviour of a hierarchy of "active" systems, must be validated. For this purpose, two applications are presented in the paper [17].

References

1. Bajic, E.: Ambient Networking for intelligent objects management, mobility and services. Seminar Institute for Manufacturing - IfM, University of Cambridge (2004)
2. Sundmaeker, H., Guillemin, P., Friess, P., Woelfflé, S. (eds.): CERP-IOT (Cluster of European Research Projects on the Internet of Things): Vision and Challenges for Realising the Internet of Things (2010)
3. Fiske, A.P.: The four elementary forms of sociality: Framework for a unified theory of social relations. Psychological Review 9(4), 689–723 (1992)
4. Främling, K., Loukkola, J., Nyman, J., Kaustell, A.: Intelligent Products in Real-Life Applications. In: Benyoucef, L., Trentesaux, D., Artiba, A., Rezg, N. (eds.) International conference on Industrial Engineering and Systems Management (IESM 2011), I4E2, Metz, pp. 1444–1453 (2011)
5. Iera, A.: The Social Internet of Things: from objects that communicate to objects that socialize in the Internet. In: Proceedings of 50th FITCE International Congress, Palermo, Italy (2011)
6. Jun, H.-B., Kiritsis, D., Xirouchakis, P.: Research issues on closed-loop PLM. Computers in Industry 58(8-9), 855–868 (2007)
7. Kawsar, F.: A document based framework for user centric Smart Object Systems, PhD in Computer Science. Waseda University, Japan (2009)
8. Kiritsis, D.: Closed-loop PLM for intelligent products in the era of the internet of things. Computer-Aided Design. 43(5), 479–501 (2011)
9. Koestler, A.: The Ghost in the Machine. Hutchinson, London (1967)
10. Kortuem, G., Kawsar, F., Fitton, D., Sundramoorthy, V.: Smart objects as building blocks for the internet of things. Internet Computing IEEE 14(1), 44–51 (2010)
11. Le Moigne, J.L.: La théorie du système général – théorie de la modélisation. Ed. Presses universitaires de France, 1ère éd. 1977, 4ème éd (1994)

12. McFarlane, D., Sarma, S., Chirn, J.L., Wong, C.Y., Ashton, K.: Auto id systems and intelligent manufacturing control. Engineering Applications of Artificial Intelligence 16(4), 365–376 (2003)
13. Meyer, G.G., Främling, K., Holmström, J.: Intelligent Products: A survey. Computers in Industry 60(3), 137–148 (2009)
14. Salkham, A., Cunningham, R., Senart, A., Cahill, V.: A Taxonomy of Collaborative Context-Aware Systems. In: Proceedings of the CAISE 2006 Workshop on Ubiquitous Mobile Information and Collaboration Systems, UMICS 2006, Luxemburg, pp. 899–911 (2006)
15. Sallez, Y., Berger, T., Deneux, D., Trentesaux, D.: The lifecycle of active and intelligent products: The augmentation concept. International Journal of Computer Integrated Manufacturing 23(10), 905–924 (2010)
16. Sallez, Y.: The Augmentation Concept: How to make a Product "Active" during its Life Cycle. In: Borangiu, T., Thomas, A., Trentesaux, D. (eds.) Service Orientation in Holonic and Multi-Agent Manufacturing Control. SCI, vol. 402, pp. 35–48. Springer, Heidelberg (2012)
17. Sallez, Y.: Proposition of an analysis framework to describe the "Activeness" of a product during its life cycle - Part II: Method and applications. In: Proceedings of SOHOMA (Service Orientation in Holonic and Multi-Agent Manufacturing Control), Valenciennes (2013)
18. Sánchez López, T., Ranasinghe, D.C., Patkai, B., McFarlane, D.: Taxonomy, technology and applications of smart objects. Information Systems Frontiers 13(2), 281–300 (2011)
19. Strohbach, M., Gellersen, H., Kortuem, G., Kray, C.: Cooperative artefacts: Assessing real world situations with embedded technology. In: Mynatt, E.D., Siio, I. (eds.) UbiComp 2004. LNCS, vol. 3205, pp. 250–267. Springer, Heidelberg (2004)
20. Wong, C.Y., McFarlane, D., Zaharudin, A.A., Agarwal, V.: The Intelligent Product Driven Supply Chain. In: IEEE International Conference on Systems, Man and Cybernetics, Hammamet, Tunisia (2002)

Proposition of an Analysis Framework to Describe the "Activeness" of a Product during Its Life Cycle

Part II: Method and Applications

Yves Sallez

Univ. Lille Nord de France, F-59000 Lille, France
UVHC, TEMPO Lab, "Production, Services, Information" team
F-59313 Valenciennes, France
yves.sallez@univ-valenciennes.fr

Abstract. This paper proposes application guidelines and two case studies for the analysis framework presented in a previous paper. This analysis framework describes the situations of interaction between a collective of "intelligent / active" products and a support system for a given function. This analysis framework contains two views (functional and organic), which can be divided into several sections that describe the different facets of the interaction situation. After presenting the application guidelines, two case studies, undertaken to test and validate the analysis framework, are reported. The first concerns the "product-driven" control of a real manufacturing cell. The resource allocation and product routing functions were more particularly studied. The second application concerns the advanced diagnosis of complex transportation systems in a railway context.

Keywords: Analysis framework, activeness concept, intelligent product, manufacturing, diagnosis.

1 Introduction

Application guidelines and two case studies for the analysis framework presented in [13] are proposed in this paper. The purpose of this framework is to describe a situation of interaction between a collective of "intelligent / active" products and a support system for a given function. The framework contains two views (functional and organic). The latter are divided into several sections and describe the different facets of the interaction situation (structure of the collective, interactions and implementation).

This paper is organized in four parts. Section 2 provides guidelines for applying the analysis framework. The recursion of the analysis approach is more particularly outlined. Section 3 describes the first application of the analysis framework in a manufacturing context. In section 4, another application is detailed in the field of railway equipment diagnosis. Concluding remarks and prospects for future research are provided in section 5.

T. Borangiu et al. (eds.), *Service Orientation in Holonic and Multi-Agent Manufacturing and Robotics*, Studies in Computational Intelligence 544,
DOI: 10.1007/978-3-319-04735-5_18, © Springer International Publishing Switzerland 2014

2   Guidelines for the Application of the Analysis Framework

This section provides guidelines for the application of the proposed analysis framework and constitutes a preliminary to the two applied studies. The analysis framework
presented in [13] is summarized in Fig. 1. The different options for each point of view
and for each section are indicated in brackets.

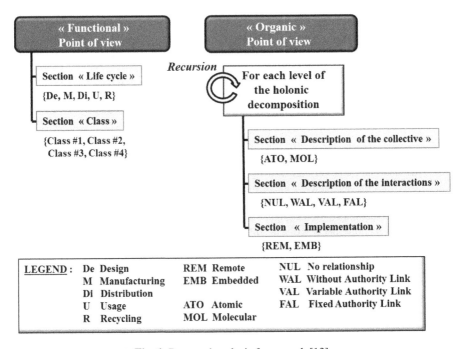

Fig. 1. Proposed analysis framework [13]

The proposed guidelines for the application of the analysis framework are described
in Fig. 2. Seven steps are considered:

— Firstly, the main function to be analysed must be identified.
— After this first step, two successive steps concern the "functional" point of view:

  • In step #2, the phase of the life cycle concerned by the function must be speci-
    fied.
  • The relationship between the "active" product and the support system must then
    be analysed in step #3 to define the class of the interaction situation.

— Step #4 constitutes a preliminary to the detailed description of the organic point of
  view. The different levels of the holonic decomposition are indicated, if they exist.

If there is a possible decomposition, the three following steps must be performed
for each level.

— Step #5 is dedicated to describing the structure of the collective. If there is a possible decomposition, then the "active" system is called Molecular (denoted MOL), otherwise it is called Atomic (denoted ATO).

— In step #6, the existing interactions between the various "active" systems of the level considered are detailed.

— Finally, in step #7 the implementation used to support the activeness is specified for the level concerned.

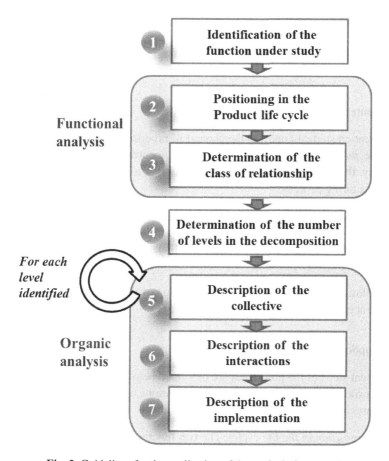

Fig. 2. Guidelines for the application of the analysis framework

For illustrative purposes, Fig. 4 in section 4 can be consulted. In this example, three levels of decomposition are considered. For more concision, the situation of interaction can be described by a shorthand notation. The example in Fig. 4 can be expressed as: [U, Class #3, [MOL, NUL, EMB], [MOL, WAL, EMB], [ATO, WAL, EMB]].

The next two sections of the paper describe the application of these guidelines to the manufacturing and in-use phases. The focus is put on the ability of the analysis framework to describe the functions addressed in the different studies. The latter are summarized and those interested can consult the references provided to obtain more details.

3   Application of the Analysis Framework in a Manufacturing Context

The first application of the proposed analysis framework in a manufacturing context, via experiments on a real flexible manufacturing cell, is presented in this section.

3.1   Context and Functions under Study

The application studied concerns the context of product-driven manufacturing control in flexible manufacturing systems [6, 8, 14]. According to this innovative control philosophy, the "active" product (AP) makes decisions concerning its "trajectory" in the manufacturing process.

According to the first step of the proposed guidelines, two main functions are considered:

— The function f_1 is responsible for resource allocation,
— The function f_2 is responsible for product routing.

The following section describes the application of the other steps of the guidelines for different experiments.

3.2   Application of the Guidelines

The different steps of the guidelines are described below for the two functions and for different experiments. For each step, differentiations are made to highlight the common features and the particularities of the experiments.

— Step #2 *(position in the life cycle)*: The two functions under study are located in the "Manufacturing (M)" phase of the life cycle,
— Step #3 *(relationship with the support system)*: The relationship differs according to the different experiments and is detailed below.
— Step #4 *(number of levels in the decomposition)*: The AP analysed does not contain any "active" sub-systems. A single level is then considered.

According to the previous step, the organic analysis is performed on a single level.

— Step #5 *(Description of the collective)*: The option "ATO" is retained because the AP cannot be decomposed into other "active" sub-systems.

— Step #6 *(Description of the interactions)*: Different options are considered according to the experiments, as detailed below.
— Step #7 *(Description of the implementation)*: To avoid the classic limitations (vulnerability to failure, difficulty in increasing scale) of a centralized "remote" implementation, an "embedded" solution was retained to support the activeness of the product [2] (see Fig. 3). For the analysis framework, the option retained was thus "EMB".

Fig. 3. "Active" product implementation

Steps #3 and #6 are detailed below for each experiment:

— In the first experiment [11], the product has a list of services it manages and it triggers the support system which controls the allocation decision process.
  • For the function f_1 (allocation), the interaction situation is clearly class #2. There is no interaction between the APs for the allocation process which leads to a "NUL" option for the interaction. The situation of interaction for the function f_1 is then characterized as: [M, Class #2, [ATO, NUL, EMB]].
  • Product routing in the cell is accomplished using a stigmergy mechanism allowing indirect exchanges of information between products (via the environment). The support system is not solicited (class #4 situation). No authority link exists, so the option retained for the interaction is "WAL". The situation of interaction for the function f_2 (routing) is then characterized as: [M, Class #4, [ATO, WAL, EMB]].

— In the second experiment [12], the allocation and routing processes were supported respectively by a traditional contract net protocol and an embedded Dijkstra algorithm that finds the shortest time-path to the chosen resource.
  • For the function f_1, the contract net protocol implies an interaction with the support system, but the decision is made by the AP thus leading to a class #3 situation. There is no interaction between the APs which leads to a "NUL" option for

the interaction. The situation of interaction for the function f_1 is then characterized as: [M, Class #3, [ATO, NUL, EMB]].

- For the function f_2, as in the first experiment, the support system is not solicited (class #4 situation). Interactions exist among APs: travel times measured by each AP are reused to update the graph used by Dijkstra's algorithm. No authority link exists leading to the "WAL" option. The situation of interaction for the function f_2 is then characterized as: [M, Class #4, [ATO, WAL, EMB]].

— In a third experiment [10, 15], the same mechanism of interaction, based on attractive potential fields, was used for the two functions. As depicted in Fig. 4, resources emit potential fields to attract APs according to the services offered. As the APs exploit the potential fields emitted by the support system's resources, the two functions are characterized by a class #3 situation. Travel times measured by each AP are reused to update the potential fields and so the "WAL" option is retained. The situation of interaction for the two functions f_1 and f_2 is characterized as: [M, Class #3, [ATO, WAL, EMB]].

— In a final experiment [1], the concept of *role* is used to support dynamic service allocation among products. A delegation mechanism introduces a master-slave relationship, and variable authority links (VAL option) are established between APs. The situation of interaction for the function f_1 is then described as: [M, Class #3, [ATO, VAL, EMB]].

3.3 Overview of the Framework

In this application field, the main contribution of the proposed analysis framework is to describe the dynamics of a collective of APs. For each experiment analysed in the previous section, the analysis framework has allowed the following relationships to be easily described:

— the relationships between the collective of APs and the support system,
— the relationships within the collective itself.

The analysis framework particularly enables self-organized systems (class #4 systems), which are rarely dealt with by typologies relating to "intelligent" products, to be described.

4 Application of the Analysis Framework in a Condition-Based Maintenance Context

This section is dedicated to the application of the analysis framework to study an AP in "Use" phase and aims to validate the holonic decomposition mechanism proposed in the framework. Contrary to the previous application in which the product was considered as atomic, the activeness of the product is here dependent on the activeness of several sub-systems.

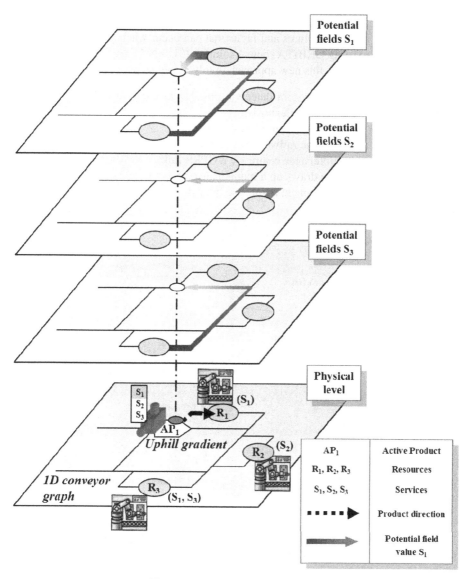

Fig. 4. 1D field emitted by resources

4.1   Context and Function under Study

One of the most promising avenues for the development of APs concerns the predictive maintenance of complex systems in use phase [3, 7]. In the context of railway transportation, the PSI team is involved in the FUI SURFER project (SURveillance active FERroviaire, translated as "active train monitoring"), launched in June 2010 and led by Bombardier Transport.

The aim of the SURFER project is to provide a more advanced solution for the online diagnosis of incipient failures and faults that can occur when a train is in service, in addition to the existing ORBITA remote solution [9].

As illustrated in Fig. 5, this new approach involves four steps:

1. An embedded diagnosis system detects a possible problem in the door operating mechanism and elaborates a diagnosis;
2. Diagnosis data are sent to the central train controller;
3. Once the train arrives at the railway station, the train controller sends the diagnosis data to the remote maintenance centre via a GPRS link;
4. The maintenance centre draws up a maintenance schedule to carry out work on the potentially defective equipment before the breakdown.

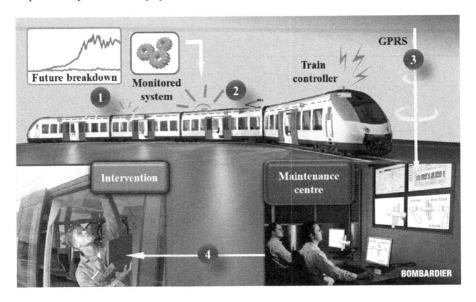

Fig. 5. "Active" train monitoring

According to the first step of the proposed guidelines, the function f_1 under study is responsible for executing the advanced diagnosis of any failures in a complex transportation system. It is the only function considered in this part.

The following section describes the application of the other steps in the guidelines.

4.2   Application of the Guidelines

The different steps of the analysis are described below for the function f_1:

— Step #2 *(position in the life cycle)*: The function under study is located in the "In-Use (U)" phase of the lifecycle,

— Step #3 *(relationship with the support system)*: The diagnostic process is performed on board the train but information needs to be exchanged with the support system (maintenance centre). The relationship is thus class #3.
— Step #4 *(number of levels in the decomposition)*: The system considered (complex transportation system) can be decomposed, through its hierarchical structure, into a set of interconnected subsystems. According to this hierarchical structure, the diagnosis is performed using decentralized recursive diagnosis structures [5]. As illustrated in Fig. 6, three levels can be distinguished: train, car and door.

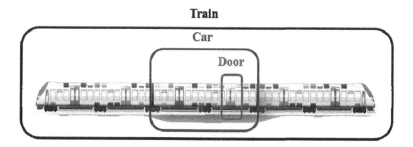

Fig. 6. The three levels of the holonic decomposition

As previously explained in the guidelines in section 2, steps #5, #6 and #7 must be performed for each level:

Level #1 (Train):

— Step #5 *(Description of the collective)*: The option "MOL" is retained because the activeness diagnosis results from the activeness of the other "active" subsystems located at "car" level.
— Step #6 *(Description of the interactions)*: Trains do not interact with each other. Consequently the "NUL" option is retained.
— Step #7 *(Description of the implementation)*: The accuracy of the diagnosis is closely linked to the quality of the system observations, so the diagnosis has to be executed as close as possible to each subsystem of the hierarchical structure. For the analysis framework, the option retained is thus "EMB".

Level #2 (Car):

— Step #5 *(Description of the collective)*: The option "MOL" is retained because the activeness diagnosis results from the activeness of the other "active" subsystems located at "door" level.
— Step #6 *(Description of the interactions)*: Two different implementations were performed in references [4, 5]:

  • In reference [5], no interaction exists among the diagnosticians at the same level. The "NUL" option is thus retained.

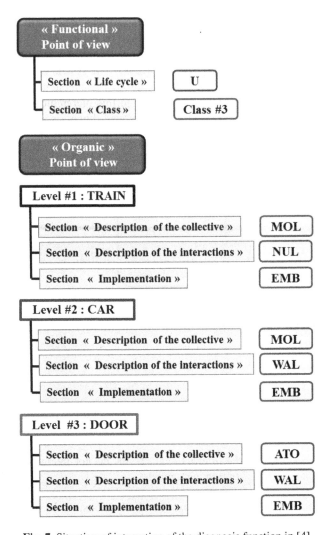

Fig. 7. Situation of interaction of the diagnosis function in [4]

- In reference [4], the diagnosticians exchange information about their respective contexts leading to a decrease in false alarms. The "WAL" option is thus retained.

— Step #7 *(Description of the implementation)*: as with the previous level, "on-board" diagnosis implies the "EMB" option.

Level #3 (Door): the analysis is the same as for level #2, except for step #5. This level is in fact the last in the decomposition and so logically the "ATO" option is chosen.

In summary, according to the two different implementations, two situations of interaction can be distinguished for the diagnosis function:

— In reference [5], the situation of interaction is characterized as follows:
  [U, Class #3, [MOL, NUL, EMB], [MOL, NUL, EMB], [ATO, NUL, EMB]].
— In reference [4], the situation is characterized as:
  [U, Class #3, [MOL, NUL, EMB], [MOL, WAL, EMB], [ATO, WAL, EMB]].
  This situation of interaction is illustrated in Fig. 7.

4.3   Overview of the Analysis Framework

This application outlines the pertinence of the analysis framework that allows:

— The different diagnosis levels to be described from a recursive point of view,
— The different interactions supporting the collaboration between "active" systems to be described.

Via the holonic decomposition principle, complex systems can be described using the analysis framework proposed. Most of the typologies associated with "intelligent" products do not enable such a detailed description.

5   Conclusion and Prospects

The aim of this paper was to demonstrate the pertinence of the analysis framework proposed in [13]. First, guidelines for the application of the analysis framework were presented. These guidelines were applied to two fields of application.

The first in a manufacturing context has exhibited the ability of the framework to describe the relationships between the collective of products and the support system and within the collective itself.

In a context of diagnosis of complex transportation systems, the second application has highlighted the ability of the framework to describe the behaviour of a hierarchy of "active" systems.

To improve our analysis framework some prospects are envisaged:

— A short-term prospect aims at developing a graphical representation, providing an efficient overview for the analysis of complex systems (with multiple decomposition levels). This type of representation can help the user manage the inherent complexity of such systems more rapidly.
— A second prospect aims at developing a dedicated software tool to help the user apply the guidelines. This software tool will exploit the previous graphical representation. Some functionalities of this software will allow a more user-friendly comparison of different situations of interaction.

Acknowledgments. This research was financed by the Inter-ministerial Fund (FUI) and the Nord/Pas-de-Calais Region, and sponsored by the i-Trans competitiveness cluster. The author gratefully acknowledges the support of these institutions.

References

1. Adam, E., Berger, T., Sallez, Y., Trentesaux, D.: Role-based manufacturing control in a Holonic Multiagent System. International Journal of Production Research 49(5), 1455–1468 (2011)
2. Berger, T., Sallez, Y., Valli, B., Gibaud, A., Trentesaux, D.: Semi-Heterarchical allocation and routing processes in FMS control: a stigmergic approach. Journal of Intelligent & Robotics Systems 58(1), 17–45 (2010)
3. Främling, K., Loukkola, J., Nyman, J., Kaustell, A.: Intelligent Products in Real-Life Applications. In: Benyoucef, L., Trentesaux, D., Artiba, A., Rezg, N. (eds.) International conference on Industrial Engineering and Systems Management (IESM 2011), I4E2, Metz, pp. 1444–1453 (2011)
4. Le Mortellec, A., Clarhaut, J., Sallez, Y., Berger, T., Trentesaux, D.: Embedded cooperative holarchy for diagnosing complex moving system. In: Borangiu, T., Dolgui, A. (eds.) IFAC Papers Online Proceedings Volume of the 14th IFAC Symposium on Information Control Problems in Manufacturing (INCOM), Bucharest, Romania (2012)
5. Le Mortellec, A., Clarhaut, J., Sallez, Y., Berger, T., Trentesaux, D.: Embedded holonic fault diagnosis of complex transportation systems. Engineering Applications of Artificial Intelligence 26, 227–240 (2013)
6. McFarlane, D., Sarma, S., Chirn, J.L., Wong, C.Y., Ashton, K.: The intelligent product in manufacturing control and management. In: 15th IFAC Triennial World Congress, Barcelona, Spain (2002)
7. Meyer, G.G., Främling, K., Holmström, J.: Intelligent Products: A survey. Computers in Industry 60(3), 137–148 (2009)
8. Morel, G., Valckenaers, P., Faure, J.M., Pereira, C.E., Diedrich, C.: Manufacturing plant control challenges and issues. Control Engineering Practice 15, 1321–1331 (2007)
9. Orbita-BT: Data monitoring Cuts rail running costs. Journal Professional Engineering. 19(20), 50–53 (2006)
10. Pach, C., Bekrar, A., Zbib, N., Sallez, Y., Trentesaux, D.: An Effective Potential Field Approach to FMS Holonic Heterarchical Control. Control Engineering Practice 20(12), 1293–1309 (2012)
11. Sallez, Y., Berger, T., Trentesaux, D.: A stigmergic approach for dynamic routing of active products in FMS. Computer in Industry 60(3), 204–216 (2009)
12. Sallez, Y., Berger, T., Raileanu, S., Chaabane, S., Trentesaux, D.: Semi-heterarchical control of FMS: From theory to application. Engineering Applications of Artificial Intelligence 23, 1314–1326 (2010)
13. Sallez, Y.: Proposition of an analysis framework to describe the "Activeness" of a product during its life cycle - Part I: Motivations and modelling. In: Borangiu, T. (ed.) Preprints of the 3rd SOHOMA (Service Orientation in Holonic and Multi-Agent Manufacturing Control) Workshop, Valenciennes, France (2013)
14. Trentesaux, D.: Distributed control of production systems. Engineering Applications of Artificial Intelligence 22(7), 971–978 (2009)
15. Zbib, N., Pach, C., Sallez, Y., Trentesaux, D.: Heterarchical Production Control in Manufacturing Systems Using the Potential Fields Concept. Journal of Intelligent Manufacturing 23, 1649–1670 (2012)

An Evolutionary Simulation-Optimization Approach to Product-Driven Manufacturing Control

Mehdi Gaham[1], Brahim Bouzouia[1], and Nouara Achour[2]

[1] Équipe Systèmes Robotisés de Production
Centre de Développement des Technologies Avancées
Baba Hassen, Alger, Algeria 16303
[2] Laboratoire de Robotique Parallélisme Électro-énergétique
USTHB, BP32 El Alia, Bab Ezzouar
Alger, Algeria
mgaham@cdta.dz

Abstract. The presently reported research proposes an adaptive manufacturing scheduling and control framework that exploits the challenging combination of the main capabilities of product-driven control paradigm and online simulation-optimization approaches. Mainly, the proposed approach employs a scheduling rule-based evolutionary simulation-optimization strategy to dynamically select the most appropriate local decision policies to be used by the agentified manufacturing system components. In addition, this approach addresses products and machines agents' local decisional efficiency issues by dynamically adapting their behaviour to the fluctuations of the manufacturing system state. The main motivation of the developed hybrid intelligent system framework is the realization of an effective and efficient distributed dynamic scheduling and control strategy, that enhances manufacturing system reactivity, flexibility and fault-tolerance, as well as maintaining global behavioural stability and optimality. In order to assess the significance of the proposed approach, a proof of proposal prototype implementation is presented and a series of numerical benchmarking experiments are discussed.

Keywords: product-driven control, agent-based manufacturing, holonic systems, simulation-optimization, dynamic scheduling, genetic algorithm.

1   Introduction

Induced by market globalization and pressure, poor demand visibility, shorter product life cycles and adoption of new consumer-driven business practices such as mass customization, are some of the innumerable factors that makes flexibility and agility the key competitiveness issues of nowadays manufacturing enterprise. However, due to their centralized or hierarchically organized planning, scheduling and control structures, legacy manufacturing control frameworks respond weakly to these challenges and behave poorly when the system is subject to internal or external unforeseen disturbances. So, distributed or "heterarchical" control paradigms, aiming to bring manufacturing enterprise more competitiveness by enhancing their ability to dexterously

T. Borangiu et al. (eds.), *Service Orientation in Holonic and Multi-Agent Manufacturing and Robotics*, Studies in Computational Intelligence 544,
DOI: 10.1007/978-3-319-04735-5_19, © Springer International Publishing Switzerland 2014

react to customer orders and changing production environments, have been proposed. Distributed manufacturing control means that each system's component representation in the control software is designed as an autonomous processing unit, which has its own goals and responsibilities and which interacts with other components for constructing overall manufacturing system dynamic behaviour [2]. With the aid of appropriate coordination strategy, decisional distribution is expected to enhance system modularity, flexibility, fault-tolerance, as well as system adaptability and reconfiguration ability. Because multi-agent technology cope efficiently with such complex distributed systems, distributed manufacturing approaches are usually put into practice using the agent paradigm [3, 4, 5, 6].

Within manufacturing dynamic scheduling and control context, full heterarchical bidding-based approaches inspired from the well-known Contract Net Protocol pioneered the usage of multi-agent local decision-making methods [7]. However, because of their inherent decisional myopia, these approaches and some of their variants are increasingly criticized for their inability to cope efficiently with the complexity of real world manufacturing dynamic scheduling and control problems [8, 9]. Actually, efficiency of dynamic scheduling process represents a vital concern for manufacturing organizations evolving within a dynamic and unpredictable environment, and although they enhance agility, adaptability and reconfiguration ability of manufacturing system, agent-based dynamic allocation approaches still incarnate immaturity facing these concerns and penalize seriously industrial adoption of this emerging paradigm. Motivating an important number of research works, hybrid hierarchic/heterarchical multi-agent decisional structures have been also investigated for the improvement of agent-based manufacturing approaches [10]. As a core line of investigation, Holonic Manufacturing Systems (HMSs) represent a major declination of distributed manufacturing dynamic scheduling and control structures where the manufacturing system components (machines, products, AGVs, etc.) feature autonomy and cooperation. Mainly, HMSs focus on decisional efficiency and are characterized by a hybrid decisional structure that combines the desirable characteristics of hierarchic and heterarchical control frameworks, which are behavioural optimality of the former and flexible strategy of the latter. Abundantly documented PROSA and ADACOR architectures exemplify the application of the holonic concept to manufacturing control [11, 12]. Another relevant characteristic of HMSs is that they promote the full integration of the manufacturing products or parts as computational control entities within the manufacturing distributed decisional system. The product becomes an active decisional and communicative entity capable of participating in, or making decisions relevant to its fabrication. Within this context, association of physical product and its informational counterpart is realized by Radio Frequency Identification (RFID)-based product identification technology (Fig. 1).

Hence, intelligent product driven manufacturing control emerges as a promising declination of multi-agent HMSs, and is actually defined by Pannequin [13] as a specialization of holonic agent-based distributed control paradigm where agent technology brings forward new fundamental insights on decentralized coordination and auto-organization, enabling new manufacturing decision-making policies and on-the-fly reconfiguration capabilities and infotronics technologies address the issue of synchronization between physical objects and their informational representation. Beside, Trentesaux and Thomas [14] arguing that product-driven control is based on the assumption that the product is the core object in the design, manufacturing, logistic and services systems, define this paradigm as *"a way to optimize the whole product life cycle by dealing with products whose informational content is permanently bound to their virtual or material content and, thus, are able to influence decisions made about them and participating actively to different control processes in which they are involved throughout their life cycle"*.

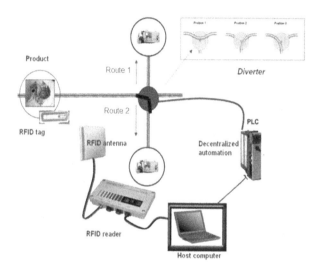

Fig. 1. Product-driven manufacturing control technological issues related to industrial implementation

Focusing on decisional efficiency, the reported research proposes an adaptive manufacturing dynamic scheduling and control framework that explores the challenging combination of main capabilities of this emerging control paradigm and simulation-optimization approaches. Relevant expected contribution of the proposed approach to multi-agent product-driven control is that it exploits a genetic algorithm simulation-optimization approach to dynamically select the most appropriate local decision policies to be used by the agentified manufacturing system components. Exactly, it addresses products and machines agents' local decisional efficiency issues by adapting dynamically their behaviour to the fluctuations of the manufacturing system state. Beside, expected to maintain the correct balance between hierarchical and heterarchical behaviour, the proposed hybrid decisional framework is mostly intended to introduce some level of optimization capabilities within product-driven manufacturing control paradigm by dynamically tuning the used local operational policies.

The rest of this paper is organized in the following way: in section 2, the adopted multi-agent control architecture is succinctly presented, section 3 is dedicated to the presentation of the genetic algorithm simulation-optimization adaptive scheduling approach, prototype implementation and results discussion are presented in section 4 and section 5 concludes the paper.

2   Presentation of the Multi-Agent Product-Driven Control System

Manufacturing facilities organized in flexible job shop production structures are the main focus of the proposed multi-agent control architecture. According to specific implementation issues related to project development perspective, the architecture is separated in two distinct and independent parts: high-level decisional and low-level system emulation part. Each agent in the multi-agent emulation part is a representation of manufacturing component - either a physical resource, a RFID reader or a product. The sited agents are mainly intended to emulate the operational and informational activities of the manufacturing facility. Informational activities such as contract-net real-time information gathering are also encapsulated in this part. Within decisional part, machine and product agents counterparts are implemented as independent agents (Decision Machine (D_Machine) and Decision Product Agents (D_Product)) encapsulating decisional capabilities of product and machine agents. According to overall system design, common manufacturing machine selection rules and job dispatching rules are used by D_Product and D_Machine agents as local decision policies. According to machine related real-time information, machines selection rules are heuristics used by D_Product agent for the selection of the next processing resource to be visited by product agent. Heuristic job dispatching rules are used by D_Machine agents to select from waiting products which one to process next. As an integral component of the decisional part of the multi-agent architecture, Operational Policies Optimization Agent (OPOA_Agent) is responsible of the online tuning of the decisional capabilities of D_Product and D_Machine agents. Instead of using a fixed decisional policy along manufacturing system operational horizon, an optimized set of decision rules is assigned to decisional agents at each occurrence of a perturbation that can affect system' stability. The architecture is shown in Fig. 2.

3   Scheduling Rules-Based Genetic Algorithm Simulation-Optimization

Although they play a significant role within practical manufacturing dynamic scheduling context, one of the commonly identified shortcomings of dispatching rules is that their relative performance depends upon the system attributes, and no single rule is dominant across all possible manufacturing system states. Addressing this issue, simulation is usually used to empirically assess the performances of various dispatching rules and to determine the best rule to be used according to the configuration of the manufacturing system. However, these approaches don't propose a clear

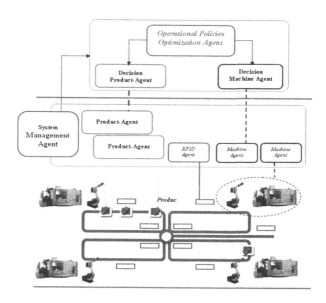

Fig. 2. Product-driven manufacturing control multi-agent architecture

optimization strategy that can guarantee the calculation of an optimal set of rules. By providing a unified integrated framework, simulation-based optimization or simply "simulation-optimization" approaches overcome this limitation. Indeed, as an increasingly investigated research topic, online scheduling rules-based simulation-optimization has been identified as offering a real efficiency perspective to practical manufacturing dynamic scheduling approaches. Within this context, optimization is used to orchestrate the simulation of a system configurations sequence (each configuration corresponds to particular settings of the decision variables) so that a system configuration, that provides an optimal or near optimal solution, is eventually obtained [15]. Decision variables correspond to the set of machines selection and jobs dispatching rules, and simulation is carried out by a simulation model that reproduces the stochastic behaviour of the modelled system.

Hence, scheduling rules-based simulation-optimization is a well suited adaptive dynamic scheduling approach as it makes possible a real-time tuning of used local operational policies according to the manufacturing system state. Moreover, scheduling rules-based simulation-optimization has been successfully applied to a number of real industrial operational management problems. Recently, in [16] a Genetic Algorithm-simulation approach to solving multi-attributes combinatorial dispatching decision (MACD) problem in a flow shop with a multiple processors (FSMP) environment is presented. This approach illustrates the effectiveness and the efficiency of that kind of methodology compared to several common industrial practices. However, as stressed by the authors, although the MACD decision is effective for a practical application, it adds complexity to the shop-floor control problem and its implementation requires supports for a sophisticated shop-floor control system that can perform the dispatching algorithms and control. This shortcoming is well addressed by the *product-driven multi-agent integration* and technological framework adopted in this research.

The system's online simulation-optimization capabilities are encapsulated in the OPOA_Agent. Triggered by the System Management Agent (SM_Agent) at each occurrence of an internal or external disturbance (product arrival, machine break-down), the OPOA_Agent uses simulation-optimization (Optimization and Simulation modules) for the calculation of the new set of decisional policies (scheduling rules) to be dispatched to the decisional agents of the system (via the communication module). The optimization is carried out using a Genetic Algorithm (GAs) as shown in Fig. 2.

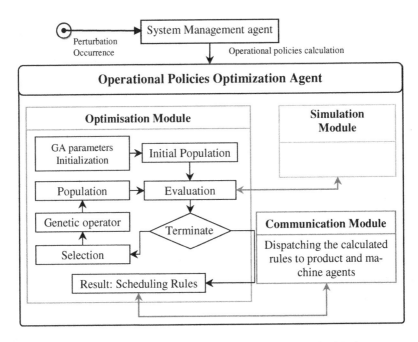

Fig. 3. Internal architecture of the Operational Policies Optimization Agent

In GAs, individuals within a population reproduce according to their fitness in an environment (optimization space). Using stochastic recombination operators, the population of individuals combines to perform an efficient domain-independent search strategy. During each generation, a new population of individuals is created from the old one via the application of genetic recombination operators (crossover, mutation), and evaluated as solutions to a given problem (the environment). Due to selective pressure (Selection operator), the population adapts to the environment over the generations, evolving better solutions. Our approach uses a real coded genetic algorithm (an individual codes a sequence of scheduling rules that equals the number of machines and products in the system) combined to classical crossover and muta-tion and selection operators.

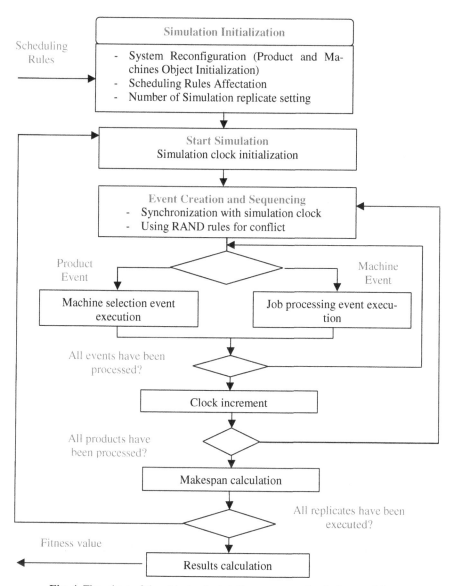

Fig. 4. Flowchart of the object oriented discrete event simulation module

Evaluation of each individual is carried out using a predefined number of simulation replicates that assesses the performances of the set of scheduling rules according to the stochastic nature of the adopted multi-agent manufacturing control framework. In charge of the simulation execution, the independent simulation module is implemented as an object oriented discrete-time simulation framework. For each simulation replicate, the simulation framework, guided by the simulation clock evolution, constructs a schedule using the manufacturing system real-time status and the set of machine selection and job dispatching rules. Stochastic transportation times and randomized synchronization of decisional time conflicts are integrated within the schedule construction and evaluation for each simulation replicate. Schedule evaluation is done using the makespan criterion. Fig. 4 illustrates the functional operation of the simulation module.

4   Prototype Development and Approach Validation

As depicted in Fig. 5, Open source NetBeans IDE and JADE (Java Agent Development Environment), have been used for the development of the prototype emulation and control system.

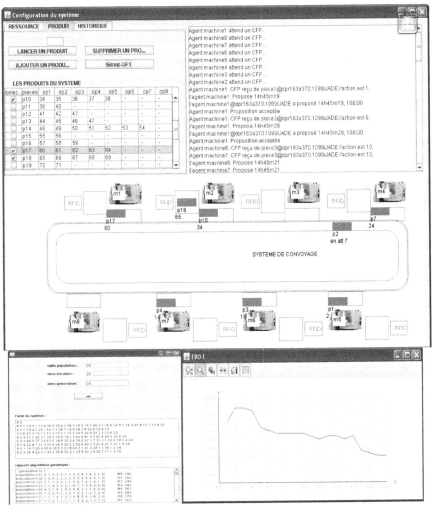

Fig. 5. Snapshots of the JADE-Based Prototype Implementation

JADE is an open source platform [14] that provides basic middleware-layer functionalities which simplify the realization of distributed applications that exploit the agent's software abstraction [15]. A multi-agent system based on Jade has amongst others the following features: fully distributed, compliant with the FIPA specifications, efficient transport of asynchronous messages, provides a library of interaction protocols. JADE also provide a runtime environment and a set of graphical tools to support programmers when debugging and monitoring applications.

In order to evaluate the pertinence of the dynamic scheduling approach in terms of solution quality and computational effort, several tests have been carried out using an instance problem of 8 machines and 15 products. Different machine selection and job dispatching rules have been used. Table 1 summarizes the set of rule.

Table 1. The set of scheduling and dispatching rules used as local decisional policies by the product and machine agent

	Rules	**Description**	
PRODUCTS	Pr1	LRW	Least remaining work
	Pr2	LRPT	Least Remaining processing time
	Pr3	LPT	Longest processing time
	Pr4	SPT	Smallest processing time
MACHINES	Mr1	SPT	Smallest processing time
	Mr2	LPT	Longest processing time
	Mr3	EDD	Earlier due date
	Mr4	LRW	Least remaining work
	Mr5	FIFO	First in first out
	Mr6	LIFO	Last in first out
	Mr7	RANDO	Random selection

The evolution of the fitness value for the conducted experiments is illustrated in Fig. 6. The GA parameters have been respectively set to 20 and 10 respectively for the population size and number of generation. The number of replicate has been empirically assessed and has been set to 10. The conducted experiments showed the effectiveness of the approach for the minimization of the makespan. Thus, simulation-optimization seems to be a well suited approach for the online determination of dynamic scheduling operational local policies, but it is still highly sensitive to the number of replicates. In fact, the dynamic nature of the simulation-based optimization problem makes this parameter a critical one. Fig. 7 shows the influence of this parameter for the conducted test problem. In this figure, the dotted curve corresponds to the evolution of solution for a number of replicates equals to 1. It can be seen that the variance of the values is clearly superior to the values variance of the other curve that correspond to a number of replicates equal to 20.

Makespan

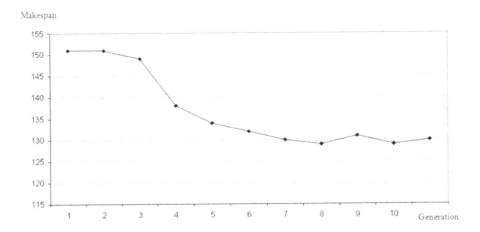

Fig. 6. Genetic algorithm Fitness Evolution

Makespan

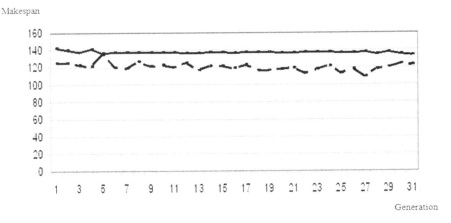

Fig. 7. Number of simulation replicates influence

4.1   Computational Results

The computational tests were conducted using flexible job shop benchmark problems (Brandimarte [17]). Six test problems have been chosen. As the problems are static, ones with no transportation time defined and with no due date, the system has been adapted according to those facts. The computational tests compare the genetic algorithm based simulation-optimization approach with some of most relevant combination of scheduling and dispatching rules. To handle stochastic nature of the system, ten realization were conducted for each rules combination and average makespan (completion time for all the tasks) are given as indicator. Genetic algorithm parameters are as follow: Population=100, Generation=50, Crossover probability=0.9, Mutation probability=0.02, Number of simulation replicates=10.

Table 2. Comparative results of the simulation approach and rules combination

	Num. of products	Num. of machines	Best rules value	Rule combination	Brandimarte values	GA values
MK 01	10	6	54	Pr 1+ Mr 1	42	47
MK 02	10	6	42	Pr 1+ Mr 1	32	38
MK 03	15	8	234	Pr 1+ Mr 5	211	222
MK 04	15	8	81	Pr 1+ Mr 1	81	77
MK 05	15	4	197	Pr 1+ Mr 5	186	188
MK 07	20	5	217	Pr 3+ Mr 1	157	168

The conducted experiments validate the proposed approach in term of computational efficiency. In fact, for that specific set of benchmark problem the approach is superior to simple combination of scheduling and dispatching rules. The approach, also, gives a very interesting results compared to those of Brandimarte, for example for the of mk04 problem.

5   Conclusions and Future Works

This paper investigates an innovative hybrid framework combining holonic product-driven manufacturing control and scheduling rules-based genetic algorithm simulation-optimization approaches for real-time adaptive dynamic scheduling of real world stochastic manufacturing systems. Both design and implementation issues related to the adopted multi-agent system have been presented, and as a core component of the overall framework, a scheduling rules-based genetic algorithm simulation-optimization approach has been described and evaluated. The applicability and effectiveness of the proposed hybrid framework has been demonstrated by the developed prototype and the conducted and succinctly presented tests.

Future research direction will deal with the investigation of a more formal approach for the determination of algorithm parameters, and particularly the number of simulation replicates.

References

1. Duffie, N.A., Piper, R.S.: Non-hierarchical control of a flexible manufacturing cell. Robotics & Computer-Integrated Manufacturing 3, 175–179 (1987)
2. Trentesaux, D.: Distributed control of production systems. Engineering Applications of Artificial Intelligence 22, 971–978 (2009)

3. Marik, V., Lazansky, J.: Industrial applications of agent technologies. Control Engineering Practice 15, 1364–1380 (2007)

4. Shen, W., Norrie, D.H.: Agent-based systems for intelligent manufacturing: a state-of-the-art survey. KAIS 1, 129–156 (1999)

5. Shen, W., Hao, Q., Yoon, H.G., Norrie, D.H.: Applications of agent-based systems in intelligent manufacturing: An updated review. Advanced Engineering Informatics 20, 415–431 (2006)

6. Deen, S.M. (ed.): Agent-based manufacturing Advances in the holonic approach. Springer (2003) ISBN 3-540-44069-0

7. Parunak, H.V.D.: Manufacturing experience with the contract net. In: Huhns, M.N. (ed.) Distributed Artificial Intelligence, pp. 285–310. Pitman (1987)

8. Maione, G., Naso, D.: A Genetic Approach for Adaptive Multiagent Control in Heterarchical Manufacturing Systems. IEEE Transactions on Systems, Man, and Cybernetics - Part A: Systems and Humans 33 (2003)

9. Zambrano Rey, G., Pach, C., Aissani, N., Bekrar, A., Berger, T., Trentesaux, D.: The control of myopic behaviour in semi-heterarchical production systems: A holonic framework. Engineering Applications of Artificial Intelligence 26, 800–817 (2012)

10. Heragu, S.S., Graves, R.J., Kim, B., Onge, A.: Intelligent Agent Based Framework for Manufacturing Systems Control. IEEE Transactions on Systems, Man, and Cybernetics - Part A: Systems and Humans 32, 560–572 (2002)

11. Van Brussel, H., Wyns, J., Valckenaers, P., Bongaerts, L., Peeters, P.: Reference architecture for holonic manufacturing systems: PROSA. Computers in Industry 37, 255–274 (1998)

12. Leitão, P., Colombo, A.W., Restivo, F.: A formal specification approach for holonic control systems: the ADACOR case. IJMTM 8, 37–57 (2006)

13. Pannequin, R., Morel, G., Thomas, A.: The performance of product-driven manufacturing control: An emulation-based benchmarking study. Computers in Industry 60(3), 195–203 (2009)

14. Trentesaux, D., Thomas, A.: Product-Driven Control: Concept, Literature Review and Future Trends. In: Borangiu, T., Thomas, A., Trentesaux, D. (eds.) Service Orientation in Holonic and Multi agent, SCI, vol. 472, pp. 135–150. Springer, Heidelberg (2013)

15. Law, A.M., McComas, M.G.: Simulation-based optimization. In: Proceedings of the 2000 Winter Simulation Conference (2000)

16. Yang, T., Kuo, Y., Cho, C.: A genetic algorithms simulation approach for the multi-attribute combinatorial dispatching decision problem. European Journal of Operational Research 176, 1859–1873 (2007)

17. Brandimarte, P.: Routing and scheduling in a flexible job shop by tabu search. Annals of Operational Research 41, 157–183 (1993)

Farm Management Information System as Ontological Level in a Digital Business Ecosystem

Luiza-Elena Cojocaru, George Burlacu, Dan Popescu, and Aurelian Mihai Stanescu

University Politehnica of Bucharest,
Faculty of Automatic Control and Computer Science, Bucharest, Romania
{luiza_cojocaru85,burlacu_george85,
dan_popescu_2002,amstanescu}@yahoo.com

Abstract. The aim of the paper is to present a methodology to develop a Farm Management Information System capable to integrate within a Digital Business Ecosystem. In additional, an ontology has been developed with the objective of representing knowledge which can be used for understanding the concepts and relationships regarding the Farm Management Information System. Also, the ontology offers decision making support to farmers in what concerns the land and crop management.

Keywords: Information System, Interoperability, Digital Business Ecosystem, Future Internet Enterprise System.

1 Introduction

In the last few years, the performance of agriculture in developed countries, the technologies brought into play, are looming the future agriculture. Precision agriculture is practiced in the most developed countries in the world on smaller surfaces, using the most modern methods of quality control for the various resources from the environment.

Sørensen et al. proposed a conceptual model to develop a Farm Management Information System (Fig. 1). The main actor in this picture is the farm manager who needs to manage the overall agricultural crop production. The managerial demands are caused not only by the internal farm production activity, but also by external entities like: government, customers, and academia, who increase the pressure on the agricultural sector to change the production methods, the quality, the price or the technological improvements [1].

The proposed **Farm Management Information System** aims to support management tasks and real-time decision making, as well as compliance management by automating data acquisition, and taking into account external parameters (e.g., regulations, best management practices (BMP), market information, etc.). In addition, the structure of the Farm Management Information System should allow the connection with external systems (e.g., market, financial, administration, etc.).

T. Borangiu et al. (eds.), *Service Orientation in Holonic and Multi-Agent Manufacturing and Robotics*, Studies in Computational Intelligence 544,
DOI: 10.1007/978-3-319-04735-5_20, © Springer International Publishing Switzerland 2014

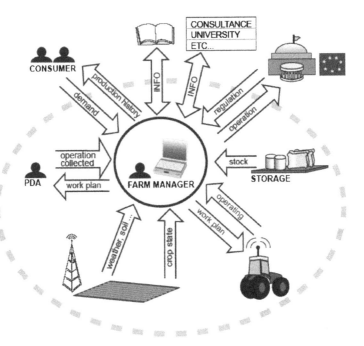

Fig. 1. Conceptual model for future Farm Management Information System [1]

T conclude, a Farm Management Information System (FMIS) is a management information system designed to assist farmers to perform various tasks from operational planning to implementation and for assessment documentation of performed field work [1].

The concept of "business ecosystem" was defined for the first time by J. F. Moore as "an economic community supported by a foundation of interacting organizations and individuals – world business organisms" [2]. This economic community produces goods and services valued by clients, and they are members of the ecosystem.

If the Farm Management Information System provides support decision inside the farm, the Digital Business Ecosystem comes to offer ITC solution in order to develop the community in which the farm is a part.

In Romania, agriculture plays an important role in our daily lives, many families produce their necessary revenues in this way. In recent years, the interest for this sector has increased and many development funds have been attracted. Even so, many farmers choose to work the land in traditional way. Their motivation is mainly related to the financial aspect necessary to implement a Farm Management Information System.

The paper proposes a methodology to develop a Farm Management Information System capable to integrate within a Digital Business Ecosystem. Also, there will be discussed an ontology developed with the objective of representing knowledge to be used for understanding the concepts and relationship related to the Farm Management Information System. This ontology offers support to farmers in taking decisions regarding the land and crop management.

2   Farm Integration within Digital Business Ecosystem

The technological achievements, as well as technological improvements that favoured the appearance of precision agriculture will be presented next:

- Ability to understand the complexity of agricultural systems (both systemic and holistic approaches), and to monitor the specific processes and phenomena, by automating data acquisition;
- Achievements in computational techniques (hardware, software and databases);
- Improvement of calculation and interpretation methods (statistics, modelling, simulation, decision support systems);
- Development of GIS (Geographic Information Systems) and improvements in the fields of Remote Sensing and GPS (Global Positioning System);
- Technical achievements regarding the improvements in agricultural machinery.

In this context, precision agriculture promotion is justified by the following reasons:

- Use of inadequate varieties to the specific conditions;
- High level of prices for fertilizer, fuel, equipment, machinery for used in agriculture works;
- Farmers do not finance the culture for an optimum production;
- Climatic mutations produced lately in our country;
- Private producer's involvement, which are not totally specialized in growing plants.

The purpose of precision agriculture is to optimize the use of resources: soil, water and fertilizer, and has the objectives: to obtain large and quality productions, to optimize economic profits, to protect the environment and to increase the sustainability of agricultural systems.

An ecosystem is viewed as a group of interacting, interrelated, or interdependent elements forming a complex whole. A system may have boundary and may interact with the surrounding environment by exchanging energy and/or materials [3].

The Digital Business Ecosystem is the result of coupling the digital ecosystem with the business ecosystem (Fig. 2).

Digital Ecosystem is a persistent digital environment populated by digital components that evolve and adapt to local conditions with the evolution of its constituents [4].

Digital Ecosystems are based on a systemic evolutionary process which consists in the following layers [5]:

- Ecosystem infrastructure is a common supportive environment including components for basic services, integrated solutions and infrastructure components;
- Specific ecosystems refers to services, solutions and components designed for a specific area;

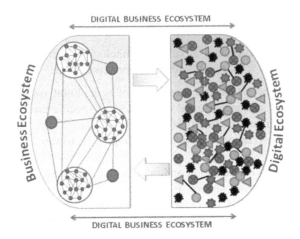

Fig. 2. Digital Business Ecosystem (Adapted from Paolo Dini [4])

- Local ecosystems represent local implementations of the specific ecosystems in nodes and networks of innovation. These instances can be linked together to form a network in order to support collaboration and knowledge sharing between members of that ecosystem.

Considering all the aspects mentioned, we believe that a farm is as part of a local ecosystem, within which it operates mainly by dynamic methods of interaction and global cooperation between farms and business communities.

3   A Methodology to Develop a Management Information System for Future Internet Based Farm

The methodology development will include four stages: requirements and analysis, modelling framework, architectural model (design) and integration.

A.   Requirements and analysis

To be able to configure the Farm Management Information System proposed, the requirements should be clearly outlined and defined.

In the following we describe the processes identified in the decision making at all levels: strategic, tactical, operational, execution and evaluation. These processes are complex and require information from multiple input sources:

- Determining the long-term vision for agricultural production on the basis of internal and external opportunities;
- Planning and setting the overall strategy based on internal and external preferences and possibilities;
- Selecting the best technologies required in the production process, taking into account the constraints in the coming years;

- Selecting the work plan for fertilizers;
- Observing and acquiring information regarding the ground conditions;
- Formulating the execution plan;
- Formulation of the optimum program in terms of the expected increases, estimated weather conditions and associated risks;
- Updating the final plan based on the latest information available;
- Determining the parameters based on all available updated information;
- Determining the progress made with respect to the planning;
- Determining the possible differences between the actual performances and the planned goal.

Well-structured decision making follows certain rules which have to be provided to the decision processes [6].

In addition to these rules, decision processes require correlating and integrating information; information decomposition requires not only the identification but also the content management of big data of dynamic and static nature [6].

In this context, the description involves the entities, the attributes related to them, the availability of these attributes at certain time moments and future requirements in considering unexpected events and uncertain availability.

The analysis phase involves the identification of the Management Information System components such as: operations, processes, conflicts and ideas underlying the design of a particular Farm Management Information System [7].

The main topics to be addressed are [7]:

- Identification of the external actors who relate with the farm (e.g., institutions or public entities, non-governmental organizations, companies, etc.);
- The way the farm should work (in terms of service delivery, operations management and monitoring, follow-up of processes and information integration);
- Information acquisition and processing, in order to help decision taking and automate specific activities, thus providing efficient operating conditions for the farm

B. Modeling framework

Fig. 3 depicts the conceptual model which is proposed for the Farm Management Information System.

This framework focuses on the way it performs *data acquisition* and *mapping* (how to make spatial distribution maps, maps of productivity and profit). The acquired information is stored on a platform together with the information from the various web services (e., meteorology data). The system provides user access to its knowledge base, from where the farmer can take a decision with respect to the land or crops.

The purpose of web services is to enable communication, interoperability and data exchange between applications. Through these services the farmer has access to: new technology, resources (human, materials), information from the market (competitors, prices, and forecasts), information from ministries and the EU legal bodies, etc.

An Agricultural Geographic Information System (AGIS) is used in order to get a better image and an easy interpretation of the collected data. So the farmer can have access to spatial distribution maps considering certain parameters (location of interest points, distances between certain parts with reference to whole land areas or only to certain areas of interest) [8].

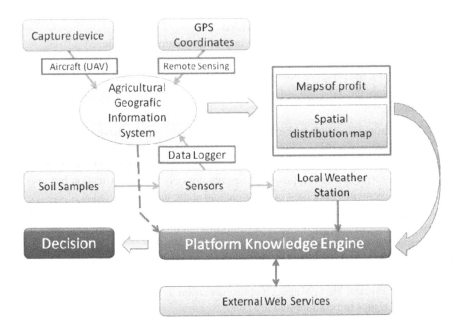

Fig. 3. Description of the conceptual model for the Farm Management Information System

The information obtained through remote acquisition play an important role in building databases of geographical data within AGIS; remote acquisition implies the existence of a capture device, attached to a platform (usually an aircraft) [9].

Most of the data in the current systems of agricultural information management are obtained through the interpretation of aerial photographs. Using a AGIS, the farmer can analyze and interpret only digital type data. So the geographic information that does not meet this requirement must be digitized. Digitizing is the process through which a physical copy of a map or a plan is converted into a digital environment using programs for objects representation in 2D or 3D. Digitizing can be done by hand and using semi-automat devices that are connected to a computer and can transmit the coordinates of a point on a 2D surface.

Topographic information from GPS have various roles: defining fields in terms of geographical boundaries, monitoring the routes of agricultural machinery in the field in order to know in real time their position, soil parameters mapping, crops mapping, maximizing yield per hectare or discovering the spread of specific disease in crops. Precision spatial dimension provided by the GPS system is used in the classical form of mechanized agriculture due to the development of new technologies and equipment including mobile devices, computing systems and distributed intelligence [10].

Presently, large amount of data from field operations are collected by agricultural machines and transmitted using various data storage and transmission media [11]. The classical sampling which was done manually by collecting soil samples from different areas of land at random or according to specific pattern has moved towards a mechanized collection by using modern agricultural machinery. This practice is known as "sampling" [12]. Sampling is used to obtain an overview distribution of chemical soil properties and / or to check for any abnormalities in normal chemical soil properties. Also physical properties (soil electro conductivity, soil pH, concentrations of NO3, K, Na) are measured directly in the field by specific laboratory techniques [13].

Data from the mapping process is used to estimate the production of farming crops in the entire area or in a specific areas of interest. In this estimation process, data from the history of agricultural land are taken into account and analyzed. A profit map can be created using inputs records in crops and outputs in harvest. So, a farmer can determine what areas of the field are not profitable.

Farmers can also make crop productivity maps using sensors that measure and record in real time the harvested crop volume. Various parameters like quantity per hectare and flow are measured. In this way, crop productivity maps are a valuable solution for technology improvement in farm management.

Spatial distribution maps of the relevant parameters are ranked based on several criteria (visualization, physical support, technical type mapping, type of variables, type of representation, etc.). One advantage of using spatial distribution maps is the avoidance of the surplus or the deficit in natural or mineral fertilizer (nitrogen, phosphorus, potassium, magnesium, nitrates).

Automated Variable Rate Technologies allow farmers to vary inputs, such as fertilizers, pesticides and seeding rates, by monitoring sample land zones and then generalizing the specific treatments decided by the information system in the entire field [14]. Varying input rates aims at either increasing crops or reducing costs, depending on the farmer's goal for the production system. Auto guidance systems, on the other hand, assist equipment operators in driving through the fields so that efforts can be focused on other important tasks. These technological tools therefore help to reduce redundancy and labour costs and to save operation times.

A set of specific sensors is used to determine soil properties (ground temperature/air temperature, humidity), hydro stress, degree and spread of crop disease (using the light reflection on the leaves to determine their level of chlorophyll), to measure various parameters (electrical conductivity, high concentrations of K, NO_3, Na, pH) or to measure the soil and crops properties.

The need for simultaneously increasing the production and the efforts for saving resources make this sensor set an important component of the information system for farm management. The usage of sensors requires means for physical storage of data with devices known as "data loggers", which are equipped with non-volatile memory. Data transmission is done either by cable (wired) or by radio signals (wireless) [15].

By using remote sensing sensors (aerial or satellite sensors) lands maps and cultures maps can be obtained. The "aerial photographer" is an optical sensor that observes the variations in soil colour, crop development and land boundaries. Images obtained by satellite or by an aircraft can provide maps of vegetative indices, which reflect also the plant health, the soil conditions or the crops status.

The obtained database is thus more accurate (in number of measured values and number of parameters) and increases the quality of decisions.

C. Architectural model

The architectural model proposed for the farm management system is organized on three levels: infrastructure layer, process layer and management layer (Fig. 4).

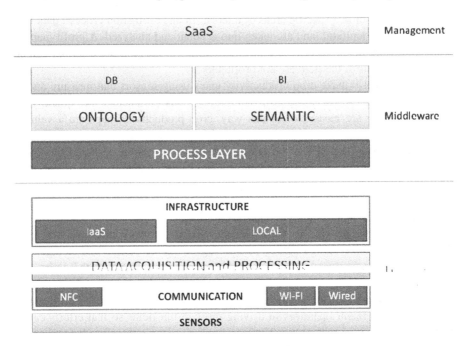

Fig. 4. Levels of architectural model of the farm management system

1. *Infrastructure Layer*: includes sensors, communication infrastructure data acquisition platform. This level will feature the data model and associated sensors.
2. *Process Layer*: includes process models, ontologies, annotations, and middleware applications database and business analysis.
3. *Management Layer*: includes services that can be composed and orchestrated according to current business needs.

Processes, production or business models are the basis for service composition. The automatic selection of necessary sensors and associated services is performed based on the process model.

Different types of associated services can be used:

- *Autonomous service* that analyzes and integrates the data directly from a type of sensors;
- *Composite service* that processes data from many types of sensors, from an intermediate level (buffer infrastructure);
- *Interoperability service* handling existing applications (legacy system) and applications from other manufacturers;
- *Collaborative service*, user interface services and external services.

The Semantic Web provides the infrastructure necessary for publication and identification of the ontological descriptions of terms and concepts. The Semantic Web Services extend web services capabilities by associating a semantic description at the web service.

In essence, the ontology provides semantics and in this context has two responsibilities: it describes the semantics of the modelling language and also the semantics of model instances.

In terms of business processes, this means that ontologies are used both for explaining the semantics of activities in general, as well as to describe the semantics of a specific activity.

D. Integration

Interoperability of systems involves that the systems are able to interact - to read and to "understand" the messages sent to each other and share the same expectations about the effect that the exchange of messages will have.

Syntactical Interoperability aims to ensure that the data contained in the messages that are exchanged between two computer systems have compatible formats. The system that transmits the message encodes database taking into account the syntax rules specified by a grammar. The receiver of the message decodes the message using the same syntax rules defined in the same grammar used by the system that has sent the message. At this level of syntactic interoperability, problems may occur when the message encoding and decoding is performed according to different rules [16].

Semantic Interoperability aims to ensure that the data transmitted by messages between two systems have the same meaning in both systems. The data transmitted between the two systems have a meaning only when they are interpreted at the level of a domain.

However, the system which transmits the message does not always know the system that will receive the message. In this case, the system will make a series of assumptions about the system's receiver domain in order to build a message. At this level of semantic interoperability, problems can occur when the system domain which transmits the message differs from the system domains which receive the message [16].

The Farm MIS implementation will take into account the implementation of different interoperability levels within the farm (Fig. 5.).

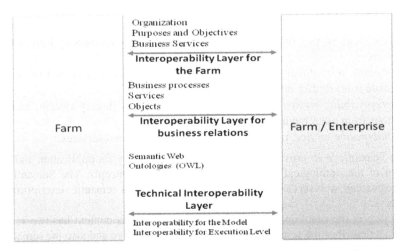

Fig. 5. Implementation of different interoperability levels within the farm

Integration of the farm within the Digital Business Ecosystems will be considered from two points of view (Fig. 6.).

From the point of view of Business Ecosystem, the integration will involve aligning the farm policies at the regional, national and European policies. Also, the integration of Business Ecosystem requires a focus on accomplishing important business-to- business (B2B) relationships between two farmers, one farmer and an enterprise or suppliers, government institutions, non-governmental institutions and customers [17].

From the perspective of the Digital Ecosystem, integrating the farm requires the use of hosted Cloud Computing based technologies: Infrastructure as a Service, Platform as a Service and Software as a Service.

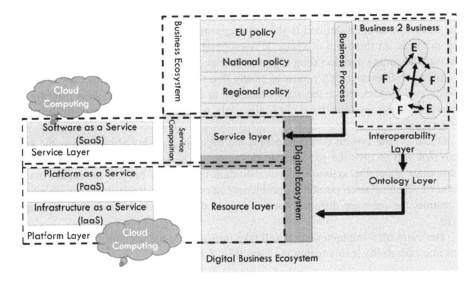

Fig. 6. Integrating the Farm within the Digital Business Ecosystem

Coupling the business ecosystem with the digital ecosystem will be made via an Interface Layer, which includes technologies like Interoperability and Ontologies.

Fig. 7 describes the farm activities and represents the connections with the entities from the Digital Business Ecosystem. The farm activities are influenced by other entities within the Digital Business Ecosystem.

Based on the information from sensors and keeping in mind the weather forecast and the available resources (water, fuels, fertilizers, human resources, equipment) the farmer establishes the work plan in terms of production.

A part of the resulting production is stored and used in the farm, whereas another part is sold. The residues from the processing activity may get biomass and fertilizer, which in turn can be used in the farm, or sold.

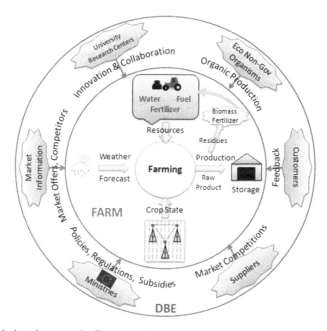

Fig. 7. Relations between the Farm and the rest of the Digital Business Ecosystem entities

In terms of market information, from here the farmer get information about offerings on the market, direct competitors or clients preferences.

The client's feedback provides the information related to the satisfaction degree as a result of using the products sold by the farmer. Based on this feedback, the farmer can guide the production respecting the client's preferences.

The farmer buys the raw materials (e.g. fuel, seeds, fertilizers, machinery and technologies) from suppliers. In the Digital Business Ecosystem he is able to choose from a competitive market those products / services that correspond to its needs.

The farmer collaborates with Universities and Research Centres, through which he may have access to the new technologies and IT solutions for his farm.

Eco Non-Governmental Organisms play an important role through campaigning for the natural production and a healthy, and unpolluted environment.

Farm activities are influenced in a decisive manner by the policies, regulations and subsidies established/offered by the Ministers (Agriculture and Environment) and by the European Community.

4   Integrating the Farm Management Information System within a Digital Business Ecosystem at Ontological Level

The proposed ontology was built using Protégé-OWL Editor, OWL Full version. Protégé is a public and free tool that uses OWL and other languages to built ontologies [18]. This OWL language can be used along with information written in RDF [19]. The Resource Description Framework (RDF) is used for conceptual description or modelling of information implemented in web resources, using syntax notations and data serialization formats [20]. OWL Full is appropriate for obtaining maximum expressiveness and the syntactic freedom of RDF with no computational guarantees [21]. Consequently, authors use OWL Full to represent knowledge in the digital business ecosystem developments.

The ontology developed with Protégé represents the knowledge that can be used to understand the concepts and relationships concerning the Farm Management Information System and its integration within a Digital Business Ecosystem (Fig. 8).

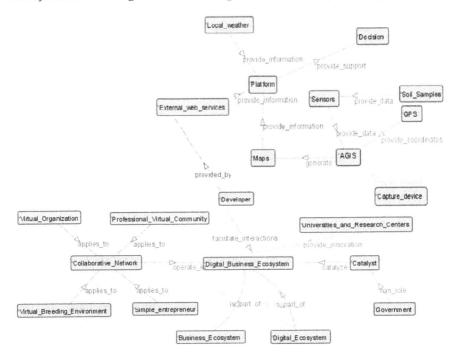

Fig. 8. General representation of Integrating Farm Management Information System within the Digital Business Ecosystem in Protégé

This ontology also provides support to a system to offer different types of information to simple entrepreneurs (e.g., farmers) as for instance, which are the necessary steps to be taken to become part (members) of the Digital Business Ecosystem, or how they can access to regional development funds.

In conclusion, the ontology helps people and machines to use the same terms for expressions and thus to achieve better mutual understanding.

Case study

In order to have a better understanding of how the Farm Management Information System can be integrated within a Digital Business Ecosystem, a case study is presented next.

Mr. Radu Ionescu, lives in the Ilfov district in Romania, has 30 years old, and wants to establish a farm for agricultural crops. So far, he cultivated only the cereals required to ensure a decent living and he was not involved in the agro-technical domain at large scale.

Mr. Ionescu has an agricultural surface of 5 hectares, but is willing to work a larger area, because in the region where he lives there are many uncultivated lands.

In the last years, he has grown only wheat and maize, but due to the fact that he was not informed and also to unfavourable weather conditions, the cereals he cultivated did not have a good quality.

The main objective of this farmer is to start a business whose activity is represented by the growth of wheat, maize and rape. He wants to sell as many products as possible and to maximize the profit. For this, he needs the most modern methods of quality control of various environmental resources, which requires that he gets all the information regarding this business.

In this context, a Farm Management Information System of the type described in this paper can provide to the farmer the technologies necessary for functional operation of his farm and the management solutions optimizing the resources and maximizing the profit.

In addition to this, the Digital Business Ecosystem can provide ICT solutions like:

- Access to a database containing information about all the suppliers, selling prices and product quality. Thus, the farmer saves time and can quickly negotiate contracts with suppliers, taking into account its preferences in terms of quality and price.
- Access to a database containing information about other sellers offering similar products in the market and their offers. Also, the farmer can publish its offer through the infrastructure provided by the Digital Business Ecosystem, and suppliers can sell goods to customers on the internet. Thus, the farmer has real possibilities to: (i) expand his activities in the virtual environment and to deliver integrated services in addition to product sales to the customer (product-service extension), and (ii) provide follow-up of delivered services and tracking of products sold to customers.
- The farmer can customize a database for himself and thus attract the customers by providing them all information related to production: variety, cultivation method, method of taking care and any other information that could differentiate him from other producers.

5   Conclusion

One of the most important outcomes of implementing Digital Business Ecosystems is the acceleration of local economic growth through new methods of dynamic business interaction and global cooperation between farmers and business communities.

The concepts related to the information systems management and design determine farmers to adopt new working methods to remain connected to the technological news. Farmers can use various services described above more efficiently and are able to outsource certain tasks that were previously carried out by them, aligning thus to the global value chain of complex services.

In this paper, the authors propose a 4-stage methodology to develop a Farm Management Information System capable to integrate within a Digital Business Ecosystem.

The ontology presented has been developed to represent the knowledge used for understanding the concepts and relation regarding the Farm Management Information and to offer support in decision making in what concerns the land and crop management and follow-up and product traceability.

The case study presented shows that the implementation of Farm Management Information Systems integrated in a Digital Business Ecosystem represents a real solution for farmers not only to make precision agriculture, but also to keep connection to business communities.

Acknowledgement. This work has been developed within the POSDRU/107/ 1.5/S/76903 project "Future expert researchers training through doctoral scholarship programs (EXPERT)", funded from the European Social Fund through the Sectoral Operational Programme Human Resources Development 2007-2013 (http://plone. cempdi.pub.ro/BurseDoctoraleID76903).

References

1. Sørensen, C., Fountas, S., Nash, E., Pesonen, L., Bochtis, D., Pedersen, S., Basso, B., Blackmore, S.: Conceptual Model of a Future Farm Management Information System. Computers and Electronics in Agriculture 72 (2010)
2. Moore, J.F.: The death of competition: leadership and strategy in the age of business eco-systems. Harperbusiness, New York (1996)
3. Stanescu, A.M., Moisescu, M.A., Sacala, I.S., Lolu, I.G.: Complex Adaptive Systems Fo-cused Intelligent Manufacturing. In: The 6th Symposium on Process Control, Ploiesti (2009)
4. Dini, P., Nachira, F.: The Paradigm of Structural Coupling in Digital Ecosystems - Toward Digital Business Ecosystems, Eds. Edward Elgar (2007)
5. European Commission, Framework Programme 7,
   http://www.digital-ecosystems.org
6. Nash, E., Dreger, F., Schwarz, J., Bill, R., Werner, A.: Development of a model of data-flows for precision agriculture based on a collaborative research project. Computers and Electronics in Agriculture 66 (2009)

7. Sørensen, C.G., Pesonen, L., Bochtis, D.D., Vougioukas, S.G., Suomi, P.: Functional requirements for a future farm management information system. Computers and Electronics in Agriculture 76(2) (2011)

8. Bachmaier, M., Gandorfer, M.: A conceptual framework for judging the precision agriculture hypothesis with regard to site-specific nitrogen application. Precision Agriculture 10 (2008)

9. Wang, N., Zhang, N.Q., et al.: Wireless sensors in agriculture and food industry - Recent development and future perspective. Computers and Electronics in Agriculture 50 (2006)

10. Lamba, D., Fraziera, W.P., et al.: Improving pathways to adoption: Putting the right P's in precision agriculture. Computers and Electronics in Agriculture 61 (2007)

11. Nikkilä, R.: Farm Management Information System Architecture for Precision Agriculture. Department of Computer Science and Engineering, Laboratory of Computer and Information Science. Helsinki University of Technology, Espoo, Finland. Master Thesis (2007)

12. Schnug, E., Panten, K., et al.: Sampling and nutrient recommendations - The future. In: 1997 International Soil and Plant Analysis Symposium. Marcel Dekker Inc., Minneapolis (1997)

13. Adamchuk, V.I., Morgan, M.T., et al.: An automated sampling system for measuring soil pH. Trans. Sensors (2011)

14. Srivastava, S.: Space Inputs for Precision Agriculture: Scope for Prototype Experiments in the Diverse Indian Agro-Ecosystems. In: Map Asia 2002, Report, Bangkok, Thailand, August 7-9, pp. 1–4 (2002) (assessed February 4, 2011)

15. Ruiz-Garcia, L., Lunadei, L., et al.: A Review of Wireless Sensor Technologies and Applications in Agriculture and Food Industry: State of the Art and Current Trends, Sensors (2009)

16. http://dublincore.org/documents.html

17. Burlacu, G., Stanescu, A.M., Sacala, I.S., Moisescu, M.A., Cojocaru, L.E.: Development of a Modelling Framework for Future Internet Enterprise Systems. In: 16th International Conference on System Theory, Control and Computing, Sinaia (2012)

18. http://protege.stanford.edu/overview/index.html

19. OWL 2 Web Ontology Language Document Overview. W3C. 2009-10-27, http://www.w3.org/TR/owl2-overview/

20. http://www.w3.org/standards/semanticweb/

21. http://www.w3.org/TR/owl-guide/

Part V
Robots for Manufacturing and Services

Vision-Guided Robot Manipulation Predictive Control for Automating Manufacturing

Corneliu Lazar[1], Adrian Burlacu[1], and Alexandru Archip[2]

[1] Dept. of Automatic Control and Applied Informatics
[2] Dept. of Computer Engineering
Technical University "Gheorghe Asachi" of Iasi, Romania
clazar@ac.tuiasi.ro

Abstract. Increasing demands of flexible automating manufacturing processes made robots with vision systems to be frequently used due to their ability to adapt to new environments and various conditions. This paper presents a visual servoing system for real time picking and placing of dynamic objects. The system is integrated into a manufacturing cell and its structure consists in a CCD camera with SIFT detector, an image based predictive controller and a robot manipulator. Using a Service Oriented Architecture (SOA), the visual servoing system is implemented and different experimental testing of the vision-guided robot predictive control architecture is conducted.

Keywords: visual servoing, automating manufacturing, predictive control, Service Oriented Architecture.

1   Introduction

In the last decades, increasing demands of automating manufacturing processes and services with larger flexibility [1] made robots with vision systems to be used frequently due to their ability to adapt to new environments and various conditions. Robot vision systems within a manufacturing production line try to emulate the person action of human vision in a general-conceptual actions of workers consisting of object and working area recognition, training and skill mastering of procedures and tool management.

Flexible vision systems are being employed more and more for visual servo control of robotic manipulators to perform complex manufacturing tasks especially for tracking and grasping dynamic objects by generating the optimum-tracking trajectory for the robot [2]. Automated manufacturing systems also require accurate positioning. Visual servoing is a very attractive technique to enhance such positioning accuracy [1, 3, 4]. The main task in visual servoing (VS) [2] is to use visual information to control the pose of the robot's end-effector relative to a target object or a set of target features. The challenge in providing visual feedback is to recover visual motion quickly and robustly.

There are two types of motion processing: full-field motion processing such as optical flow and feature-based motion processing such as key-point tracking. Feature

T. Borangiu et al. (eds.), *Service Orientation in Holonic and Multi-Agent Manufacturing and Robotics*, Studies in Computational Intelligence 544,
DOI: 10.1007/978-3-319-04735-5_21, © Springer International Publishing Switzerland 2014

tracking concentrates on spatially localized areas of the image and because image processing is local, the amount of data that must be processed is relatively low and thus, this type of motion processing is more suitable for real time implementation.

In comparison with *look-and-move* systems, which perform the control of the robot by using firstly the vision system to provide the input to robot controller and then with joint feedback to internally stabilize the robot, the *visual servo* systems use a visual controller to compute directly the input to the robot joints. Thus, if the look-and-move open-loop approach assumes that the environment including the object remains static after the robot has started to perform blind movements, the visual servoing is a closed-loop approach which gives more flexibility and can be used for moving objects. Currently automated visual feedback robot systems use vision and robot system as separate tools.

In this paper, we propose a service oriented architecture-based visual servoing system for the picking and placing of dynamic objects in real time. Considering service orientation of robot vision tasks in the manufacturing context, the VS system is treated as a composite process consisting in recognition, tracking, picking and handling parts from static or moving scenes. For each component of the composite VS system, service with IT support [5, 6, 7] was associated. Our architecture integrates a CCD camera with a Scale Invariant Features Transform (SIFT) detector, an image based predictive controller and a robot manipulator into a single entity. The visual features chosen for object description and recognition were introduced by [8] and are invariant to scale, orientation, illumination and affine transformation. An important stage in visual servoing is finding the correspondence between features from different frames. The descriptor attached to any SIFT feature is a magnitude orientation based representation of the surrounding region, which makes it suitable for visual servoing [9].

The visual servoing system developed for automating manufacturing is based on a predictive control architecture. Employing a point features motion predictor, a multivariable image based predictive controller was designed for vision-guided robot manipulation control. In order to have a more accurate control over the dynamic of the servoing system, a reference trajectory of the point features in the image plane was considered. Using a cost function based on errors between the reference trajectory and the predicted feature point trajectories, the convergence of robot motion has been obtained through nonlinear constraint optimization, which takes into consideration the visibility constraints due to sensor characteristics. A manufacturing system consisting of a work cell with a manipulator robot, a numerical control machine, a conveyor, computers and sensors was used for experimental testing of the vision-guided robot predictive control architecture.

The paper is organized as follows. In Section 2 the SIFT features detection and tracking are analyzed. Section 3 contains the design details of the image based predictive controller. The SOA based visual servoing implementation and experimental results are discussed in Section 4, while conclusions are revealed in Section 5.

2   Visual Features Recognition and Tracking

The choice of visual features has a strong influence on the performance of the image-based servoing systems used to control robot manipulators. A frequent approach is image-point-based visual servoing that uses simple features. During the pose control of the robot's end-effector, two types of transformations must be taken into account: change of scale and change of viewpoint. A change of scale implies the change of the size of the object in the image and a change in viewpoint implies that the object is being viewed from a different location. For these reasons we chose for visual features extraction the SIFT algorithm [8] which gives features (named *key points*) that are invariant to image scaling and rotation, and partially invariant to change in illumination and 3D camera viewpoint.

2.1   Key Point Extraction and Description

The SIFT detector was developed following the stages of computation from [8].

Scale-Space Extreme Detection. In this first stage, a search in all the image scales is made in order to find scale and orientation invariant features. For this purpose, the scale space theory is applied. This approach states that the scale invariant locations in an image can be found by making the convolution product between the image $I(x, y)$ and the Gaussian function $G(x, y, \sigma)$, dependent on scale σ

$$L(x, y, \sigma) = G(x, y, \sigma) * I(x, y), \tag{1}$$

where the $*$ operator denotes the convolution. In order to detect the stable locations in scale space, the Difference of Gaussians (DOG) function $D(x, y, \sigma)$ is used, which computes the difference between two Gaussian filtered images, whose scale differs with a multiplicative constant k :

$$D(x, y, \sigma) = L(x, y, k\sigma) - L(x, y, \sigma). \tag{2}$$

Now the DOG space is employed to find local extrema: each image pixel, placed on the current scale point is compared to its eight neighbors in the current image and nine neighbors in the scale above and below, to find if it has an extreme value. Having this step accomplished, the next stage follows.

Key Point Localization. For each extrema feature, its localization and scale is determined and the points that are placed on edges or that have low contrast are rejected. The first three terms of a Taylor expansion for DOG function $D(x, y, \sigma)$ are considered

$$D(\mathbf{x}) = D + \frac{\partial D^{T}}{\partial \mathbf{x}} \mathbf{x} + \frac{1}{2} \mathbf{x}^{T} \frac{\partial^{2} D}{\mathbf{x}^{2}} \mathbf{x} \tag{3}$$

where D and its derivatives are evaluated in each extremum point, and $\mathbf{x} = (x, y, \sigma)^T$ is the offset related to this point. By setting (3) to zero the true extremum location is obtained:

$$\hat{\mathbf{x}} = -\frac{\partial^2 D^{-1}}{\partial \mathbf{x}^2} \frac{\partial D}{\partial \mathbf{x}}. \tag{4}$$

If $\hat{\mathbf{x}}$ is greater than 0.5 in any dimension, the point is rejected because it is close to another extremum point. In the end, the offset is added to the feature's coordinates in order to establish a range for the extremum location. In order to discard points with low contrast, that are unstable, the value of DOG function in the extremum $D(\hat{\mathbf{x}})$ must be less than a threshold. This can be obtained by substituting equation (4) into (3) giving:

$$D(\hat{\mathbf{x}}) = D + \frac{1}{2}\frac{\partial D^T}{\partial \mathbf{x}}\hat{\mathbf{x}}. \tag{5}$$

Orientation Assignment. For this purpose, the scale of the feature is used to select the image with the nearest scale from the Guassian pyramid. This ensures that the future calculations are made in a scale invariant space. For each selected image $L(x, y)$ found with that scale, the magnitude $m(x, y)$ and the orientation $\theta(x, y)$ are computed:

$$m(x, y) = \sqrt{(L(x+1, y) - L(x-1, y))^2 + (L(x, y+1) - L(x, y-1))^2}$$
$$\theta(x, y) = tan^{-1}(\frac{L(x, y+1) - L(x, y-1)}{L(x+1, y) - L(x-1, y)}) \tag{6}$$

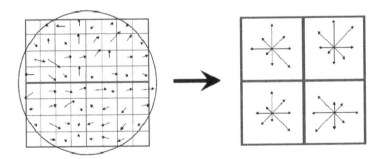

Fig. 1. Key point descriptor construction

Key Point Descriptor. In this final stage, for each chosen feature, a region centred in that point is selected, using a Gaussian window and for all the points localized in that region, the gradient magnitude and the orientation are determined. In Fig. 1, on the left side, the gradients are represented with arrows, and the Gaussian window with a circle. In the right side the keypoint descriptor is presented. First, the entire region is partitioned in four sub-regions, and for each one, an eight direction histogram is created.

2.2   Visual Features Tracking

The SIFT features

$$s = (s_p \ s_d)^T = (s_{p1} .. s_{pm} \ s_{d1} .. s_{dm})^T, \qquad (7)$$

contain both information about the key point position s_{pi} and the descriptor s_d. The keypoint position is stuctured as a vector $s_{pi} = (u_i, v_i)$, where u_i, v_i are feature point coordinates in the image plane. The robot motion and implicitly the camera motion is obtained by minimizing the error in the image plane between the reference of the feature positions s_p^* and the current feature positions $s_p(k)$ at discrete-time k. In order to correctly compute the error $e(k)$ between the feature points s_{pi}^* and $s_{pi}(k)$, the features descriptor s_d is used to establish the corresponding points between the desired image $I^*(x, y)$ and the current image $I_k(x, y)$. Using the descriptor s_d for tracking the visual features, the computation time is greatly reduced because the system no longer uses a matching algorithm and thus the task of locating features in sequential scenes is easier.

3   Image-Based Predictive Controller

The predictive controller developed for image-based visual servoing (IBVS) combines concepts from model based predictive control theory (MPC) [15] with results obtained by IBVS in motion control problems for autonomous robot manipulators. The strategy of predictive control based on receding horizon principle has been adapted for IBVS, as depicted in Fig. 2. Thus, the distance between the camera and the object is represented on the z axis. In the desired position for object grasping the distance z^* is minimum and the image $I^*(x, y)$ is obtained with the corresponding feature points s_p^*. In the followings $I_k(x, y)$ will be denoted by $I(k)$.

Thus, in IBVS control systems, the set point is the desired features s_p^* obtained from a reference image of the desired grasping when the end-effector is correctly positioned relative to the target object. Starting from the current position of the camera with the distance $z(k)$ and the features $s_p(k)$, a reference trajectory is necessary in visual predictive control to define the way how to reach the desired

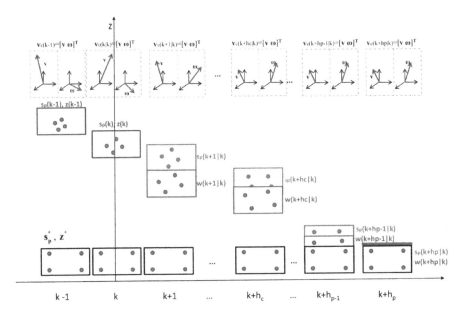

Fig. 2. Predictive control strategy for IBVS

features s_p^* over the prediction horizon. Such a reference trajectory, denoted with $w(k+i\,|\,k)$ in Fig. 2, starts at the current discrete time k with the current image $I(k)$ having the feature points $s_p(k)$.

The reference trajectory is designed from the image sequences, $I(k+i), i = \overline{1, h_p}$ over the prediction horizon h_p, with the feature points so that to obtain $w(k+h_p\,|\,k) = s_p^*$. The notation $w(k+i\,|\,k)$ specifies that the reference trajectory depends on the initial conditions at discrete time k. In order to predict the visual features evolution in the image over the horizon h_p, $s_p(k+i\,|\,k), i = \overline{1, h_p}$, a nonlinear global model or a local model based on the interaction matrix is used together with the past values of the plant input and output and the feature control sequence $v_c^*(k+i\,|\,k), i = \overline{1, h_p}$. For the IBVS control systems, the plant is the robot-visual sensor ensemble, so the input is the camera velocity reference $v_c^*(k)$ with its components, linear velocity v and angular velocity w and the output is represented by the feature points $s_p(k)$. In Fig. 2 the camera velocity is represented through its components. The future control sequence is obtained via the optimization of a cost function, generally defined as a quadratic function of the predicted control error and the control input.

3.1   Reference Trajectory Generation

In order to generate a reference trajectory in the image plane, we have chosen the method developed in [10] based on the 3D motion planning approach for image-based visual servoing task from [11].

We consider that the initial image is $I(k)$ with feature points $s_p(k)$, the final one is $I(k + h_p)$ with the desired feature points s_p^* and the fixed object is described by four feature points assumed to be coplanar, but not collinear. The reference trajectory varies gradually over the prediction horizon from the current feature points $s_p(k)$ at discrete time k to the desired feature points $s_p^*(k)$ at discrete time $k + h_p$. Taking into account the collineation matrix $\mathbf{G}$ representing the projective homography between the initial image $I(k)$ and the final image $I(k + h_p)$, the homogeneous coordinates of the four feature points from the final image $\tilde{s}_{pi}^* = (u_i^*, v_i^*, 1)^T$ can be expressed in relation to the coordinates of points from the initial image $\tilde{s}_{pi} = (u_i(k), v_i(k), 1)^T$, resulting:

$$\tilde{s}_{pi}^* = \mathbf{G} s_{pi}(k), i = \overline{1,4}. \tag{8}$$

In order to obtain a path between the current point feature positions $s_p(k)$ at time $t = KT_s = 0$ and the desired one s_p^* at time $t = (k + h_p)T_s = t_{h_p}$ it is necessary to build a sequence of collineation matrices that will be correlated with a time variation law. The sequence can be obtained using the parameter-dependent matrix

$$\mathbf{G}_d(q) = \mathbf{K}\mathbf{H}_d(q)\mathbf{K}^{-1}, \tag{9}$$

where $q = q(\tau)$ is a monotonic function of the non dimensional time $\tau = t / t_{h_p}$ ranging from $q(0)$ at $t = kT_s = 0$ to $q(1)$ at $t = (k + h_p)T_s = t_{h_p}$ and the matrix $\mathbf{K}$ with the intrinsic camera parameters. The matrix $\mathbf{H}_d(q)$ must fulfill the conditions: $\mathbf{H}_d(q(0)) = \mathbf{I}, \mathbf{H}_d(q(1)) = \mathbf{H}$. In the general motion case, when both a rotation and a translation of the camera are needed in order to reach the final view, the time dependent Euclidean homography $\mathbf{H}_d(q)$ is given by

$$\mathbf{H}_d(q) = \mathbf{R}_d^T(q)(\mathbf{I} - \frac{\mathbf{t}_d}{d_k}(q)\mathbf{n}_k^T)$$

$$\text{with } \mathbf{R}_d(0) = \mathbf{I}, \mathbf{R}_d(1) = \mathbf{R}_{khp}, \frac{\mathbf{t}_d}{d_k}(0) = \mathbf{0}, \frac{\mathbf{t}_d}{d_k}(1) = \frac{\mathbf{t}_d}{d_k}, \tag{10}$$

where $\mathbf{R}_{khp}$ is the rotation matrix between the desired camera frame F_{k+h_p} and the initial camera frame F_k , $\mathbf{t}_{kh_p}$ is the translation between F_{k+h_p} and F_k expressed in

F_{k+h_p}, d_k is the distance from point features plane Π and the origin of F_k and $\mathbf{n}_k$ is the unitary normal to Π expressed in F_k [10]. Starting from the initial conditions $s_p(k)$, the reference trajectory is generated using the above algorithm in order to obtain $w(k+h_p \mid k)$ at the end of the prediction horizon.

3.2   Local Model Based Prediction

The plant model consists of the robot model combined with the camera model, generating a nonlinear global model. This complex model, used to predict the evolution of the visual features with respect to a control sequence over the horizon h_p generates difficulties in the predictor development. To overcome these problems due to the complexity of the nonlinear global model, a local model was proposed in [12] based on the relation between the camera velocity $\mathbf{v}_c$ and the time variation of the visual features $\dot{s}_p$ given by the interaction matrix.

Let us consider that the object from a visual servoing application is characterized by m point features defined by

$$\mathbf{s}_p = [\mathbf{s}_{p1}^T ... \mathbf{s}_{pm}^T]^T , \tag{11}$$

where feature point s_{pi} has the coordinates (u_i, v_i) in the image plane. Considering the position and the orientation of the object and the camera to be $x_o \in \mathbb{R}^6$ and $x_c \in \mathbb{R}^6$, respectively, the point features are achieved employing the camera model

$$s_p = i(x_c, x_o), \tag{12}$$

where i is the mapping $i : \mathbb{R}^6 \times \mathbb{R}^6 \to \mathbb{R}^{2m}$. The relation between the camera velocity $\mathbf{v}_c$ and the velocity of the point features $\dot{s}_p$ is obtained by computing the derivative of (12) and for static objects, one obtains

$$\dot{s}_p = L(s_p, d_c)\mathbf{v}_c, \tag{13}$$

where L is the interaction matrix and $d_c = [z_1 ... z_m]^T$ is a vector with the distances z_i from the object points to the camera frame. The interaction matrix is defined in [2] as

$$L = [L_1 ... L_m]^T , \tag{14}$$

where L_i is the interaction matrix of the ith feature point of coordinates (u_i, v_i) in the image plane

$$L_i = \begin{bmatrix} \dfrac{-\lambda}{z_i} & 0 & \dfrac{u_i}{z_i} & \dfrac{u_i v_i}{\lambda} & -\dfrac{u_i^2 + \lambda^2}{\lambda} & v_i \\[2.5ex] 0 & \dfrac{-\lambda}{z_i} & \dfrac{v_i}{z_i} & \dfrac{v_i^2 + \lambda^2}{\lambda} & -\dfrac{u_i v_i}{\lambda} & -u_i \end{bmatrix}, \tag{15}$$

and λ is the focus distance.

Starting from the relation between the camera and point features velocities given by (13) and considering the robot model as a virtual Cartesian motion device (VCMD) with camera velocity reference $\mathbf{v}_c^*$ as input and camera velocity $\mathbf{v}_c$ as output, the plant model for image prediction from Fig. 3 was developed in [10]. The VCMD is considered an inner velocity loop which controls the camera velocity and the outer one implements the image based control loop. The inner loop is regarded as an analogue system because the sampling period of the inner loop is very short (usually 1 ms) and it is described by the transfer matrix $\mathbf{G}(s)$. This transfer matrix approximates the nonlinear robot dynamics using different approaches [13, 14] and typically has a diagonal form obtained through a suitable design of the multivariable inner velocity control loop.

The input of the VCMD is the camera velocity reference $\mathbf{v}_c^*$ which is applied to analog velocity control loops described by the transfer matrix $\mathbf{G}(s)$ through a zero-order holder (ZOH). Thus, the discrete time VCMD model is

$$\mathbf{G}(z) = (1 - z^{-1})Z\{\mathbf{G}(s)/s\}, \tag{16}$$

where Z symbolizes the z transform. Having $\mathbf{v}_c(k)$ as output of the VCMD discrete time model, the discrete time relation between camera and point features velocities is obtained by discretization of (13) using Eulers' method

$$s_p(k+1) = s_p(k) + T_e L_k \mathbf{v}_c(k), \tag{17}$$

where L_k is the interaction matrix computed with the point features $(u_i(k), v_i(k))$ acquired at the current discrete time k with a camera and an appropriate point feature detector and T_e is the sampling period.

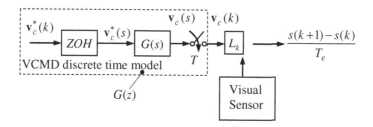

Fig. 3. Local model based predictor

It is assumed that it is possible to compute at every sampling period the depth $z_i(k)$ of the current point features with respect to the camera frame.

The one-step ahead prediction of the image point features evolution can now be calculated using (17) and the discrete model (16) of the VCMD, resulting

$$s_p(k+1|k) = s_p(k|k) + T_e L_k \mathbf{G}(z)\mathbf{v}_c^*(k|k), \tag{18}$$

where the notation $s_p(k+1|k)$ indicates that the prediction is computed at the discrete time k. Shifting the one-step ahead prediction model (18) by recursion, the i-step ahead predictor $s_p(k+1|k)$ is obtained:

$$s_p(k+i|k) = s_p(k+i-1|k) + T_e L_{k+i-1}\mathbf{G}(z)\mathbf{v}_c^*(k+i-1|k). \tag{19}$$

Next, the i-step ahead predictor will be used to design an image-based predictive control law.

3.3   Image-Based Predictive Control Law

The control law is obtained through the minimization of a quadratic objective function so that future system outputs converge for a desired reference trajectory $w(k+i|k), i = \overline{1,h_p}$. The error in image space over the prediction horizon h_p is given by:

$$e(k+i|k) = s_p(k+i|k) - w(k+i|k), i = \overline{1,h_p} \tag{20}$$

and the objective function to be minimized is defined by

$$J = \frac{1}{2}\sum_{i=1}^{h_p} e^T(k+i|k)\mathbf{Q}e(k+i|k) + \sum_{i=0}^{h_c-1} v_c^{*T}(k+i|k)\mathbf{R}v_c(k+i|k), \tag{21}$$

where $\mathbf{Q}$ and $\mathbf{R}$ denote positive definite, symmetric weighting matrices and h_c is the control horizon.

The main constraints are associated to the limits of the image called the *visibility constraint*, ensuring that all the features are always visible:

$$(u_i(k), v_i(k)) \in [(u_{min}, v_{min}), (u_{max}, v_{max})]. \tag{22}$$

Other two constraints related to the robot are frequently used, respectively the torque constraints, the joint boundaries and the camera velocity constraints.

4   SOA Based Visual Servoing Implementation and Experimental Results

In this section a design method for a visual predictive control scheme is analyzed in a manufacturing system application.

4.1   SOA Based Manufacturing System

As part of the CIM Laboratory, the manufacturing system consist of two industrial robots, both being produced by by Asea Brown Boveri Sweden, namely IRB 1400 and IRB 2400. In the working space of the IRB 1400 there are: a storage device, a machine tool PC Mill 55 and a pallet conveyor, which was built using conveyor components from Flex Link. In the working space of the IRB 2400 there are: a work table, a storage device which contains pallets (both the empty ones and those filled by the robot) and the conveyor.

The goal for the manufacturing system is to achieve a desired number of pallets, each of them with a specified layout of processed parts. The IRB 1400 robot must solve the manipulation tasks regarding the loading and unloading of the machine tool. The main task of the IRB 2400 is to place the processed parts on the right positions on the pallets from the storage device S2. The flow of the manufacturing process which involves IRB2400 is composed from a sequence of steps. First, a processed part is transported by the conveyor into an area were it can be picked up by the manipulator robot. Next, the part is visually analyzed and classified according to a predefined parts database. For part recognition, the vision system takes an image of the part from the top and the complete segmentation of the image into object and background is done. For the object, using SIFT detector, the key points position s_p and the descriptor s_d are determined. The SIFT descriptor of a part is compared with descriptor models from a database and using a minimum distance classifier one obtaines how close each model matches with the part. After the recognition phase, in order to grasp the part, the IBVS control system is started. The IBVS control system for the picking and placing of dynamic objects in real time was achieved by integrating the CCD camera with SIFT detector, the image based predictive controller and the robot ABB into a Service Oriented Architecture (SOA).

The robot-vision tasks consisting in recognition, tracking, picking and handling parts from the work table is a composite process that consists of the four basic steps: image acquisition, image processing using the SIFT algorithm, image based control and robot execution. For each of these steps a corresponding service is defined. The choice of the service type to be used (for instance, SOAP/XML or RESTful web services) and its corresponding behaviour (either stateless or stateful) is dependent on the actual production environment. Our prototype assumes a closed environment and uses stateless services. Thus, a service is developed for each of the previously defined steps. The image acquisition service is designed to interact with the remote camera and provides the required images for analysis. The service accepts a request for image acquisition and the type of image to be provided and returns an image of the scene represented according to the desired format. The image processing service contains the SIFT algorithm. This second service relies on the image previously acquired and returns the key point position s_p and the determined descriptor s_d. The image based control service deals with establishing the trajectory to be followed by the video camera in order to fulfil the servoing task. The input message for this service consists of the SIFT descriptors of the current and target positions, while the resulting output of this third service represents the desired camera trajectory. Finally, the robot execution service is connected to the robot controlling equipment and is responsible for the actual movements of the robot. All four services create a SOA based visual sevoing prototype.

4.2   Visual Servoing Implementation

The flexibility of the manufacturing system presented in 4.1 was increased by a visual servoing scheme (Fig. 4). This scheme is composed from three different modules: Image Based Control Strategy, Robot Driver Interface (RDI) and Robot Controller, each one having a different execution time. The following notions are considered: T_{ap} the execution time for the Image Based Control module, T_c the RDI module execution time and T the execution time for Robot Controller module. The sampling period (T_e) of the entire architecture is obtained from:

$$T_e = T_{ap} + T_c + T. \tag{23}$$

The Image Based Control Strategy module implements an image based predictive control algorithm using SIFT features. The RDI module has three blocks: I/O, Robot Interface and User Interface. Robot Controller is a module composed from all the entities that ensure the correct execution of the desired robot motion.

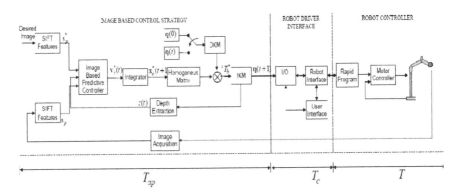

Fig. 4. Visual servoing structure

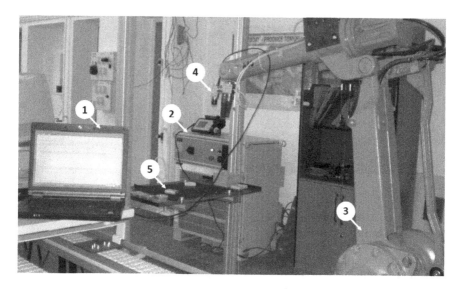

Fig. 5. Visual servoing implementation

Instructions from the RAPID program are used to generate trajectories for the internal arm-level. Next, the motor controllers are responsible for the low-level motion control of each joint.

The visual servoing architecture proposed in the present paper was implemented, tested and validated using the IRB 2400 robot (Fig. 5). A visual servoing system was developed and it is composed from: a master PC (label 1), the S4CPlus ABB controller (label 2), the IRB 2400 manipulator robot (label 3), a visual sensor mounted on the gripper (label 4) and the work table (label 5). The link between the visual sensor (label 4) and the master PC (label 1) is realized by an IEEE-1394b serial bus interface. The robots controller (label 2) communication with the master PC is performed over an Ethernet network, and images were acquired using a Sony XCD-V60CR visual sensor.

This type of industrial camera is based on a serial communication interface IEEE-1394b (FireWire). The IEEE-1394b interface allows a transfer rate of 800 Mbit/s and use a 9-pin connector. The intrinsic camera parameters used are : the unit cell size 7.4 $\mu m \times 7.4 \mu m$, the depth resolution was set to RAW16: 10bits/pixel and the focal length was set to $4.5\,mm$. Since an eye-in-hand camera configuration is used, the extrinsic camera parameters are related to the robot effector frame.

The visual sensor is able to acquire 90 fps with 640×480 image resolution. In order to optimize the computational time for the point features extraction, images with 320×240 resolution were considered.

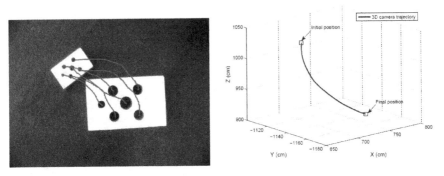

Fig. 6. a) Features trajectories between start and desired configurations b) 3D camera trajectory

4.3   Case Study

Next, the grasping of processed part using visual servoing is presented. Considering an initial state and desired state of the processed part image (Fig. 6), key points were extracted using the SIFT detector.

The first step was to establish the correspondence between the initial and the desired feature points. This step is done using the SIFT descriptor, which allows a correct assessment of correspondence between the two views. Having established the correspondence for each iteration, the image based predictive control law was implemented in order to ensure the smooth behavior of the camera while the visibility and robot constraints are considered.

The image based predictive controller was tunned manually and the parameters were set on: $h_p = 5$, $h_c = 2$, $\mathbf{R} = \mathbf{I}_6$ and $\mathbf{Q} = e^{1-i}\mathbf{I}_6, i = \overline{1, h_p}$. This structure of the $\mathbf{Q}$ matrix ensures weighting the error at each sampling instant more and more over the prediction horizon, thus stressing the error at the first step in the predictive sequence. For the image based predictive controller, the minimization of the cost function was done using the *fmincon* function from Matlab. The *fmincon* parameters were set to keep the camera velocity between $[-0.5; 0.5]\frac{m}{s}$ and $[-11.25; 11.25]\frac{deg}{s}$.

The experimental results show that the servoing task is fulfilled, the feature point trajectories in the image plane are depicted in Fig. 6a, while the camera trajectory in Cartesian space is given in Fig. 6b.

5   Conclusions

This paper has focused on manufacturing automation, the main issues being the visual feature tracking, the object recognition and the image-based predictive control of a manipulator used for picking and placing of the dynamic objects in real time in a robot work-cell. A SOA based visual servoing implementation prototype was detailed and the advantages of such an approach were highlighted. The vision-guided robot system is able to recognize general 2D and 3D objects to be manipulated based on SIFT key points. For grasping the objects after the recognition phase, an IBVS control system was developed by integrating a CCD camera with SIFT detector, an image based predictive controller and the robot ABB into a single application. A manipulator robot based manufacturing system was used for experimental testing of the vision-guided robot predictive control architecture in an application of objects manipulation.

References

1. Blomdell, A., Bolmsj, G., Brogrdh, T., Cederberg, P., Isaksson, M., Johansson, R., Haage, M., Nilsson, K., Olsson, M., Olsson, T., Robertsson, A., Wang, J.: Extending an industrial robot controller implementation and applications of a fast open sensor interface. IEEE Robotics and Automation Magazine 12(3), 85–94 (2005)
2. Chaumette, F., Hutchinson, S.: Visual Servo Control, Part I: Basic Approaches. IEEE Robotics and Automation Magazine 13(4), 82–90 (2006)
3. Xie, W.F., Li, Z., Tu, X.W., Perron, C.: Switching control of image based visual servoing with laser pointer in robotic assembly systems. IEEE Transactions on Industrial Electronics, 520–529 (2009)
4. Bilen, H., Hocaoglu, M., Unel, U., Sabanovic, A.: Developing robust vision modules for microsystems applications. Machine Vision and Applications 23(1), 25–42 (2012)
5. Zhao, Y.Z., Zhang, J.B., Zhuang, L.: Service-oriented architecture and technologies for automating integration of manufacturing systems and services. In: Proceedings of the 10th IEEE conference on Emerging Technologies and Factory Automation, Catania, pp. 349–355 (2005)
6. Deng, W., Yang, X.: Flexible Application System Integration Based on Agent-Enabled SOA in the Industrial Field. In: Proceedings of Second International Symposium on Electronic Commerce and Security, Nanchang, pp. 120–124 (2009)
7. Mendes, J.M., Restivo, F., Leitão, P., Colombo, A.W.: Injecting Service-Orientation into Multi-Agent Systems in Industrial Automation. In: Rutkowski, L., Scherer, R., Tadeusiewicz, R., Zadeh, L.A., Zurada, J.M. (eds.) ICAISC 2010, Part II. LNCS, vol. 6114, pp. 313–320. Springer, Heidelberg (2010)
8. Lowe, D.G.: Distinctive Image Features from Scale-Invariant Keypoints. Journal of Computer Vision 60(2), 91–110 (2004)
9. Maxim, A., Lazar, C., Burlacu, A., Copot, C.: Robotic visual servoing system based on SIFT features. In: 16th International Conference on Systems Theory, Control and Computers, Sinaia, pp. 1–6 (2012)
10. Lazar, C., Burlacu, A., Copot, C.: Predictive Control Architecture for Visual Servoing of Robot Manipulators. In: Proceedings of the 18th IFAC World Congress, pp. 9464–9469 (2011)
11. Allotta, B., Fioravanti, D.: 3D Motion Planning for Image-Based Visual Servoing Tasks. In: Proceedings of the IEEE International Conference on Robotics and Automation, Barcelona, pp. 2173–2178 (2005)
12. Lazar, C., Burlacu, A.: Visual Servoing of Robot Manipulators using Model-Based Predictive Control. In: Proc. of the 7th IEEE Int. Conference on Industrial Informatics, Cardiff, pp. 690–695 (2009)

13. Gangloff, J., de Mathelin, M.: High speed visual servoing of a 6 DOF manipulator using multivariable predictive control. Advanced Robotics 21(10), 993–1021 (2003)
14. Fujimoto, H.: Visual servoing of 6 dof manipulator by multirate control with depth identification. In: Proceedings of the 42nd IEEE Conference on Decision and Control, Maui, Hawaii, pp. 5408–5413 (2003)
15. Rawlings, J., Mayne, D.: Model Predictive Control: Theory and Design. Nob Hill Pub Press (2009)

Integration of Visual Quality Control Services in Manufacturing Lines

Florin D. Anton, Silvia Anton, and Theodor Borangiu

Dept. of Automatic Control and Applied Informatics,
University Politehnica of Bucharest, Bucharest, Romania
{florin.anton,silvia.anton,theodor.borangiu}@cimr.pub.ro

Abstract. The paper discusses a new method for integrating visual quality control in highly dynamic manufacturing lines where new products are added in the production inventory and older products are removed from production. In such dynamic lines a new approach for visual inspection is required, a method which requires that the new tasks for visual quality control should be included in the product data and not in the vision/robot programs. The paper proposes a system for visual quality control which allows to define new inspection tasks and to modify the old ones without intervening in the production process. The visual inspection task is defined, linked with a product data and stored on the production server; when a product must be created the task definition is retrieved from the server, a special program on the robot station is parsing the data and executes the vision checks, after which the results are reported. The paper is structured in five chapters presenting the motivation of the paper and trends in visual quality inspection, the product data and visual inspection task definition, the vision systems used, the manufacturing line and ends with experimental results and conclusions.

Keywords: Visual quality control, Assembly robots, Industrial robots, Robotic manipulators, Production data.

1 Introduction

In manufacturing lines an important role is held by quality control operations. Quality control is the operation which gives an important feedback about the processes involved in creating the products and the problems which alter the final product. Depending on the type of the product, quality control is done by hand, by human operators specialized in quality control using different tools (different types of light, magnification tools, etc.). If the product and the manufacturing process permit, an automatic visual quality control can be implemented and this will boost the productivity of the line; the employees in quality control can be redirected to other activities, and the cost of production decreases.

Different approaches have been proposed during the years: Skinner et. al. [1] proposed a method which simulates the processes of adaptation, fixation, and feature extraction in the visual system, with an original algorithm for optimizing convolution

T. Borangiu et al. (eds.), *Service Orientation in Holonic and Multi-Agent Manufacturing and Robotics*, Studies in Computational Intelligence 544,
DOI: 10.1007/978-3-319-04735-5_22, © Springer International Publishing Switzerland 2014

masks to distinguish between acceptable and unacceptable images. In [2] is presented an automatic visual inspection of the surface appearance defects of bearing roller, where they applied an analogue thresholding technique for improving inspection reliability. In [3] an automated system for design-rule-based visual inspection of printed circuit boards is proposed; the visual inspection is based on processing the image using morphological tools and checking a set of geometric design rules - subject discussed also in [4]. Yang et. al., [5] push the visual inspection a step forward by using CAD data in the visual inspection process.   In [6-10], feature extraction and processing in automated visual inspection systems are introduced, the solution being tested for visual quality control in the electronics industry. In Zang and Hashimoto [11-12], a system framework realizing the automatic visual inspection with random initial object poses   is presented; the system uses a visual servo technique in order to accurately place the camera so that a reliable visual inspection can be carried out. In general colour image processing is used in special applications, [13-15] the majority of visual quality control tasks being executed on gray level images.

This paper presents a method and strategy for integrating automated visual control in highly dynamic manufacturing lines by creating an inspection template and associating this template to the product data. In this way, the necessity to implement a different visual inspection application for a new product is eliminated, the inspection application will be a generic application, but parameterized with different sets of characteristics for each product / inspection process.

2   Holonic Manufacturing Control with SOA Knowledge Base

The proposed system is integrated in a holonic manufacturing line which uses KB (Knowledge Base) SOA for customer request management. This approach consists in four information areas (Fig. 1.):

• Offer Request Management
• Customer Order Management
• Order and Supply Holon Management
• Order and Supply Holon Execution

Offer request management is responsible for publishing the offers and allowing the service requestors to select an offer or a set of offers and request execution of these offers. The orders are published in a service registry from where the service requestors can search and find the available services and can make requests are then sent to the customer order management as online product orders.

The *customer order management* module receives online or offline product orders which are sent to a customer order interpreter and also receives updates from the service provider about supplier data and engineering, updates which are stored in a production database along rules and production strategies, resource holons (RH) and product holons (PH). The information from the database is interconnected with the customer order interpreter which processes product orders by creating an aggregate list of product orders that is then sent to a job-shop scheduler (in hierarchical planning mode) for batch production planning.

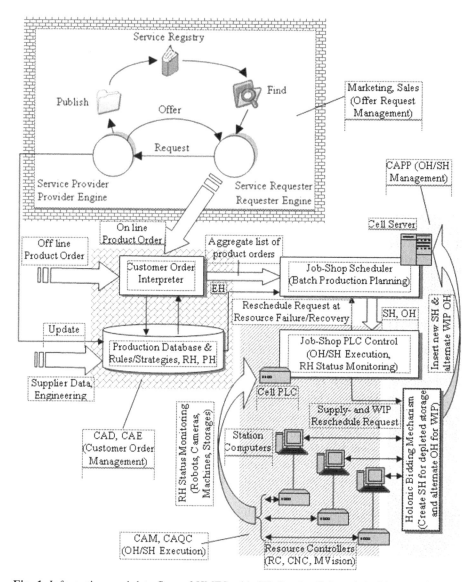

Fig. 1. Information- and data flow of HMES with KB Service Oriented Architecture for customer request management

The job-shop scheduler is executed on an IBM x3500 server which is the production cell server and is included in the order holon (OH) / supply holon (SH) management module. The scheduler takes as input the list of product orders and supply holons for the depleted storage of raw materials associated with each assembly station and creates order holons and supply holons which are then sent for execution in the "production with on line supply" execution module.

The order holon/supply holon execution module is composed by the resources which forms the assembly cell: a cell PLC which executes the order holons and supply holons and monitors the resources holons. The resources holons are resources controllers (controllers of the robots inside the assembly cell), CNC machines and machine vision systems. The resources holons are executing different operations sent by the PLC and also send status information about the current operation and overall status of the resource back to the PLC. In this way a faulty operation or a malfunction inside the cell can be detected and proper measures can be taken. Also the resource holons can generate supply holons in order to request that a depleted storage for raw materials used in production should be refilled.

3   Product Data

In order to create an command (order) which will be processed by the manufacturing line a set of product data must be associated with each product to be manufactured. The product data describes the manufacturing process (the operations and the execution order), the raw materials needed, the resources used for the manufacturing process, the visual quality inspection tasks, etc. The data associated with the product has the following structure:

1. Product code: #IDpr$_k$;
2. Associated set of operations: [o_k1, o_k2, ... , o_kf]
3. Operations order: [p(o_k1), p(o_k2), ... , p(o_kf)] where p(o_ki) $\in$ {$\emptyset$, ... , (f − 1)} are operations executed before the current operation
4. Necessary materials (components) (array of materials [type; number] / operation): [m (t, nm)] (o_k1), [m (t, nm)] (o_k2), ... , [m(t, nm)] (o_kf), where [m (t, nm)] (o_ki) = [t_1i | nm_1i, t_2i | nm_2i, ... , t_fi | nm_fi]
5. Necessary tools (tools array [type; number] / operation): [s (t, ns)] (o_k1), [s (t, ns)] (o_k2), ... , [s (t, ns)] (o_kf), where [s [t, (ns)] (o_ki) = [t_1i | ns_1i, t_2i | ns_2i, ... , t_ui | ns_ui]
6. Necessary programs (array of #ID programs / operation): IDprg(o_k1), IDprg(o_k2), ... , IDprg(o_kf), where IDprg(o_ki) = IDprg_1i, IDprg_2i, ... , IDprg_vi

For the visual inspection tasks (VIT) an image must be acquired and the pallet/object must be detected and located, then a set of tools are applied in order to make measurements, features detection and localizations, etc. The VIT is returning a FAILED/PASS result for each visual tool (telling whether a feature has been detected, the measurements are in the defined tolerances, etc.) applied in order to assess the quality of the inspected product.

For example, in order to measure an object the measurement task (MSO) is inserted after a detection/locating object task; the measurement operation requires a program identified by IDprg( ) and having the parameters:

- MSO(*o*) [object_type, meas_1 (), ... , meas_k (), validation (), message (), action()]

  - *objec_typet*: blob
  - *meas_i* (meas_type, AOI (meas_i), crit[type_cr, reference, val_std] (meas_i), ...
    , res[type_cr, reference, val_crt] (meas_i))

    meas_type: the type of the measurement:

    <u>GLOBAL</u>:

    Shape descriptor:

    Scalar:
    - ♣ perimeter length
    - ♣ convex hull, convex deficiency
    - ♣ minimal bounding rectangle
    - ♣ compactness, roundness, eccentricity
    - ♣ diameter of the boundary and basic rectangle
    - ♣ mass distribution as a function of distance measured from the object's centroid (cent)
      - ○ dist(cent, furthest point on boundary)
      - ○ dist(cent, closest point on the boundary)
      - ○ number of protuberances
      - ○ ratio of object's projection onto the major and minor axes of the best-fit ellipse

    Vector:
    - ♣ polar radii signature
    - ♣ linear offset signature
    - ♣ slope of the boundary
    - ♣ curvature

    Region descriptor:

    Scalar:
    - ♣ standard moments of (p + q) order
    - ♣ central moments of (p + q) order
    - ♣ area
    - ♣ 7 invariant moments (to rot., trans., and mirroring)
    - ♣ principal axes of object
    - ♣ area of object / area of convex hull
    - ♣ area of object / area of its circumcircle
    - ♣ area of object / area of the min. bounding rectangle
    - ♣ number of holes in the object's body
    - ♣ Euler number of the region

    Vector:
    - ♣ skeleton of object

LOCAL:

Anchor point existence:
- ♣ "corner" type
- ♣ "colour contrast" type
- ♣ "hole centre" type
- ♣ "mass centre" type
- ♣ "contour point" type

Straight line existence

Arc shape existence

Length of line:
- ♣ between 2 anchor points
- ♣ between an anchor point and a line
- ♣ between an anchor point and an arc

Angle:
- ♣ between 3 anchor points
- ♣ between two lines

Length of an arc (distance)

Collinearity of 3 anchor points

In order to execute the measurement the area of interest AOI(meas_i) should be defined. The AOI is the area where the measurement should be executed and is defined by shape, dimensions and location specified relative to the measured object (axis of minimum inertia of the object and the centre of the object).

Also the following information should be available:

- crit[cr_type, mark, std_val] (meas_i): standard structure of the measurement, where:

- cr_type (meas_i) = cr_type_i1, cr_type_i2, ... , cr_type_if is an array of criteria which form a measurement meas_i of the type meas_type

- mark (cr_type_ij) = mark_j1, mark_j2, ... , mark_jm is an array of marks (for example points, coordinates, distances, angles, graph nodes, ...) which mark individual measurements

- std_val (mark_jk) = std_val$_{\_k11}$, ε_{k11} ... , std_val$_{\_k1x}$, ε_{k1x} |

std_val$_{\_k21}$, ε_{k21} ... , std_val$_{\_k2y}$, ε_{k2y} |

....

std_val$_{\_ke1}$, ε_{ke1} ... , std_val$_{\_kez}$, ε_{kez} is the set of arrays of variable length of the of the standard measurement values and admissible errors, associated to each mark of the current criteria of the measurement

- res[cr_type, mark, crt_val] (meas_i): the structure of the result of the measurement, where crt_val (mark_jk) = crt_val$_{\_k11}$, ... , crt_val$_{\_k1x}$ |

crt_val$_{\_k21}$, ... , crt_val$_{\_k2y}$ |

....

crt_val$_{\_ke1}$, ... , crt_val$_{\_kez}$ is the set of arrays of variable length of the current measured values, associated of each mark of the current measurement criteria.

A component $meas\_j(),1 \le j \le k$ of the measurement MSO(o) is corresponding if for all measurements that have the type crit_type_jh, $1 \le h \le f$, all components of the array of standard values [std_val ( )] are different with at most the components associated from the array of admissible errors, [$\varepsilon()$]

validation ([MSO($o\_$i)]): is a function returning a Boolean value (TRUE / FALSE), which interconnects through relational ($\le,\ge,==,<,>$,) and logic (AND, OR, NOT, ...) operators the components of the array of executed measurements, $meas\_i()$

message (valid (MSO(o)), [T/F[cr_type(meas_j)]], [[crt_val (meas_j)]): data file containing the global result, on sets of measurements, and all values measured / evaluated (for each type of criteria of each measurement meas_j ($o\_$i), $1 \le j \le k$):

- valid (MSO(o)): logical value (T/F) indicating the global result of product measurement (acceptable/not acceptable)

- [T/F[cr_type(meas_j)]]: logical value (T/F) indicating the result of each criteria cr_type_jm, $1 \le m \le f$ for each measurement meas_j, $1 \le j \le k$

- [[crt_val (mark_jk)](meas_j)]: the set of all current values (result) for all criteria types cr_jm, $1 \le m \le f$ of a measurement and for all measurements $meas\_j$ of MSO(o), $1 \le j \le k$

-     action(act_type,   [SIGNAL(adr)],   [MP_i(o)],   [MG_i(o)],   dest) (valid(MSO(o)),

... $1 \le i \le p$: action initiated depending of the global result (T/F) of the executed measurement MSO(o); the parameters are:

- act_type = {none I digital output command I pick-and-place object}: the action which will be initiated depends on the global result of the measurement; any combination of the three types of actions is admitted, for example:

$$\text{act\_type} = \begin{cases} \text{pick - and - place,} & \text{valid(MSO}(o)) = \text{TRUE} \\ \text{none,} & \text{valid(MSO}(o)) = \text{FALSE} \end{cases}$$

or

$$\text{act\_type} = \begin{cases} \text{pick - and - place,} & \text{valid(MSO}(o)) = \text{TRUE} \\ \text{sig(\#1002, -\#1003),} & \text{valid(MSO}(o)) = \text{FALSE} \end{cases}$$

- [SIGNAL(adr)] = [sig(adr_1), ..., sig(adr_m)]: the list of the digital output signal addresses, with sign (+ for SET, – for RESET), which will be commanded depending on the logical value of the global result of the measurement, valid(MSO(o)), for example:

$$\text{sig(adr\_}k) = \begin{cases} \text{on,} & \text{sign(adr\_}k) = \pm \text{ and valid(MSO}(o)) = \text{TRUE/FALSE} \\ \text{off,} & \text{sign(adr\_}k) = \mp \text{ and valid(MSO}(o)) = \text{TRUE/FALSE} \end{cases}$$

- [MP_i(o)], $1 \le i \le p$: is the set of grasping models robot-object
- [MG_i(o)], $1 \le i \le p$, is the set of gripper models, associated with grasping models robot-object, in order to avoid collisions
- dest: is the destination of the measured object (where the object will be placed), for "pick-and-place" actions:

$$dest = \begin{cases} dest\_1, & valid(MSO(o)) = TRUE \\ dest\_2, & valid(MSO(o)) = FALSE \end{cases}$$

As an example let's take the shift cam represented in Fig. 2. An 8-element vector $[D_1, D_2,...., D_8]$ is defined, where $D_i, i = 1,....8$ is the twice the distance from the minimum inertia axis (AIM) to the part's edge, measured respectively at $d_i, i = 1,....8$ mm from the " d_0 small contour's edge.

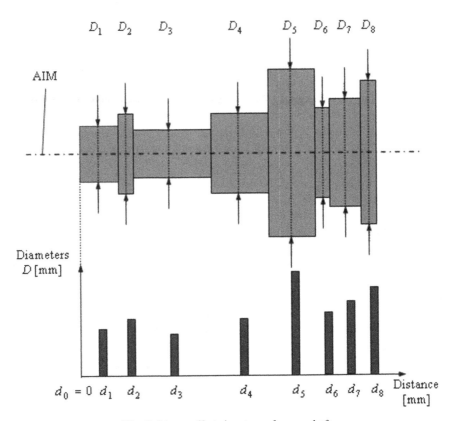

Fig. 2. Linear offset signature of a cam shaft

The product data for the object in respect with the measurement will contain the following information:

- *object_type*: blob
- *number of measurements*: 1 (meas_1( )):
  - ♣ meas_type: GLOBAL, vector edge descriptor of the type "linear offset signature "
  - ♣ AOI: rectangular (minimum rectangle box – MRB); $wd_{MRB} \times ht_{MRB}$;

    $(x_{c\_AOI} = x_{c\_MRB}, y_{c\_AOI} = y_{c\_MRB})\big|_C$, orient $= 0$ (C = mass centre,

    $(x_{c\_MRB}, y_{c\_MRB}) =$ centre of MRB

- ♣ crit [ ]:
  - □ crit_type: coincidence of the measured values with the standard ones (real values array) using an array of admissible errors
  - □ mark:

    [mark_0 $=$ d_0 (AIM$\cap$contour$=d_0$,dist(C,d_0)$=init\_val$, mark_1 $= d_1$, .., ... , mark_8 $= d_8$, means 9 points on AIM, trained and detected (identified) as follows:
    - Training: (i) mouse click on the contour (P_m) at the intersection with AIM; (ii) the system detects the point d_0 by searching: $d_0 \in$ contour AND $d_0 \in$ AIM and computes dist(C,d_0) $= init\_val$; (iii) the user specifies the number of marks for measurement: marks number $= 8 - 8$ bars are automatically displayed with $(\text{dist}(d_{i-1},d_i) = \text{dist}(d_i,d_{i+1}) \forall 1 \leq i \leq \text{marks no})$, normal to AIM, and their distances from d_0, $\text{dist}(d_0,d_i),1 \leq i \leq 8$, measured along AIM; (iv)   the user can modify the position of the bars which defines the measurement marks by following the online display of the d_i positions.
    - Detection: (i) intersection of the AIM with the contour of the current object, selection of one of the 2 points P found $-\varepsilon_{\text{init}} \leq |\text{dist}(P,C) - \text{init\_val}| \leq \varepsilon_{\text{init}}$, so $d_0 \equiv P$; (ii) measuring the distances $D_i = \text{dist}(d_i,\text{contour}),1 \leq i \leq 8$.
  - □ [std_val]: array of standard values [mm] of the diameters of the object in the established measurement marks $d_i, 1 \leq i \leq 8$ and of the admissible errors [%] for each diameter: [$D_{\text{std}\_i},\pm\varepsilon_i;...,D_{\text{std}\_8},\pm\varepsilon_8$]
- ♣ res [ ]:
  - □ [crt_val]: array of current measured values [mm], $D_{\text{mas}\_i},1 \leq i \leq 8$
- • validation ( ): $(-\varepsilon_1 \leq |D_{\text{std}\_1} - 2 \cdot D_{\text{mas}\_1}| \leq \varepsilon_1)$ AND ... ($-\varepsilon_8 \leq |D_{\text{std}\_8} - 2 \cdot D_{\text{mas}\_8}| \leq \varepsilon_8)$
- • message ( ):
  - ♣ valid(MSO(shift cam): T/F
  - ♣ [T/F[cr_type(meas)]: [T/F (meas_ D_1), ..., T/F (meas_ D_8)]
  - ♣ [crt_val(meas_1)]: array of current measured values, [$D_{\text{mas}\_1},...,D_{\text{mas}\_8}$]
- • action ( ):
  - ♣ act_type(valid(MSO("R087")): sig(#1002) = on if (valid (MSO) == T), respectively sig(#1003), −sig(#1004) if (valid(MSO) == F)
  - ♣ SIGNAL (adr): #1002, #1003, #1004

4   Automated Visual Quality Control

The Visual Quality Control (VQC) process is designed to be fully automated and to be modified *on the fly* during the execution of the production. The artificial vision systems used are produced by the company Adept Technology Inc. and there are two types of systems from the point of view of architecture:

- The first system (VS1) is integrated in the controller of the robot and is programmed using the high level language used also for the robot control (the vision system uses a dedicated set of instructions which are used for placing on the processed image a set of visual tools which are returning a set of information of interest regarding the features of the objects in the image)
- The second system (VS2) (is a new generation system), is a computer based system. The video cameras are mounted on the robot arms or on a fixed position top-down looking and are connected to the computer using high speed FireWire IEEE 1934 (High Performance Serial Bus) connections. The programming of the system is done in two steps:
  - **Creating a vision project for image processing**. The vision project is composed by two sections: a section which describes the connections between the cameras and the robot controllers, the setup of the cameras (vision calibration, vision-robot calibration) and a section which contains the vision sequence. The vision sequence represents a set of visual tools connected in a sequence for processing the image acquired from the camera.

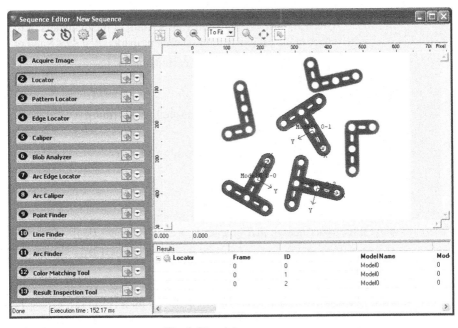

Fig. 3. The vision sequence

— **Creating the robot program** which connects to the vision system and triggers the acquisition of the image and execution of the vision sequences. The same program is able to modify the parameters of the visual tools and to retrieve the results obtained after the visual tools are executed.

Comparing the two vision systems VS1 and VS2 we can see that both of them are using visual tools and also both systems are accessed by the robots through special vision programming instructions.

Both systems are using a program which retrieves the visual inspection task data from the production server, parses the data and creates the program used for visual inspection. The VS1 system, depending on the task definition selects and executes directly the vision instructions and obtains the results. The second system VS2 has a generic vision sequence which is loaded on the image processing system (Adept Sight) and the program on the robot, depending on the visual task definition only configures the vision sequence (see Fig. 3.)

5   The Manufacturing Line

The system was tested on a flexible manufacturing line which uses 6 Adept Technology Inc. industrial robots (see Fig. 4): a Cartesian robot (R1) which communicates with the work cell's PLC in order to: (1) feed the production line with pallet carriers transporting pallets on which during the assembly processes there will be progressively created create products on the pallets, (2) check the quality of the final products and (3) store the pallets carrying the finally assembled products in a local buffer.

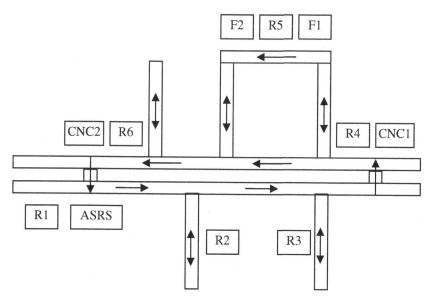

Fig. 4. The manufacturing line

Two SCARA robots (R2, R3) are used for standalone and collaborative assembly / disassembly tasks, two vertically articulated robots (R4, R6) are used for assembly / disassembly tasks but also for CNC machine feeding and discharge of manufactured assembly components. The last SCARA robot (R5) is used for feeding the stocks of all the other robots; when one such storage is low, the robot makes a request to the SCARA robot and this latter one robot fills a pallet with components using two feeding devices and send the pallet to the robot that has issued the request.

In this flexible cell each robot is using an artificial vision system (Adept Sight) with two video cameras. The SCARA robots are using the vision in order to verify the current status of the assembly and to recognize, localize and compute the grasping position of the objects the position of which is not known. The vertically articulated robots are using the vision for the same tasks as the SCARA robots, and also for verifying the shape of the objects created by the CNC machines. The Cartesian robot uses the vision in order to verify the integrity and the quality of the final product which is exiting the cell; if the product passes the quality check will be stored on the ASRS system of robot 1 and if the product is not passing the checks it will sent back in order to be disassembled and then reassembled.

6   Experimental Results

The proposed system for VQC has been tested and a set of preliminary results have been obtained in different visual inspection tasks. In Fig. 5 a set of measurements for VQC are presented (measuring a cam shaft and measuring angles and circle dimensions defined by anchor points); the computations use algorithms for object recognition (based on geometric model) and measurements (linear and circular measurements, circle detection, etc.).

The results obtained using this system are the same as using a regular visual inspection, the only drawback is the delay added by reconfiguring multiple times a tool, for example if a "Locator" tool is used multiple times in the same inspection the tool must be reconfigured each time the parameters must be changed.

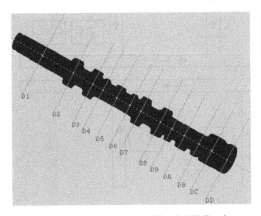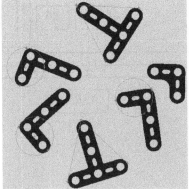

Fig. 5. VQC using measurements

Table 1 presents the measurement results for the cam shaft in Fig. 5; the first column represents the AIM angle and D1 from DD are the measurements in mm for each mark placed on the object. In the table you can see that for the 11 different orientations of the AIM the dimensions of the object in D1 to DD are varying so the tolerances and the array of acceptable errors must be selected accordingly.

Table 1. Linear offset signature results

AIM	D1	D2	D3	D4	D5	D6	D7	D8	D9	DA	DB	DC	DD
39.73	20.25	30.37	30.37	23.21	23.41	22.37	31.01	31.01	24.88	25.72	24.88	31.65	31.00
13.85	19.60	31.14	31.14	23.28	23.21	22.64	31.15	31.15	24.44	25.67	25.09	31.81	31.80
2.00	20.24	30.92	30.94	22.62	23.21	22.60	30.94	30.94	25.01	25.61	24.42	31.55	30.96
27.03	20.65	30.62	31.28	22.54	23.11	23.21	30.73	31.30	24.44	25.10	24.45	31.32	30.66
43.35	20.61	31.52	31.01	22.61	22.62	22.62	30.88	31.04	24.11	25.60	24.93	31.72	31.72
86.23	20.17	31.11	31.11	22.69	23.53	22.69	31.52	31.32	24.37	25.21	25.21	31.97	30.49
124.50	19.66	31.48	31.48	22.98	23.65	23.65	31.48	31.48	24.97	24.97	24.18	31.48	30.82
190.00	20.21	30.92	31.52	23.20	23.20	23.17	31.50	30.90	24.97	25.53	24.96	31.48	31.48
220.35	19.84	30.86	30.85	22.93	22.93	22.85	31.49	31.41	24.71	25.85	25.19	31.46	31.38
267.42	19.75	30.66	31.08	23.51	23.51	23.92	32.31	31.06	24.75	25.59	24.74	31.87	31.43
293.22	20.36	30.60	30.59	23.01	22.48	23.13	31.09	30.57	25.11	25.11	24.96	31.21	31.87

7 Conclusion

The described VQC system has a major advantage over other visual inspection tasks because there is no need to stop or intervene in the production flow. The task definition is done on an isolated vision system and then the task description is saved on the server. The system has also a drawback: when the number of objects/vision tools executed is very large and the required execution time is below 30ms per visual tool then the system is no longer able to execute all inspection tasks. This is happening because the generic program for visual inspection cannot be optimized. Normally if a vision sequence is executed the necessary time for each visual tool to be executed varies between 8 to 20 ms, depending on the tool and the associated parameters.

In our case, because we are using a generic sequence, the tools inside the sequence may be reconfigured multiple times during the execution, and that is the cause of the remaining time up to 30ms. For example if an object must be inspected using five arc detectors, the arc detector tool will be configured five times for each arc during a single inspection. This problem cannot be solved using the current vision system.

References

1. Skinner, D.R., Benke, K.K., Chung, M.J.: Application of Adaptive Convolution Masking to the Automation of Visual Inspection. IEEE Transactions on Robotics and Automation 6(I), 123–127 (1990)

2. Xian, W., Zhang, Y., Tu, Z., Hall, E.L.: Automatic Visual Inspection of The Surface Appearance Defects of Bearing Roller, CH2876-1/90/0000/1490$01.00 0 1990 IEEE, 1490-1494 (1990)
3. Oguz, S.H., Onural, L.: An Automated System for Design-Rule-Based Visual Inspection of Printed Circuit Boards. In: Proceedings of the 1991 IEEE Int. Conference on Robotics and Automation Sacramento, California, pp. 2696–2701 (April 1991)
4. Hassan, M.H., Diab, S.L.: Visual inspection of products with geometrical quality characteristics of known tolerances. Ain Shams Engineering Journal 1, 79–84 (2010)
5. Yang, C.C., Marefat, M.M., Kashyap, R.L.: Active Visual Inspection Based on CAD Models, 1050-4729/94 $03.00 0 1994 IEEE, 1120-1125 (1994)
6. Garcia, H.C., Villalobos, J.R., Runger, G.C.: An Automated Feature Selection Method for Visual Inspection Systems. IEEE Transactions on Automation Science and Engineering 3(4), 394–406 (2006)
7. Garcia, H.C., Villalobos, J.R.: Automated Feature Selection Methodology for Reconfigurable Automated Visual Inspection Systems. In: Proceedings of the 3rd Annual IEEE Conference on Automation Science and Engineering Scottsdale, AZ, USA, September 22-25, pp. 542–547 (2007)
8. Garcia, H.C., Villalobos, J.R.: Automated Refinement of Automated Visual Inspection Algorithms. IEEE Transactions on Automation Science and Engineering 6(3), 514–524 (2009)
9. Wu, H., Feng, G., Li, H., Zeng, X.: Automated Visual Inspection of Surface Mounted Chip Components. In: Proceedings of the 2010 IEEE International Conference on Mechatronics and Automation, Xi'an, China, August 4-7, pp. 1789–1794 (2010)
10. Mar, N.S.S., Yarlagaddan, P.K.D.V., Fookes, C.: Design and development of automatic visual inspection system for PCB manufacturing. Robotics and Computer-Integrated Manufacturing 27(2011), 949–962 (2011)
11. Zang, C., Hashimoto, K.: A flexible visual inspection system combining pose estimation and visual servo approaches. In: 2012 IEEE International Conference on Robotics and Automation River Centre, Saint Paul, Minnesota, USA, May 14-18, pp. 1304–1309 (2012)
12. Zang, C., Hashimoto, K.: A Flexible Camera Positioning Strategy for Robot-based Visual Inspection Applications. In: Proceedings of 2012 IEEE International Conference on Mechatronics and Automation, China, August 5-8, pp. 527–532 (2012)
13. Wu, H., Zhang, X.M., Hong, S.L.: A Visual Inspection System for Surface Mounted Components Based on Color Features. In: Proceedings of the 2009 IEEE International Conference on Information and Automation, Zhuhai/Macau, China, June 22-25, pp. 571–576 (2009)
14. Hata, S., Suezawa, K., Hayashi, J.: Evaluation of Sensitivity Strength for Visual Inspection System of Color Display Device. In: 19th IEEE International Symposium on Robot and Human Interactive Communication Principe di Piemonte - Viareggio, Italy, September 12-15, pp. 773–778 (2010)
15. Wu, D., Sun, D.W.: Colour measurements by computer vision for food quality control - A review. Trends in Food Science & Technology 29, 5–20 (2013)

Optimized Location Discovery Algorithm for Cooperative Activities of a Robot Team

Radu Dobrescu[1], Matei Dobrescu[1], Gheorghe Florea[2], and Victor Purcarea[3]

[1] Politehnica University of Bucharest, Bucharest, Romania
[2] Society of Systems Engineering, Bucharest, Romania
[3] University of Medicine and Pharmacy "Carol Davila", Bucharest, Romania
rd_dobrescu@yahoo.com

Abstract. This paper presents a novel location discovery algorithm used in collaborative work of a team of mobile mini-robots, which are connected as nodes of an inter-robot communication network. The location discovery scheme is a bio-inspired solution for collective decision making based on Outside of Transmission Range information from some reference nodes. The results obtained by simulation show that the proposed scheme can locate more nodes than the basic location discovery scheme by multi-lateration.

Keywords: cooperative work, robot team, location discovery, transmission range, decision making.

1 Introduction

Recent technological advances in mechanics, sensing, and control technologies have made the design of autonomous mobile robots feasible, yet extremely challenging. The demand for such robots has grown, mainly in the dynamic environments, where uncertainty and unforeseen changes can happen. It was realized that a single robot is not the best solution for many application domains. Instead, teams of robots are called upon to work in an intelligently coordinated fashion for achieving efficiency and reliability via redundancy. Multi-robot systems can be more effective than a single robot in many areas. However, when designing such systems it should be noticed that simply increasing the number of robots assigned to a task does not necessarily improve the system's performance - multiple robots must intelligently cooperate to avoid disturbing each other's activity and achieve efficiency. Among the solutions for such cooperative work, one can note some success in analysing bio-inspired rules related to distributed consensus, data aggregation, and collective decision making in multi-robot formation (MRF) tasks [1]. Many of these systems deal with a large number of robotic agents and the MRF is thus viewed as a *swarm* [2], or a *colony* [3] or, more generally, as a *robot collective* [4].

There are several key advantages to the use of such intelligent swarm robotics. First, such systems inherently enjoy the benefit of parallelism. In task-decomposable application domains, robot teams can accomplish a given task more quickly than a single robot, by dividing the task into sub-tasks and executing them concurrently. Second, decentralized systems tend to be much more robust than centralized systems. A team of robots may provide a more robust solution by introducing redundancy, and by eliminating any single point of failure. Third, another advantage of the decentralized swarm approach is the ability of dynamically reallocating sub-tasks between the swarm's units, thus adapting to unexpected changes in the environment due to the benefit of locality. Finally, by using heterogeneous swarms, even more efficient systems could be designed, thanks to the utilization of different types of agents whose physical properties enable them to perform much more efficiently in certain special tasks.

2   Related Work

Various works in the field proposed many different system models, including continuous-time dynamic systems [5]. Typical problems studied under the continuous model include flocking, moving formation, and pursuit, but the continuous model is less suitable for algorithm research, where agent actions are inherently discrete events. Others, such as Marchese [6], investigated swarm algorithms in discrete spaces, inspired by cellular automata models. The first models proposed by Suzuki and Yamashita [7] and Prencipe *et al.* [8] were adopted and developed by other researchers, including Peleg *et al.* [9], Defago *et al.* [10]), Lin *et al.* [11], and us [12]. As in the above works, we concentrate on the algorithmic aspects of the problem, such as agent communication and coordination, correctness, and termination, while abstracting away all physical aspects, such as dynamics of mechanics, kinematics, collisions, sensor implementation, obstacles in the environment, noise and failures and so on.

Coordination can be challenging in large multi-agent systems where agents are spatially distributed or heterogeneous, as some agents will be able to obtain important information not available to others. One approach that can incorporate this privileged information is to designate a centralized coordinator to collect information from all agents, then disseminate a decision to the whole group [13].

While most systems assume leaderless scenarios, some recent studies have shown how to control the whole group's behaviour by controlling specific agents designated as leaders, whether a single agent [14], multiple agents [15], [16], or even virtual leaders not corresponding to actual agents [17].

In our work, inspired by recent biological studies [18], [19], and looking for an improvement of previous results of our research in this field [20], we consider a team of mobile mini-robots (MMRs) as a particular form of Multi Agent System (MAS), by specifically addressing reactivity. In this aim we examine the use of a decentralized group of extremely simple robotic agents for cooperatively accomplishing missions of global properties. We show that using only local interactions, such simple MMRs can cope with the lack of a central supervisor, communication resources and large memory resources.

3   Material and Methods

A key principle in the notion of swarms, or multi agent robotics, is the simplicity of the agents. As a result, the capabilities and the resources of such simple agents are assumed to be very limited, with respect to the following aspects: memory resources; sensing capabilities; computational resources. Communication is very limited too; distinctions between implicit and explicit communication are usually made. Implicit communication occurs as a side effect of other actions, whereas explicit communication is a specific act intended solely to convey information to other robots on the team [21].

3.1   Hardware Structure Details

This paper gives an overview of some technological problems in the development of mobile mini-robots, designed as remote oriented vehicles (ROVs), while trying to take into consideration space environment requirements. Remote Oriented Vehicles (ROVs) extend the reach of human activities and exploration, and are currently of utmost interest to difficult operations. ROVs critically depend on communication to report their observations to human experts or to data repositories, to obtain high-level direction concerning their tasks and objectives and to coordinate among multiple units. Communication is currently implemented with proprietary, application-specific, and inflexible protocols that use point-to-point, fixed-bandwidth channels. At present, fully autonomous machines are poorly equipped to survive in realistically complex and uncertain environments. However, incremental progress and useful intermediate results can be achieved through collaboration between local autonomy and remote supervision.

The mobile mini-robot (MMR) is a four wheel drive vehicle of ROV (remote oriented vehicle) type, driven by one D.C. motor. The motor disposes of an integrated gear and of incremental encoders used to actuate the chain drives. The incremental encoders have a resolution of 128 pulses per revolution. The control algorithms for motor speed and acceleration profiles are implemented on this basic platform, whereas more complex control algorithms for path-planning have to be implemented on the hardware of the extension Printed Circuit Board (PCB).

The MMR mechanical structure is shown in Fig. 1. This design allows to integrate information processing based on a micro-controller on the chassis of the robot, as well as to create the interconnect structure without using additional PCB. Choosing an appropriate connector for the backplane, it is possible to keep the extended information processing module exchangeable to leave it out to save energy and still be able to perform tasks that are not computational expensive. For more details on the size and mechanical characteristics see [22].

The chassis, built as a Moulded Interconnect Device (MID), provides the interconnection between the information processing modules and the peripheral devices, sensors and actuators, IR (Infrared devices), US (Ultrasonic devices), human interface devices and allows communication to extension modules, plugged on top of the robot like GPS (Global Position System), Remote Control, High Resolution Video Camera and all kinds of wireless communications: WiFi, ZigBee and antennas.

Fig. 1. Mobile mini-robot prototype

The design of MMR's mechanics allows to integrate information processing based on a micro-controller on the chassis of the robot, as well as to create the interconnect structure without using additional PCB. Choosing an appropriate connector for the backplane, it is possible to keep the extended information processing module exchangeable to leave it out to save energy and still be able to perform tasks that are not computational expensive.

3.2 Details on Data Acquisition

Sensing the environment, MMR can navigate and find and examine objects. It is necessary to integrate several different sensors on the robot, as every sensor has its own advantages or disadvantages. This leads to the integration of short-range sensors as a minimum requirement (US, IR and Human Interface Devices) and to the (optional) integration of long-range sensors.

In the basic design, the vehicle has three kinds of sensors for environment interactions: infrared proximity sensors, Ultra Sonic Sensors, Human Interface Devices, a High Resolution Video Camera and a Differential Global Positioning System (DGPS) device to determine MMR location in a specified area. For short-range sensing and simple navigation tasks, two IR sensors are placed around the robot's chassis in a height of 35mm. Short IR pulses are emitted cyclically around the robot and received by the IR sensors. The analogue sensor signals are processed by a DSP soldered on the inner side of the wheel house and can be requested via the I2C link.

For more detailed environment sensing a high resolution video camera system is integrated into the robot's body. The integrated image device incorporates a CMOS sensor with VGA resolution, on-chip image processor and JPEG codec for optional image compression.

3.3   Details on Communication Characteristics

The communication solution allows MMR to participate in two networks: an internal network that connects on-board sensors, actuators and other devices (intra-robot communication) and an external network that enables the coordination and communication with other entities (inter-robot communication in a group of robots using Zig-Bee). IP support enables the logical and programmatic unification of the external and internal network, ensures remote supervision of manipulation tasks and achieves the ability to interface with remote control components or human supervisors in the presence of time delays. Supplementary communication offers the possibility to control resolvability in terms of rapid reprogramming (addition of new functionality after hardware deployment), dynamic reconfiguration (creation of new collections of tasks) and extensibility (growth through modular incorporation of additional assets).

Communication among a group of robots can be achieved by ZigBee communication protocol. We already have implemented routines on the ZigBee chip to achieve communication in WSN. It is also possible to create point-to-point communication links between robots or to use the USB connector of the extension module for additional communication hardware e.g. another wireless LAN.

4   A Method for MMR Localisation Based on Cooperative Work

A very interesting development is the implementation of algorithms to coordinate several MMRs in the same area, allowing units to find each other in a distributed environment even with no prior knowledge of their respective existence. We will discuss in the following the proposed solution for robot localization in a specified area. A MMR is considered to be a node in the inter-robots communication network.

4.1   Node Location through Multi-lateration

In location discovery using multi-lateration, a node obtaining its location through external devices (in our case by GPS) is called a *reference node*. Unknown nodes exchange messages with neighbouring reference nodes and can find out where these neighbouring reference nodes are. They can also compute the distances to the reference nodes through received signal strength or time of transmission. As shown in Fig. 2, in a two dimensional space, given the locations of three neighbouring reference nodes and distances to these reference nodes, an unknown node can compute its own location if at least one of these reference nodes is not collinear with the rest of the reference nodes. If fewer than three reference nodes exist in the neighbourhood, an unknown node cannot uniquely determine its own position.

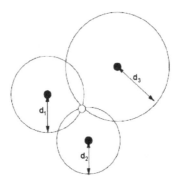

Fig. 2. Location through trilateration (○ node with unknown position; ● reference node)

Assume that we are given a node with unknown position $n_i(x,y)$. If we assume that there are three reference nodes as shown in Fig. 2 having the coordinates (x_1, y_1), (x_2, y_2) and (x_3, y_3), then the coordinate of node $n_i(x, y)$ is computed by

$$x=\frac{(y_2-y_3)(d_1^2-x_1^2-y_1^2)+(y_3-y_1)(d_2^2-x_2^2-y_2^2)+(y_1-y_2)(d_3^2-x_3^2-y_3^2)}{2(y_2-y_1)(x_3-x_1)-2(y_3-y_1)(x_2-x_1)}$$

$$y=\frac{(x_3-x_2)(d_1^2-x_1^2-y_1^2)+(x_1-x_3)(d_2^2-x_2^2-y_2^2)+(x_2-x_1)(d_3^2-x_3^2-y_3^2)}{2(y_2-y_1)(x_3-x_1)-2(y_3-y_1)(x_2-x_1)}$$

(1)

where d_1, d_2 and d_3 mean the Euclidean distances from $n_i(x,y)$ to the three nodes, respectively, which are given by

$$d_i^2 = (x-x_i)^2 + (y-y_i)^2 .. i=1,2,3$$

(2)

Given that each reference node can cost much more than a normal node and consumes significantly more energy in obtaining its location through external devices, it is preferable to use as few reference nodes as possible. It is necessary then to look at possible ways to minimize the number of initial reference nodes in location discovery using multi-lateration schemes. A method that implies cooperation between nodes and allows a minimum of reference nodes for localisation will be discussed in the following. The method uses the communication characteristics to determine the location when the nodes are placed outside of the transmission range (OTR) and was described in [23]. The definition for OTR information is based on the following observation: if two sensor nodes, N_1 and N_2, cannot hear from each other and so are not neighbouring nodes, then the distance between them $dist(N_1,N_2)$ must be larger than $r = \max(\min(r_1); \min(r_2))$, where r_1 (respectively r_2) is the minimum range over all directions that N_1 (respectively N_2) signal propagates.

We will discuss two scenarios to illustrate how to utilize OTR information to determine an unknown node's position (see Fig.3). In both scenarios it is assumed that a reference node N is out of the transmission range of node with unknown position U. It is assumed that the network is connected and so N can reach U through multi-hop flooding, but for simplicity, the other nodes in the network are not shown.

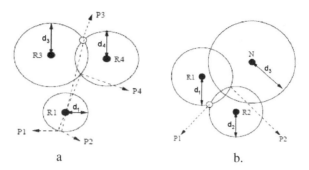

Fig. 3. Two scenarios to determine the position of an unknown node

The scenario illustrated in Fig. 3a corresponds to the situation when the unknown node has two neighbouring reference nodes, $R1$ and $R2$. The distance measured from the unknown node to $R1$ and $R2$, is d_1 and d_2 respectively. There are two possible positions that the U node might be, $P1$ and $P2$. If another reference node, N, exists in the network and the unknown node cannot hear from N, considering that $P2$ is in N's transmission range while $P1$ is not, it can be inferred that the unknown node can only reside in $P1$ because it would hear from N if it were at $P2$. So, if (x_N, y_N) is the coordinate of the reference node N, (x_1, y_1) and (x_1', y_1') the coordinates of the two possible positions of the unknown node U, and r the minimum transmission range of N, N can determine U's location if:

$$\sqrt{(x_N - x_1)^2 + (y_N - y_1)^2} > r \ \& \ \sqrt{(x_N - x_1')^2 + (y_N - y_1')^2} \leq r \qquad (3)$$

The scenario illustrated in Fig. 3 b corresponds to the situation when N is an unknown node with one neighbouring reference node and U has two neighbouring reference nodes. Node U is located at either $P3$ or $P4$. N is located at d_1 away from the reference node $R1$. $P1$ is the farthest point from $P4$ among all the possible locations N might be. If the distance between $P4$ and $P1$ is smaller than r, it can be concluded that U must reside at $P3$, because otherwise U would be a neighbouring node of N.

Generally, let consider an unknown node U, located at either (x, y) or (x', y'). An unknown node N, which is d_1 away from its neighbouring reference node $R1$, can determine U's position under the following condition:

$$\sqrt{(x - x_{R1})^2 + (y - y_{R1})^2} > r - d_1 \text{ or } \sqrt{(x' - x_{R1})^2 + (y' - y_{R1})^2} > r - d_1 \qquad (4)$$

4.2   The Algorithm for Location Discovery

In the proposed location discovery scheme, reference nodes disseminate their positions to neighbouring unknown nodes. If more than three neighbour nodes are reference nodes, an unknown node estimates its own location using trilateration. Otherwise, the unknown sensor node sends messages to non-neighbouring nodes to check whether they can help to resolve its location using OTR information. Once its location is resolved, an unknown node becomes a reference node and disseminates its position to other unknown nodes in the network to enable the continuation of the location discovery process.

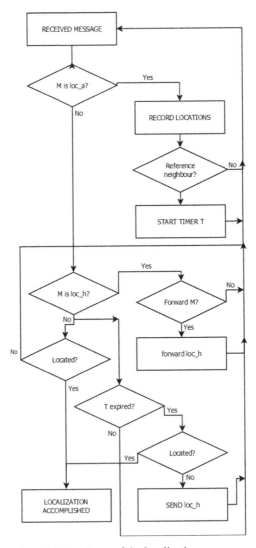

Fig. 4. The scheme of the localization process

Fig. 4 presents the major steps of the localization process executed at a reference node R. The reference node starts the localization process by announcing its location to neighbouring nodes using *"loc_a"* (Location announce) message. It then keeps waiting for messages from other nodes. U also starts a timer T if *loc_a* is the first announce message it receives. Based on the type of message received, the reference node, R, responds as follows:

- If a *"loc_h"* (Location help) message for U is received, R simply discards this message if it has already processed the help request from U. Otherwise, R checks condition (3) and sends "Location help reply" to U if it can determine the location of U using OTR information.
- If R cannot utilize its OTR information to uniquely locate U position, R decreases the value T of the timer of the "Location help" message by one and forwards the "Location help" message to its neighbours if T is still bigger than zero.
- If the timer T expires and U still cannot resolve its location, U initializes the timer of the "Location help" message to be H and sends it out to its neighbouring nodes.

5   Simulation Results

The main metric checked in the simulation is the number of nodes for which the position was established after the location discovery process completes. The initially configured reference nodes start to broadcast location information at the beginning of simulation. Each node maintains a neighbour table to know if it is out of another's node range. Various levels of densities were simulated by modifying the number of nodes in a field of 50x50m. A transmission range of 12 m was considered. For simplicity, it is assumed that all MMRs have the same value of transmission range.

Let note that the value of H is critical to the performance of the proposed scheme. On the one hand, a large value of H allows an unknown node to reach more MMRs for help and thus have a higher chance of resolving its location ambiguity. However, on the other hand, it also leads to a high level of communication overhead. Therefore, care must be taken in choosing the value of H to ensure a high possibility of location discovery at a low cost of communications.

Fig. 5 presents the number of resolved nodes after the location discovery completes in networks with various levels of densities (split in decades from 40 to 80). In these scenarios, three nodes are initially configured as reference nodes.

The results show that the number of resolved nodes remains the same in most cases. It slightly increases in the network of 50 and 80 nodes when H increases from 2 to 3. The reason is that nodes multi-hop away may be geographically too far away from the unknown node U to provide any useful OTR information. The value of H is set to be 2 in the rest of the simulations. The number of resolved nodes after the location discovery is completed is used to measure the effectiveness of the proposed scheme in comparison to the basic multi-lateration scheme.

When the average node degree is small, no OTR information can be used due to the lack of connectivity in the network. On the other hand, when the average node degree is large, no OTR information is needed since normal MMRs can estimate their locations using reference nodes. In the other cases, the proposed scheme can locate more nodes than the basic multi-lateration scheme. The results show that as the inter-robot network connectivity starts to decrease, the OTR information can be used to locate more nodes in the network

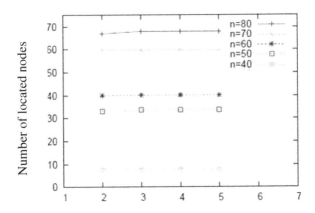

Number of   hops to send "loc_h" message

Fig. 5. The effect of *H* in unknown nodes localization

It is worth noting that the average number of resolved nodes does not always increase as the density increases in the network. This is due to the fact that nodes are uniformly placed over the entire area. Nonetheless, a relative big increase of network density does result in an increase of number of resolved nodes after the location discovery process.

In high density network, there is no need to use OTR information for location discovery since one can gain sufficient information about reference nodes and compute their locations using multi-lateration.

6   Conclusions

This paper presents a location discovery scheme used in the framework of a cooperative work of a team of mobile mini-robots connected as nodes of an inter-robot communication network. The location discovery algorithm is based on Outside of Transmission Range information and it is shown how this information can be used to resolve MMRs location ambiguities when combined with multi-lateration. The simulation results show that with OTR information, fewer reference nodes are needed to locate nodes in the network, which in turn reduces cost and energy consumption of the whole network since reference nodes are usually much more expensive and consume more energy.

The algorithm used in localization is one among several utilised for collective decision making for MMRs activities. These algorithms allow MMRS to self-organize to identify, collect, and deliver the information specific to each decision maker efficiently. Therefore it is possible to reduce the overload of less relevant information and time required for information processing and to facilitate rapid absorption of critical information by decision makers.

Future work will be oriented in substantial development of the middleware that is available on a MMR platform. Important open issues include the limit to software agent mobility, which is imposed by the low bandwidth and unreliability of communication links and from the MMR spatial mobility. A very interesting development will be the implementation of a methodology to achieve objectives such as survivability and evolvability. We consider that eventually we will be able to coordinate several MMRs in the same area, allowing units to find each other in a distributed environment even with no prior knowledge of their respective existence.

Acknowledgements. This work was supported by FP7 project REGPOT - 2010 - 1, ERRIC – Empowering Romanian Research on Intelligent Information Technologies.

Authors contributions: Location discovery algorithm – Radu Dobrescu (25%), Node location through multi-lateration algorithms – Gheorghe Florea (25%), Simulation and tests - Matei Dobrescu (25%), References to related work and mathematical review - Victor Purcarea (25%).

References

1. Cao, M., Morse, A.S., Anderson, B.D.O.: Reaching a consensus in a dynamically changing environment: A graphical approach. SIAM Journal on Control and Optimization 47(2), 575–600 (2008)
2. Gazi, V., Passino, K.M.: Stability analysis of social foraging swarms. IEEE Transactions on Systems, Man, and Cybernetics, Part B: Cybernetics 34(1), 539–557 (2004)
3. Olfati-Saber, R., Fax, J., Murray, R.: Consensus and cooperation in networked multi-agent systems. Proceedings of the IEEE 95(1), 215–233 (2007)
4. Couzin, I.D.: Collective cognition in animal groups. Trends in Cognitive Sciences 13(1), 36–43 (2009)
5. Boccara, N.: Modeling complex systems. Graduate Texts in Contemporary Physics. Springer, New York (2004)
6. Marchese, F.: MRS Motion Planning: the Spatiotemporal MultiLayered Cellular Automata Approach. In: Chugo, D., Yokota, S. (eds.) Introduction to Modern Robotics. iConcept Press (2011)
7. Suzuki, I., Yamashita, M.: Distributed anonymous mobile robots. Formation of geometric patterns. SIAM Journal on Computing 28(4), 1347–1363 (1999)
8. Prencipe, G., Santoro, N.: Distributed algorithms for autonomous mobile robots. In: 5th IFIP Int. Conference on Theoretical Computer Science (TCS 2006), pp. 47–62. Springer (2006)
9. Peleg, D.: Distributed coordination algorithms for mobile robot swarms: New directions and challenges. In: Pal, A., Kshemkalyani, A.D., Kumar, R., Gupta, A. (eds.) IWDC 2005. LNCS, vol. 3741, pp. 1–12. Springer, Heidelberg (2005)

10. Défago, X., Gradinariu, M., Messika, S., Raipin-Parvédy, P.: Fault tolerant and self-stabilizing mobile robots gathering. In: Dolev, S. (ed.) DISC 2006. LNCS, vol. 4167, pp. 46–60. Springer, Heidelberg (2006)
11. Lin, Z., Broucke, M.E., Francis, B.A.: Local control strategies for groups of mobile autonomous agents. IEEE Transactions on Automatic Control 49(4), 622–629 (2004)
12. Dobrescu, R., Popescu, D., Florea, G.: Cellular Automata Models for Cooperation in Multirobot Systems. In: Niola, V., Bojkovic, Z., Garcia-Planas, I. (eds.) Recent Researches in Circuits, Systems, Multimedia and Automatic Control, Proceedings of the 12th WSEAS International Conference on Robotics, Control and Manufacturing Technology (ROCOM 2012), pp. 124–129. WSEAS Press (2012)
13. Tanner, H., Kumar, V.: Leader-to-formation stability. IEEE Transactions on Robotics and Automation 20(3), 433–455 (2004)
14. Ji, M.: Graph-Based Control of Networked Systems. PhD thesis, Georgia Tech (2007)
15. Altshuler, Y.: Multi Agents Robotics in Dynamic Environments, Ph.D. Thesis, Technion, Israel Institute of Technology (2010)
16. Gordon, N.: Fundamental Problems in the Theory of Multi-Agent Robotics, Ph.D. Thesis, Technion, Israel Institute of Technology (2010)
17. Zheng, X., Koenig, S.: Reaction functions for task allocation to cooperative agents. In: Proceedings of the 7th International Joint Conference on Autonomous Agents and Multiagent Systems, pp. 559–566 (2008)
18. Yu, C.-H.: Biologically-Inspired Control for Self-Adaptive Multiagent Systems. Ph.D. Dissertation, Harvard University (2010)
19. Bhatt, R.M., Tang, C.P., Krovi, V.N.: Formation optimization for a fleet of wheeled mobile robots a geometric approach. Robotics and Autonomous Systems 57(1), 102–120 (2009)
20. Dobrescu, R., Dobrescu, M., Mocanu, S.: Convergence of Communication with Computing for a remote oriented mobile minirobot. In: Dolgui, A., Morel, G., Pereira, C. (eds.) Information Control Problems In Manufacturing, Elsevier Science, pp. 179–185 (2006)
21. Yamamoto, K., Izumi, T., Katayama, Y., Inuzuka, N., Wada, K.: Convergence of mobile robots with uniformly-inaccurate sensors. In: 16th International Colloquium on Structural Information and Communication Complexity, pp. 309–322 (2009)
22. Dobrescu, R.: Cooperative control for groups of autonomous mobile minirobots. In: Vladareanu, L., Chiroiu, V., Bratu, P., Magheti, I. (eds.) Automation & Information: Theory and Advanced Technology, Proc. of the 9th WSEAS Int. Conf. on Automation & Formation (ICAI 2008), pp. 402–407 (2008)
23. Hui, L.: A framework for enabling energy efficient semantic views in wireless sensor networks for data intensive applications. Doctoral Dissertation, University of Pittsburgh (2010)

A Study of Feasibility of a Human Finger Exoskeleton

Daniele Cafolla and Giuseppe Carbone

LARM: Laboratory of Robotics and Mechatronics – DICEM –
University of Cassino and South Latium,
Via Di Biasio 43, 03043
Cassino (Fr), Italy
{cafolla,carbone}@unicas.it

Abstract. Finger impairment following stroke results in significant deficit in hand manipulation and the performance of everyday tasks. Recent advances in rehabilitation robotics have shown improvement in efficacy of rehabilitation. Current devices, however, lack the capacity to accurately interface with the human finger at levels of velocity and torque comparable to the performance of everyday hand manipulation tasks. This paper tries to fill this need with a newly designed system intended to aid in hand rehabilitation. A 3D CAD model and simulations have been developed for verifying the engineering feasibility.

Keywords: robotics hands, exoskeleton, hand rehabilitation, robot services.

1   Introduction

The hand is a very useful and fragile organ of the human body, it can suffers injuries or at worst have them since the day of birth. The hand is the tool with which the human being interacts with the environment and, furthermore, it has the function of data acquisition system. The structure of the hand is able to adapt to a very high number of configurations that allow to grasp and manipulate objects of different size and shape. The hand also has a high concentration of nerve endings and receptors, whose signals are processed by a large area of the brain, [1]. Cerebral vascular accident, or stroke, remains the leading cause of adult disability and it is estimated that there are nearly 800,000 stroke incidents in the United States annually [2]. Though stroke causes deficits in many of the neurological domains, the most commonly affected is the motion system [3]. Nearly 80% of stroke survivors suffer hemiparesis of the upper arm [3] and impaired hand function is reported as the most disabling motion deficit [4]. Currently, even following extensive therapeutic interventions in acute rehabilitation, the probability of regaining functional use of the impaired hand is low [5]. Adequate hand function/prehension is vital for many activities of daily living including feeding, bathing and dressing. Accordingly, attention has been addressed to both understanding the mechanisms for hand motion impairment and optimizing hand therapy techniques. A number of factors that contribute to hand impairment have been investigated.

T. Borangiu et al. (eds.), *Service Orientation in Holonic and Multi-Agent Manufacturing and Robotics*, Studies in Computational Intelligence 544,
DOI: 10.1007/978-3-319-04735-5_24, © Springer International Publishing Switzerland 2014

Evidence indicates that hypertonia in finger flexor muscles [6] and weakness in both finger extensor and flexor muscles [7] impairs voluntary hand function. Muscle weakness is not uniform between the extensor and flexor muscles [6], and stroke survivors generally tend to regain functional flexion with minimal recovery of extension. These imbalances are related to altered muscle activation patterns where elevated levels of flexor activity occur during extension movements [6]. However, studies have shown that repetitive training paradigms such as simple flexion and extension finger movements can improve grasp and release function [8, 9]. The use of rehabilitation robotics to provide motion therapy has shown great potential. Some of the benefits of rehabilitation robotics include introducing the ability to perform precise and repeatable therapeutic exercises, reduction of the physical burden of participating therapists, incorporation of interactive virtual reality systems, and collection of quantitative data that can be used to optimize therapy sessions and assess patient outcomes. Many investigators have focused on developing devices designed to retrain an impaired upper limb [10, 11, 12], and robot-assisted therapy is proven to significantly improve proximal arm function [13, 14, 15]. This paper tries to fill this need with a newly designed system intended to aid in hand rehabilitation. The motion assistance equipment consists of two parts: mechanisms for the fingers and support of these mechanisms.

2   The Attached Problem

The rehabilitation of a body segment as the hand seeks to achieve multiple purposes: Arc of motion recovery; Muscle function recovery; Recover hand skills for daily activities; Proprioceptivity recovery; Retrieving sensibility. The traumatized hand must be placed in splint with either metacarpophalangeal joints flexed, the interphalangeal joints and thumb extended abducted. The mobilization must start as soon as possible, [1]. There are a number of exercises to be carried out as soon as possible to resume the activity of the hand. Those commonly prescribed are illustrated in Fig. 1, [15]. The rehabilitation procedure is as follows: Pain reduction; Control and treatment of oedema; Active mobilization; Passive mobilization; Orthosis, in the first stage, if necessary, must be static. Very useful are the dynamic splint, designed and manufactured directly on the patient's hand, frequently changed according to the rule enunciated by Brand: "moderate traction applied over a long period". The dynamic component has a flexible and progressive traction on the coated tendon and is modified based on the improvement of articular range, [15].

Active mobilization is intended to be when an external force is applied to the finger for its rehabilitation. This active mobilization can be achieved either manually by a physician or by means of special springs or dynamic splits. Passive mobilization is intended to be when no external force is applied to the finger. An example of passive mobilization is a special bandage that allows passive motion of the finger.

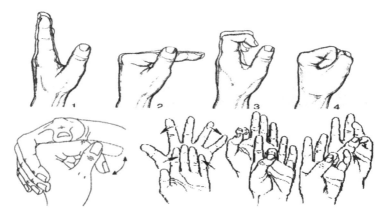

Fig. 1. Commonly prescribed exercises to resume the activity of the hand, [15]

3   The Proposed Exoskeleton

LARM Hand III has been the third robotic hand prototype that has been designed and built at LARM in Cassino. It has one active degree of freedom for each finger and a specific cross four bar driving mechanism.

Using Solid Works, starting from the design of LARM Hand III, shown in Fig. 2, single robotic finger has been isolated in order to study a way to interface it to the human finger. To make the prototype more ergonomic and in order to use it as an exoskeleton, a design solution shown in Fig. 3 has been studied to move the articulated quadrilateral, necessary for the movement of the finger, outside the finger body. In this way, the human finger can be placed where the mechanism was. The next problem to solve was that the original frame shape was too bulky for the purpose. The dimensions of the frame have been decreased and it has been modified so that the prototype can be fastened to the palm of the hand using a small belt. In this way, the frame has been adapted as the exoskeleton was not sufficiently ergonomic so the problem was solved starting from the LARM Hand IV, shown in Fig. 4. Using the same procedure shown previously the result of the new prototype is shown in Fig. 5.

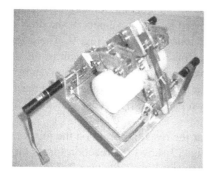

Fig. 2. LARM Hand III

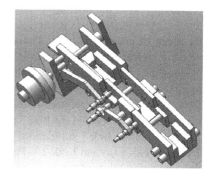

Fig. 3. First prototype

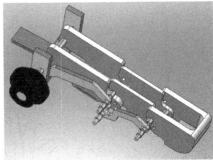

Fig. 4. -LARM Hand IV   **Fig. 5.** -Final Prototype

4   Numerical Simulations

The proposed exoskeleton is operated with an engine attached to a shaft that is stuck to the finger.

Fig. 6 shows the torque output needed to move the prototype, calculated by using the software MSC ADAMS. In the obtained results of the simulation, the transient phase was mostly due to the kinematic model approximations and interpolation errors. It can be noticed that the motor at the beginning reaches the torque necessary to generate the mechanism of 0,020 Nm. Then the torque begins to decline: this is due to the fact that the engine is going to stop and it is aided by the contribution of gravitational force that is considered in the simulation process. The transmission mechanism is based on articulated quadrilaterals. Thanks to it by a single input more links can be moved with a path as the desired movement for rehabilitation exercises. Contributions of the latters are studied on each component to better understand the accelerations and velocities to which the entire model is subject.

Fig. 7 shows the results of the simulation of the Phalanx 1 stuck on the motor shaft. The speed of the centre of mass (CM) has a constant value of 5.0 deg/s. As the Phalanx is directly stuck in the motor shaft it receives a constant motion until the end of the simulation.

Moving on to the next body, in Fig. 8 is shown the angular velocity of Phalanx 2 that is constant and tends to increase as influenced by the movement of the body and the force of gravity. The last Phalanx 3 behaviour is shown in Fig. 9. The angular velocity increases steadily until the end of the simulation, the movement of the third phalanx may be regarded as being fully developed when the contribution given by the force of gravity is fully developed. Of particular interest are the graphs of the reaction forces, the plots being referred to the axes shown in Fig. 10.The red axis is the X axis, the green - the Y one and the blue - the Z one.

Fig. 11 shows link 1 simulation, in particular the reaction forces on the joints Link 1-Frame (Fig. 11a) and Link 1 – Phalanx 1. Fig. 12 shows the reaction forces on the two remaining joints of the main moving mechanism: Link 2-Phalanx 2 (Fig. 12a); b) Link 2-Phalanx 3 (Fig. 12b).This reaction forces show that the movement is transmitted perfectly to each part of the body through the forces acting on the joints. It has

been studied that the arcs made by the trajectory of each body, approximate accepta-
bly the needed movement maintaining a low-cost device. The simulation sequence in
Fig. 13 shows how the finger properly follows the natural trajectory that would follow
without any trauma.

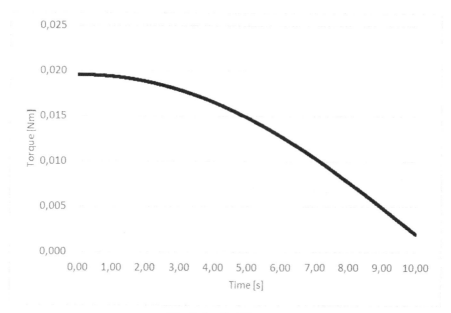

Fig. 6. Applied Torque

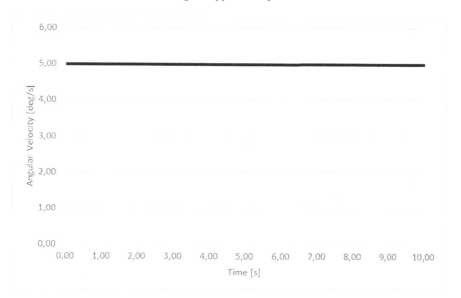

Fig. 7. Phalanx 1 simulation results: CM Angular Velocity

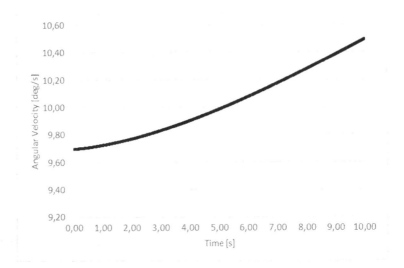

Fig. 8. Phalanx 2 simulation results: CM Speed

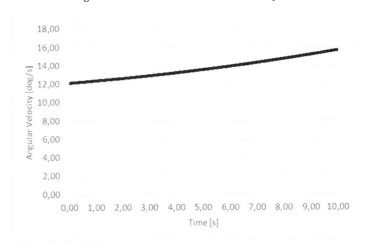

Fig. 9. Phalanx 3 simulation results: CM Speed

All the simulation results have shown proper smooth behaviour feasible with operation ranges. The results show that this prototype can be used in post-operative rehabilitation care or light trauma to get the finger used again to the movement without initial effort on the part of the patient. The motor that shall be used is a Nema 8HS11-0204S, Bipolar Stepper, Step Angle 1.8°, Holding Torque 1.6Ncm, Rated Torque 1.28Ncm(1.84oz.in) Rated Speed 200rpm, Maximum Speed 600rpm, Recommended Voltage 12-24V, shown in Fig. 14. It has the proper size and weight in order to avoid uncomfortable wearing of the exoskeleton as well as its suitable operation and can be controlled easily through an Arduino board. As engineering advantage, the proposed project has the availability of the material, the low-cost of construction, and the easy assembly and use.

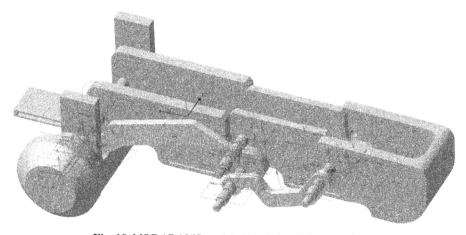

Fig. 10. MSC ADAMS model: Simulation References Axes

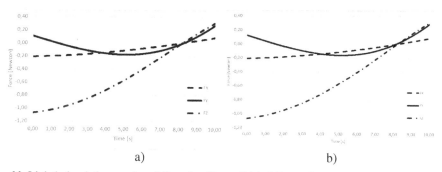

a)   b)

Fig. 11. Link 1 simulation results: a) Reaction Forces Link 1-Frame b) Reaction Forces Link 1-Phalanx 2

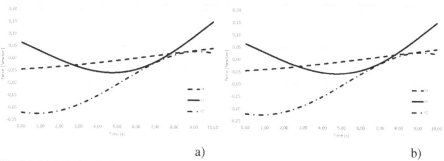

a)   b)

Fig. 12. Link 2 simulation results: Reaction Forces Link 2-Phalanx 2 b) Reaction Forces Link 2-Phalanx 3

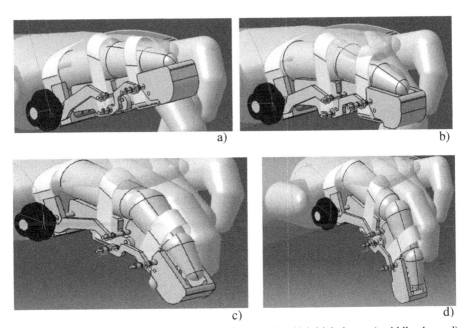

Fig. 13. Rehabilitation Movement simulation: a) no motion, b) initial phase: c) middle phase, d) final phase

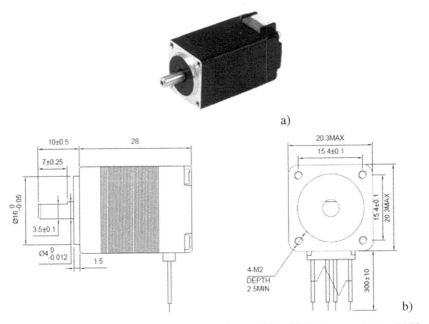

Fig. 14. Nema 8HS11-0204S Stepper Motor: a) Nema 8HS11-0204S Stepper Motor, b) Nema 8HS11-0204S Stepper Motor Scheme

5   Conclusions

In this paper, a finger exoskeleton has been proposed, based on the finger design of LARM Hand Version III and IV and 3D CAD model that have been developed in SolidWorks. In MSC ADAMS a Dynamic Simulation has been done to check the feasibility of the model and the important parameters for a future realization, choosing a commercial low cost stepper motor to be used for the real prototype.

Finally a hand model has been developed in the CATIA environment to test the wearability of the Exoskeleton. Results of numerical simulations have been reported to show the engineering feasibility with satisfactory results. The aim of this paper is to detail a preliminary study of feasibility in order to achieve a low-cost and easy - operation solution for an exoskeleton for rehabilitation purposes. Many of the mentioned aspects (such as the tightening of belts and finger comfort) shall be addressed during the construction of a prototype.

References

1. AA.VV. The Merck manual of diagnosis and therapy. In: Berkow, R. (ed.) Merck Research Laboratories, Rahway, N.J., XVI ed. (1992)
2. Stroke Statistic: 2009 Update. Centennial, CO: National Stroke Association (2009)
3. Kwakkel, G., Kollen, B.J., Van der Grond, J., Prevo, A.J.: Probability of regaining dexterity in the flaccid upper limb. The Impact of Severity of Paresis and Time Since Onset in Acute Stroke, Stroke 34, 2181–2186 (2003)
4. Carbone, G., Kwakkel, G., Kollen, B.J., Van der Grond, J., Prevo, A.J.: Probability of regaining dexterity in the flaccid upper limb. Impact of Severity of Paresis and Time Since Onset in Acute Stroke, Stroke 34, 2181–2186 (2003)
5. Ceccarelli, M.: Experimental Tests on Feasible Operation of a Finger Mechanism in the LARM Hand. Mechanics Based Design of Structures and Machines An International Journal 36(1), 1–13 (2008)
6. Heart Disease and Stroke Statistics: Update, Dallas, Tex: American Heart Association (2005)
7. Duncan, P.W., Bode, R.K., Min Lai, S., Perera, S.: Rasch analysis of a new stroke specific outcome scale: the Stroke Impact Scale. Arch. Phys. Med. Rehabil. 84(7), 950–963 (2003)
8. Carey, J.R., Durfee, W.K., Bhatt, E., Nagpal, A., Weinstein, S.A., Anderson, K.M., Lewis, S.M.: Comparison of finger tracking versus simple movement training via telerehabilitation to alter hand function and cortical reorganization after stroke. Neurorehabil Neural Repair 21(3), 216–232 (2007)
9. Carey, J.R., Kimberley, T.J., Lewis, S.M., Auerbach, E.J., Dorsey, L., Rundquist, P., Ugurbil, K.: Analysis of fMRI and finger tracking training in subjects with chronic stroke. Brain 125(pt 4), 773–788 (2002)
10. Hesse, S., Schulte-Tigges, G., Konrad, M., Bardeleben, A., Werner, C.: Robot assisted arm trainer for the passive and active practice of bilateral forearm and wrist movements in hemiparetic subjects. Arch. Phys. Med. Rehabil. 84(6), 915–920 (2003)
11. Kamper, D.G., Harvey, R.L., Suresh, S., Rymer, W.Z.: Relative contributions of neural mechanisms versus muscle mechanics in promoting finger extension deficits following stroke. Muscle Nerve 28(3), 309–318 (2003)

12. Dovat, L., Lambercy, O., Salman, B., Johnson, V., Milner, T., Gassert, R., Burdet, E., Leong, T.C.: A technique to train finger coordination and independence after stroke. Disabil. Rehabil. Assist. Technol. 5(4), 279–287 (2010)
13. Krebs, H.I., Hogan, N., Volpe, B.T., Aisen, M.L., Edelstein, L., Diels, C.: Overview of clinical trials with MIT-MANUS: a robot-aided neuro-rehabilitation facility. Technol. Health Care 7(6), 419–423 (1999)
14. Lum, P.S., Burgar Reinkensmeyer, D.J., Kahn, L.E., Averbuch, M., McKenna-Cole, A., Schmit, B.D., Rymer, W.Z.: Understanding and treating arm movement impairment after chronic brain injury: progress with the ARM guide. J. Rehabil. Res. Dev. 37(6), 653–662 (2000)
15. Jones, C.L., Wang, F., Osswald, C., Kang, X., Sarkar, N., Kamper, D.G.: Control and Kinematic Performance Analysis of an Actuated Finger Exoskeleton for Hand Rehabilitation following Stroke. In: Proceedings of the 2010 3rd IEEE RAS & EMBS International Conference on Biomedical Robotics and Biomechatronics, September 26-29. The University of Tokyo, Tokyo (2010)
16. Takahashi, C., Der-Yeghiaian, L., Le, V., Cramer, S.C.: A robotic device for hand motor therapy after stroke. In: Proceedings of IEEE 9th International Conference on Rehabilitation Robotics: Frontiers of the Human-Machine Interface Chicago, Illinois, pp. 17–20 (2005)
17. Shor, P.C.: Evidence for strength imbalances as a significant contributor to abnormal synergies in hemiparetic subjects. Muscle Nerve 27(2), 211–221 (2003)
18. Ceccarelli, M., Carbone, G.: Design of LARM Hand: problems and solutions. In: IEEE-TTTC International Conference on Automation, Quality and Testing, Robotics, AQTR 2008, Cluj-Napoca, pp. 298–303 (2008); (best paper award): in Journal of Control Engineering and Applied Informatics 10(2), 39-46 (2008)
19. Ceccarelli, M., Nava Rodriguez, N.E., Carbone, G.: Optimal Design of Driving Mechanism in a 1-d.o.f. Anthropomorphic Finger. In: International Workshop on Computational Kinematics, paper 03CK 2005, Cassino (2005)

Author Index

Printed in the United States
By Bookmasters